Fermentation Processes

Fermentation Processes

Emerging and Conventional Technologies

Edited by

Mohamed Koubaa
ESCOM, UTC, EA 4297 TIMR, Compiègne, France

Francisco J. Barba
Faculty of Pharmacy, Preventive Medicine and Public Health, Food Science, Toxicology and Forensic Medicine Department, Universitat de València, València, Spain

Shahin Roohinejad
Burn & Wound Healing Research Center, Shiraz University of Medical Science, Shiraz, Iran

WILEY Blackwell

This edition first published 2021

Registered Offices
John Wiley & Sons, Inc., 111 River Street, Hoboken, NJ 07030, USA
John Wiley & Sons Ltd, The Atrium, Southern Gate, Chichester, West Sussex, PO19 8SQ, UK

Editorial Office
The Atrium, Southern Gate, Chichester, West Sussex, PO19 8SQ, UK

For details of our global editorial offices, customer services, and more information about Wiley products, visit us at www.wiley.com.

Wiley also publishes its books in a variety of electronic formats and by print-on-demand. Some content that appears in standard print versions of this book may not be available in other formats.

Library of Congress Cataloging-in-Publication Data

Names: Koubaa, Mohamed, editor. | Barba, Francisco (Francisco J.), editor. | Roohinejad, Shahin, editor.
Title: Fermentation processes : emerging and conventional technologies / edited by Mohamed Koubaa, Francisco J. Barba, Shahin Roohinejad.
Description: Hoboken, NJ : Wiley-Blackwell, [2020] | Includes bibliographical references and index.
Identifiers: LCCN 2020031691 (print) | LCCN 2020031692 (ebook) | ISBN 9781119505853 (cloth) | ISBN 9781119505846 (adobe pdf) | ISBN 9781119505839 (epub)
Subjects: LCSH: Fermentation.
Classification: LCC QR151 .F449 2020 (print) | LCC QR151 (ebook) | DDC 572/.49–dc23
LC record available at https://lccn.loc.gov/2020031691
LC ebook record available at https://lccn.loc.gov/2020031692

Cover Design: Wiley
Cover Image: Courtesy of Mohamed Messaoud

Set in 9.5/12.5pt STIXTwoText by SPi Global, Pondicherry, India

10 9 8 7 6 5 4 3 2 1

Contents

About the Editors

Mohamed Koubaa

Dr. Mohamed Koubaa is an assistant professor in process and chemical engineering at ESCOM Chimie (Ecole Supérieure de Chimie Organique et Minérale) in Compiègne (France). He obtained an engineering diploma in biological sciences in 2007 from the National School of Engineers of Sfax (Tunisia), followed by a master's degree from the University of Technology of Compiègne (France). In July 2008, he received a full doctoral scholarship award from the Ministry of Higher Education, Research, and Innovation (France), and graduated in February 2012. He has performed different postdoctoral research stays as a research assistant in the Department of Industrial Process Engineering, University of Technology of Compiègne (France), the Department of Biological Engineering, National School of Engineers of Sfax (Tunisia), and as a postdoctoral research assistant in the prestigious Department of Molecular Genetics, The Ohio State University, Columbus, Ohio (USA). His research focus is on the use of nonconventional processing for the preservation and/or the extraction of bioactive compounds from liquid and solid foods and more recently in microbial growth stimulation by using emerging technologies. He has more than 100 publications (book chapters and articles in peer-reviewed international journals) to his credit.

Francisco J. Barba

Dr. Francisco J. Barba is an associate professor in Nutrition and Food Science and Technology, Faculty of Pharmacy, University of Valencia, Spain. He holds a European PhD (with distinction) from the University of Valencia and has obtained degrees in Pharmacy, Food and Technology. He performed postdoctoral stays in the Université de Technologie de Compiègne (UTC), Département de Génie des Procédés Industriels, Laboratoire Transformations Intégrées de la Matière Renouvelable (Compiegne, France), and Marie Curie IEF in the Department of Food Chemistry (University of Copenhagen) to explore different non-thermal applications for preserving and extracting bioactive compounds from plant food materials and by-products. Prior to his current appointment, he was also engaged as a visiting researcher in the Department of Food Biotechnology and Food Process

Engineering in Technological University of Berlin, Germany. His research focus is on innovative food processing technologies such as high-pressure processing, supercritical fluids, electrotechnologies, ultrasound, and microwaves for preservation and/or extraction of bioactive compounds from liquid and solid food. He has more than 300 publications to his credit, including more than 250 published or accepted peer-reviewed papers in international journals in the food science and technology area (h-index = 53, SCOPUS). He has been included in the Highly Cited Researchers 2019, 2020 list, the latest classification of the Clarivate Analytics bibliometric data provider. Dr. Barba currently serves as associate editor for prestigious journals such as *Food Research International*, *Journal of Food Composition and Analysis*, *Journal of Food Processing and Preservation*, *Antioxidants*, *Foods*, *Molecules*, and *Frontiers in Nutrition*, among others.

Shahin Roohinejad

Dr. Shahin Roohinejad obtained his BSc in 2000 in the field of food science and technology from the Islamic Azad University, Iran. He completed his MSc in food biotechnology at the University Putra Malaysia (UPM) in 2009. In July 2011, he received a full doctoral scholarship award from the Department of Food Science at the University of Otago in New Zealand and graduated in December 2014. In 2015, he received the Georg Forster Research Fellowship Award granted by the Alexander von Humboldt Foundation to pursue his postdoctoral research at the Department of Food Technology and Bioprocess Engineering, Max Rubner-Institut (MRI), the German Federal Research Institute of Nutrition and Food. He joined the Department of Food Science and Nutrition, University of Minnesota as a Postdoctoral Research Associate in 2017. Followingly he started to work as a Research Scientist at Tillamook County Creamery Association in Oregon, USA in 2018. Currently, he is a Senior Scientist - Next Generation Oral Products at Reynolds American Inc., USA. He is a professional member of the Institute of Food Technologists (IFT), a member of IFT Press Advisory Group, and a Global Harmonization Initiative Ambassador in the USA. In the last 15 years, he has been working on different food areas such as fermentation, emerging food processing, emulsion-based systems, nanotechnology, and functional foods. His research activities have resulted in more than 100 original papers in peer-reviewed journals, book chapters, abstracts, and short papers in congress proceedings.

List of Contributors

Marina Al Daccache
Faculté des Sciences, Centre d'Analyses et de Recherche, UR TVA, Laboratoire CTA, Université Saint-Joseph, Beyrouth, Lebanon

Francisco J. Barba
Faculty of Pharmacy, Preventive Medicine and Public Health, Food Science, Toxicology and Forensic Medicine Department, Universitat de València, València, Spain

Ivonne Delgadillo
Escola Superior de Biotecnologia, Universidade Católica Portuguesa, Porto, Portugal

LAQV-REQUIMTE, Department of Chemistry, University of Aveiro, Aveiro, Portugal

Ana M. Gomes
Escola Superior de Biotecnologia, Universidade Católica Portuguesa, Porto, Portugal

Seyedeh-Sara Hashemi
Burn & Wound Healing Research Center, Shiraz University of Medical Science, Shiraz, Iran

Mahdi Irani
Department of Food Science and Technology, Ferdowsi University of Mashhad (FUM), Mashhad, Iran

Sucheta Khubber
Food Engineering and Nutrition, Center of Innovative and Applied Bioprocessing, Mohali, India

Mohamed Koubaa
ESCOM, UTC, EA 4297 TIMR, Compiègne, France

Rita P. Lopes
LAQV-REQUIMTE, Department of Chemistry, University of Aveiro, Aveiro, Portugal

Jose M. Lorenzo
Centro Tecnológico de la Carne, Parque Tecnológico de Galicia, Ourense, Spain

Nicolas Louka
Faculté des Sciences, Centre d'Analyses et de Recherche, UR TVA, Laboratoire CTA, Université Saint-Joseph, Beyrouth, Lebanon

Richard G. Maroun
Faculté des Sciences, Centre d'Analyses et de Recherche, UR TVA, Laboratoire CTA, Université Saint-Joseph, Beyrouth, Lebanon

Krystian Marszałek
Department of Fruit and Vegetable Product Technology, Prof. Wacław Dąbrowski Institute of Agricultural and Food Biotechnology, Warsaw, Poland

Department of Food Technology and Human Nutrition, Institute of Food Technology and Nutrition, College of Natural Science, University of Rzeszow, Rzeszow, Poland

Francisco J. Marti-Quijal
Faculty of Pharmacy, Preventive Medicine and Public Health, Food Sciences, Toxicology and Forensic Medicine Department, Universitat de València, València, Spain

Maria J. Mota
LAQV-REQUIMTE, Department of Chemistry, University of Aveiro, Aveiro, Portugal

Maryam Nejadmansouri
Department of Food Science and Technology, College of Agriculture, Shiraz University, Shiraz, Iran

Mehrdad Niakousari
Department of Food Science and Technology, College of Agriculture, Shiraz University, Shiraz, Iran

Alireza Rafati
Division of Pharmacology & Pharmaceutical Chemistry, Sarvestan Branch, Islamic Azad University, Sarvestan, Iran

Maryam Razmjooei
Department of Food Science and Technology, College of Agriculture, Shiraz University, Shiraz, Iran

Fabienne Remize
UMR QualiSud, Université de La Réunion, CIRAD, Université Montpellier, Institut Agro, Université d'Avignon, Sainte Clotilde, France

Shahin Roohinejad
Burn & Wound Healing Research Center, Shiraz University of Medical Science, Shiraz, Iran

Dominique Salameh
Faculté des Sciences, Centre d'Analyses et de Recherche, UR - EGP, Laboratoire E2D, Université Saint-Joseph, Beyrouth, Lebanon
dominique.salameh@usj.edu.lb
Faculté des Sciences, Centre d'Analyses et de Recherche, UR - EGP, Laboratoire E2D, Université Saint-Joseph, Beyrouth, Lebanon

Jorge A. Saraiva
QOPNA, Chemistry Department, University of Aveiro, Aveiro, Portugal
LAQV-REQUIMTE, Department of Chemistry, University of Aveiro, Aveiro, Portugal

Sylwia Skąpska
Department of Fruit and Vegetable Product Technology, Institute of Agricultural and Food Biotechnology, Warsaw, Poland

Justyna Szczepańska
Department of Fruit and Vegetable Product Technology, Institute of Agricultural and Food Biotechnology, Warsaw, Poland

Igor Tomasevic
Faculty of Agriculture, University of Belgrade, Belgrade, Serbia

Artur Wiktor
Faculty of Food Sciences, Department of Food Engineering and Process Management, Warsaw University of Life Sciences (WULS-SGGW), Warsaw, Poland

Dorota Witrowa-Rajchert
Faculty of Food Sciences, Department of Food Engineering and Process Management, Warsaw University of Life Sciences (WULS-SGGW), Warsaw, Poland

Łukasz Woźniak
Department of Fruit and Vegetable Product Technology, Institute of Agricultural and Food Biotechnology, Warsaw, Poland

Preface

Bioprocesses find many traditional or new applications in the agri-food, chemical, pharmaceutical, and environmental industries. Enhancing these processes for better production of microbial biomass and/or products has interested many scientists in the last two decades. One of the strategies consists of changing the medium composition or the fermentation parameters (e.g. oxygenation, agitation, temperature, etc.), most of the time via a design of experiment approach. Besides, some emerging technologies (e.g. pulsed electric fields, ultrasounds, high hydrostatic pressure, microwaves, etc.) have shown their efficiency to enhance the fermentation processes. These technologies when applied at high intensities cause cell disintegration and find their applications in bioprocesses, for example, to produce sugar monomers from lignocellulosic biomass. Nonetheless, their application at sublethal levels may induce stress of microorganisms and affect the microbial growth and the formation of the products during fermentation. The beneficial effects of microbial stimulation by emerging technologies include mainly the shortening of the fermentation time, the acceleration of the substrate consumption, and the increase of the microbial biomass.

This book covers the principles of conventional fermentation processes, the major microorganisms used in bioprocesses, their implementation in industrial fermentation processes, the medium condition changes, and the use of emerging technologies for enhancing the fermentation processes. Besides, the mechanisms of action of the above-mentioned emerging technologies are discussed.

This book is designed to assist scientists working on fermentation processes as well as those working in the food, nutraceutical, pharmaceutical, and beverage industries. The topics covered in this book are suitable for teaching in courses such as bioprocess technology, microbiology, new product development, and food processing.

We gratefully acknowledge the contribution of colleagues from all around the world, the cover designer Mohamed Messaoud, and the professional assistance provided by the staff of Wiley.

Mohamed Koubaa, Francisco J. Barba, and Shahin Roohinejad

1

Introduction to Conventional Fermentation Processes

Mohamed Koubaa

ESCOM, UTC, EA 4297 TIMR, Compiègne, France

1.1 Bioprocesses

Bioprocesses represent all the methods and techniques that use microbial, plant, or animal cells or their components (e.g. enzymes, proteins, genes, etc.) for the production of goods and services (Sindhu et al. 2017). Bioprocess technology is, in fact, an extension of the ancestral techniques used at the time to develop useful products (Kalaichelvan and Pandi 2019). Nowadays, microbial cells are not only used in common processes, such as for the production of alcoholic beverages (e.g. wine, beer, etc.) or dairy products (e.g. yogurt, cheese, etc.), but also to produce a wide diversity of complex molecules. In this sense, bioprocesses find many traditional or new applications in the following industries:

- Agri-food industry: production of animal proteins, amino acids, fermented foods and beverages, vitamins, enzymes, etc.
- Chemical industry: production of organic acids, ethanol, solvents, polymers, biogas, etc.
- Pharmaceutical industry: production of antibodies, vaccines, hormones, plasmids, steroids, etc.
- Environmental industry: decontamination of wastewater, air, and soil; development of agricultural and industrial by-products, etc.

In this respect, bioprocesses are exploited in three specific fields: fermentation processes, animal and plant cell cultures, and environmental bioprocesses. This chapter will mainly focus on conventional fermentation processes.

Most of the methods and techniques used in bioprocesses are based on fermentation technology. This is not surprising since the first ancestral processes were based on microbial fermentation. For most people, fermentation simply refers to the production of alcohol (beer and wine) or the deterioration of food by microorganisms (curd). Nevertheless, the word fermentation takes on a broader common industrial meaning. It is any process for producing a substance or biomass of cells on a large scale by using the culture of a microorganism, in aerobic or anaerobic conditions.

Fermentation Processes: Emerging and Conventional Technologies, First Edition.
Edited by Mohamed Koubaa, Francisco J. Barba, and Shahin Roohinejad.

To be able to carry out these fermentations, it is imperative to cultivate microorganisms in tanks equipped with a certain number of more or less sophisticated systems; these tanks are called *fermenters* or *bioreactors*. Their role is to provide a controlled environment for optimal growth of microbial cells throughout the culture by constantly stirring the medium, infusing sterile air – in the case of aerobic fermentation – and controlling the temperature and pH of the fermentation broth. By using these tanks, contamination by other microorganisms is avoided by constantly maintaining asepsis conditions.

The first modern fermenters were designed in the 1950s to support the industrial production of penicillin and other newly discovered antibiotics. Since then, they have been able to control several other types of crops and to substantially increase the quantity of products marketed in each of the three fields of application of industrial bioprocesses mentioned above. Six major groups of products could then be obtained by fermentative processes, namely the production of (i) microbial biomass, (ii) microbial metabolites, (iii) microbial enzymes, (iv) recombinant proteins, (v) microbial plasmids, and (vi) bioconversion.

1.1.1 Production of Microbial Biomass

Commercial production of microbial biomass can be divided into two major processes: the production of viable microorganisms used primarily for fermentative applications (Vitorino and Bessa 2017) and the production of microbial cells, usually dead, that can serve as protein-rich supplements (Matassa et al. 2016).

In the first case, we can cite several examples: the production of bakery yeasts for the production of bread, the production of yeasts to perform alcoholic fermentation (e.g. beers, wines, spirits, etc.), and the production of lactic acid bacteria for the manufacturing of cheese, yogurt, fermented meats (i.e. sausages), or fermented vegetables (e.g. sauerkraut, marinated pickles, etc.). Some food supplements composed of live lactic acid bacteria, also called *probiotics*, are produced by fermentation. They can be defined as live microorganisms, and the adequate amounts of them supply a health benefit to the host (Otles and Ozyurt 2019). Their role is to exert a beneficial effect by improving the quality of the intestinal flora. These microorganisms are usually supplied as a lyophilized powder in hermetically sealed sterile bags or containers. Generally, the name of ferments is given to microorganisms that serve to start a fermentation process (Koutinas 2017). Some microbial strains such as the bacterium *Bacillus thuringiensis*, whose spores produce a very effective toxin against pest larvae (biological insecticide), are also grown.

In the second case, it is a question of producing microbial biomass to exploit the nutritional potential of the proteins that it produces (Matassa et al. 2016). This biomass is incorporated into prepared foods to increase their protein content without significant fat intake, which improves their nutritional quality. The yeast *Candida utilis* is mostly used as a dietary supplement because of its exceptionally high protein content (50–55% of dry weight). This yeast can be used as a valuable raw material to produce various preparations enriched with valuable bioelements (e.g. selenium, magnesium, etc.). The use of such preparations in the human diet provides an interesting alternative to classical, pharmacological supplementation and prevents deficits of important elements, while their addition to feedstock significantly improves the results of animal production (Kieliszek et al. 2017).

1.1.2 Production of Microbial Metabolites

Microorganisms are characterized by a variety of metabolic pathways that allow them to synthesize a host of organic compounds, called *metabolites*, many of which are potentially useful. In this type of fermentation, it is sought to produce by the metabolic activity of a microorganism a substance that is too complex to be chemically synthesized at a reasonable cost (Jeandet et al. 2013).

Metabolites are generally divided into two categories depending on whether they are produced in relation to growth or not. The first ones are called *primary metabolites* and are produced in large enough quantities during the exponential growth phase by essential metabolic pathways that are common to many microorganisms. Several primary metabolites produced by fermentation are the residues of the energetic catabolism of microorganisms. These are mainly alcohols, solvents, and organic acids used in food or the chemical industry. Others are derived from cellular anabolism. These are mostly amino acids and vitamins produced for food or pharmaceutical purposes (Sanchez and Demain 2009).

The second ones are called *secondary metabolites* and are usually produced in small quantities during the stationary phase, and sometimes even during the decline phase, by particular metabolic pathways that are exclusive to a few species and usually give them a survival advantage in the wild. Secondary metabolites form an extremely heterogeneous group of compounds, derived from anabolism, whose main uses are in pharmaceuticals (e.g. antibiotics, growth factors, enzyme inhibitors, etc.). Although not essential for microbial growth, secondary metabolites are very important for health, nutrition, and the economics of our societies (Bérdy 2005).

1.1.3 Production of Microbial Enzymes

Enzymes are proteins that act as catalysts in the biochemical reactions of metabolism (Cooper 2000). When purified, they make it possible to carry out these reactions under controlled conditions. They can be produced from animal, plant, and microbial cells. Nevertheless, microbial enzymes stand out in large quantities and often at low cost by fermentation processes (Raveendran et al. 2018). Most enzymes produced by fermentation are associated with primary metabolism and are primarily used in the agri-food industry to process many foods; however, more and more enzymes associated with secondary metabolism are produced for pharmaceutical purposes.

1.1.4 Production of Recombinant Proteins

Nowadays, the advances in genetic engineering techniques allow introducing genes from animal and plant cells into microorganisms. These genetically modified cells will produce the so-called *recombinant proteins* because their synthesis relies on the recombination of microbial DNA with foreign DNA (Griffiths et al. 2000). Several microbial species have been selected as hosts for such productions (e.g. *Escherichia coli*, *Saccharomyces cerevisiae*, *Yarrowia lipolytica*, etc.). The research and development effort required to develop such strains is, however, colossal, and the recombinant proteins produced by fermentation are therefore almost all dedicated to pharmaceutical uses (e.g. human insulin, human growth hormone, etc.).

1.1.5 Production of Microbial Plasmids

There has been a marked interest in the production of plasmids by fermentation (Carnes and Williams 2014; Carnes et al. 2006). Plasmids are self-replicating extrachromosomal DNA molecules found in Gram-negative and Gram-positive bacteria as well as in some yeast and other fungi (Actis et al. 1999). To produce appreciable amounts, it is first necessary to introduce a plasmid of interest into microbial cells such as *E. coli* or *S. cerevisiae*. Subsequently, culturing these microorganisms in a bioreactor develops significant biomass. The plasmids then replicate independently in the new cells produced, and at the end of the fermentation, they are recovered and purified (Carnes and Williams 2014).

First developed in research programs in molecular biology and genetic engineering, plasmids are todays used in new applications of high technology such as in gene therapy (Sousa et al. 2009). Indeed, plasmids obtained by fermentation may contain therapeutic genes derived from the recombination of DNA that will be used to produce previously defective or nonexistent proteins to correct a genetic abnormality in a human organ. These plasmids are inserted into synthetic vectors and injected into the target cells of the affected organ by using particular techniques.

1.1.6 Bioconversion

A microbial cell can be used to convert or transform any substance into a value-added product (Garlapati et al. 2016), a bit like conventional conversions of grape must into wine, wine into vinegar, or milk into yogurt. These transformations contribute to producing very valuable compounds in the pharmaceutical industry, such as antibiotics, vitamins, steroids, and prostaglandins. These conversions are based on the biochemical reactions of microorganisms used such as hydroxylation, dehydroxylation, O-methylation, O-demethylation, glycosylation, deglycosylation, dehydrogenation, hydrogenation, C-ring cleavage of the benzo-γ-pyrone system, cyclization, and carbonyl reduction (Cao et al. 2015).

The bioconversion of compounds by microorganisms is much more advantageous than the conventional chemical transformation because the reactions can occur at low temperature and low pressure and without the addition of catalysts. To understand the different steps required to carry out a fermentation process, it is of paramount importance to understand first the microbial metabolism and how a substrate could be transformed by a microorganism to maintain its growth and the production of targeted compounds.

1.2 Energetic Metabolism

Microorganisms need energy and carbon for their metabolism and are classified as autotrophs and heterotrophs depending on the sources of energy and nutrients (Misra 2011). There are only two sources of energy metabolizable by the cells: light energy captured during photosynthesis and energy from the oxidation of organic and inorganic molecules. Nevertheless, cells can be categorized into nutritional categories depending on how they meet these needs. In bioprocesses, three of these classes are potentially exploited on an industrial scale: phototrophs, chemolithotrophs, and chemoorganotrophs (Jurtshuk 1996).

Phototrophs use light as a source of energy and carbon dioxide (CO_2) as a source of carbon. They include photosynthetic bacteria (cyanobacteria), algae, and green plants. Chemolithotrophs rely on electrons from reduced inorganic compounds, such as iron, nitrogen, or sulfur, as a source of energy (oxidation of the inorganic material) and CO_2 as a carbon source. They include several bacterial species that are primarily used in environmental bioprocesses, particularly in aerobic wastewater treatment. The chemoorganotrophs use, as a source of energy, electrons from hydrogen atoms that are part of organic compounds (oxidation of organic matter), which also serve as a carbon source. They are the ones who enter the vast majority of bioprocesses, particularly in fermentation processes (bacteria, yeasts, and molds) and in animal cell cultures. The next sections will discuss the metabolism of this class of cells.

All living organisms need energy to grow and reproduce. In chemoorganotrophs, this energy is obtained during the degradation of organic compounds. Mainly, carbohydrates, lipids, and proteins are oxidized to release the chemical energy they contain. This energy will be then transferred to adenosine diphosphate (ADP) and inorganic phosphate (P_i) molecules before being stored in the form of adenosine triphosphate (ATP) (Figure 1.1). ADP and ATP correspond to molecules of adenosine monophosphate (AMP) plus one and two high-energy phosphates (AMP ~ P and AMP ~ P ~ P, respectively). The energy is stored in these compounds as high-energy phosphate bonds. All living cells must maintain steady-state biochemical reactions for the formation and use of such high-energy compounds.

Biochemical assimilation (anabolism) and dissimilation (catabolism) of nutrients by a chemoorganotroph are mediated by a network of enzymatic reactions perfectly synchronized and regulated. Anabolic pathways consist mainly of the reductive processes that lead to producing new cellular molecules, while catabolic pathways include the oxidative processes involved in removing electrons from substrates or intermediates that are used to generate energy.

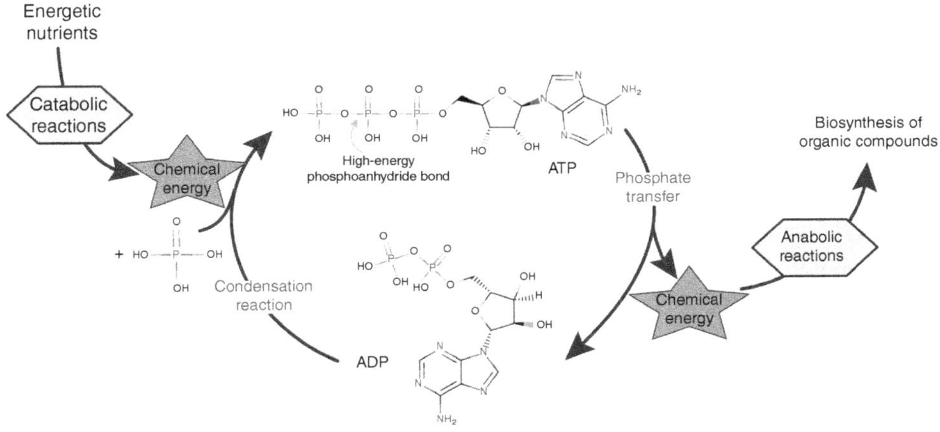

Figure 1.1 Energy coupling and the role of ATP in microbial metabolism. ADP: adenosine diphosphate, ATP: adenosine triphosphate

1.2.1 Energy Transfer and Redox Reactions

The energy released by the catabolic reactions is associated with the electrons of molecules that are degraded during these reactions. This energy is transferred to a molecule of ADP to form ATP. More specifically, a phosphate group is added to the ADP molecule, with an energy investment, to form an ATP molecule (Figure 1.1). The energy contained in organic molecules cannot be released all at once; otherwise, it would be practically all lost in the form of heat, as during combustion. It must, therefore, be gradually transferred to the ATP molecules via the cascades of redox reactions.

First, the energetic nutrients are oxidized, which allow them to behave as electron donors (e^-). During this reaction, two electrons and two protons (H^+) are transferred to a coenzyme known as nicotinamide adenine dinucleotide (NAD^+) to be reduced in the form of $NADH + H^+$. A part of the chemical energy contained in nutrients is then transferred to the NAD^+ by the electrons, according to the reduction reaction in Eq. (1.1).

$$NAD^+ + 2H^+ + 2e^- \rightarrow NADH + H^+ \tag{1.1}$$

Overall, each time an organic molecule is oxidized (the loss of electrons and H^+ ions), there is simultaneously a reduction of NAD^+ taking place. This is why we talk about "redox reactions." The newly formed NADH will then undergo oxidation, in turn, to release stored energy (Eq. (1.2)), which will eventually be transferred to ATP molecules by various chemical processes.

$$NADH + H^+ \rightarrow NAD^+ + 2H^+ + 2e^- \tag{1.2}$$

At the end of the energy transfer process, the released electrons and protons (H^+) must be picked up by a final acceptor. This acceptor will vary according to the preferred catabolic pathway: aerobic respiration, anaerobic respiration, or fermentation.

1.2.2 Aerobic Respiration

In this catabolic pathway, the final electron acceptor is molecular oxygen (O_2), and the organisms using it are, therefore, dependent on air for their survival. It takes place in three stages, each involving a series of chemical reactions: glycolysis, citric acid cycle, and electron transport chain.

1.2.2.1 Glycolysis Pathway

In the glycolytic pathway, occurring in all tissues, glucose is oxidized to provide energy (i.e. ATP) and intermediates for other cellular metabolic pathways. Glycolysis is at the hub of carbohydrate metabolism where glucose is converted to pyruvate following a series of 10 enzymatic reactions (Figure 1.2). This metabolic pathway is known as *aerobic glycolysis*, as the reoxidation of the NADH formed during the oxidation of glyceraldehyde 3-phosphate requires O_2. Aerobic glycolysis sets the stage for the oxidative decarboxylation of pyruvate to acetyl coenzyme A (acetyl-CoA), a major fuel of the tricarboxylic acid cycle (Ferrier 2017).

During glycolysis, two ATP molecules are initially consumed to phosphorylate glucose, which thus receives an essential energy supply to continue the catabolic pathway.

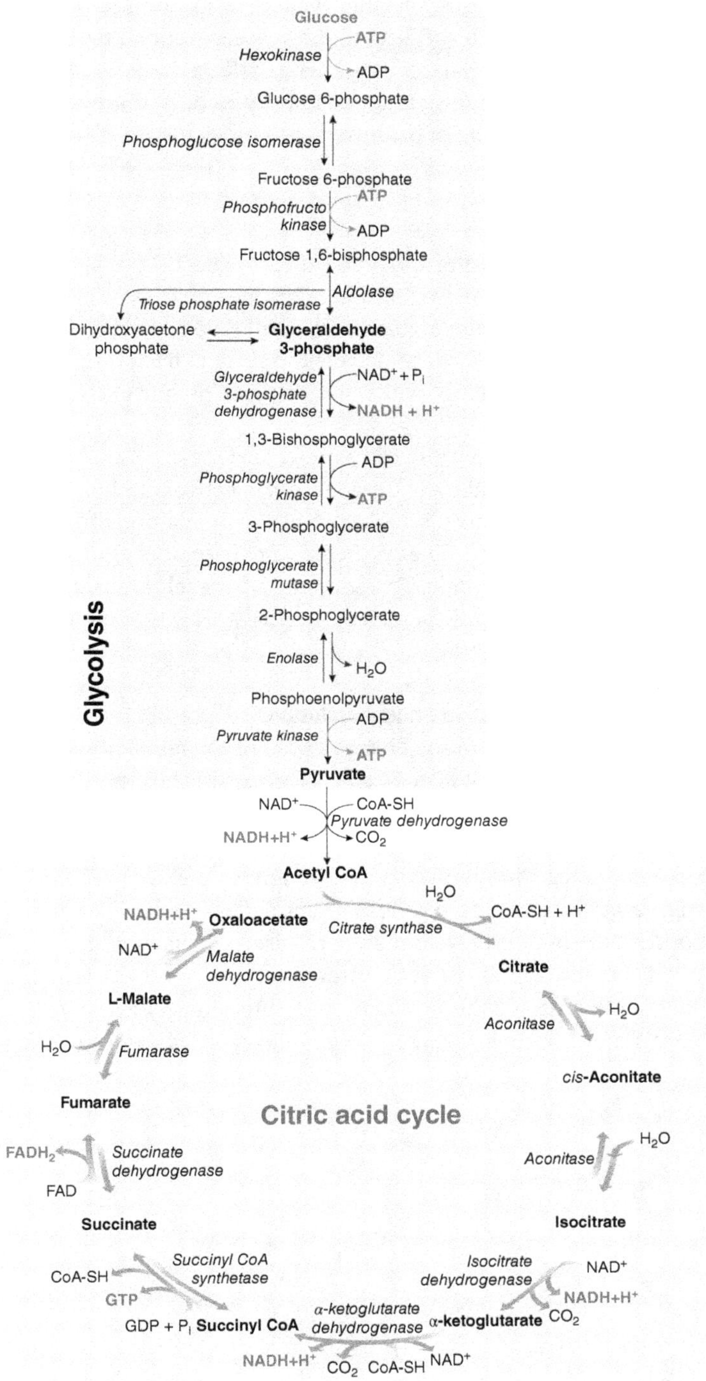

Figure 1.2 Glycolysis and citric acid cycle pathways.

Subsequently, two molecules of NAD^+ oxidize phosphorylated sugar and are reduced to $NADH + H^+$. During these reactions, which will lead to the synthesis of two molecules of pyruvic acid from a glucose molecule, a part of the energy released allows the direct synthesis of four molecules of ATP. Considering that two molecules of ATP are consumed and four are produced, glycolysis presents a net balance of two molecules of ATP for each molecule of oxidized glucose.

1.2.2.2 Citric Acid Cycle

Pyruvic acid is first decarboxylated (release of CO_2) and then oxidized by a molecule of NAD^+ (reduced to $NADH + H^+$) to form the two carbon molecules of acetyl-CoA. As two molecules of pyruvic acid are produced from a glucose molecule, two molecules of CO_2, $NADH + H^+$, and acetyl-CoA are produced. In prokaryotic cells (e.g. bacteria), the citric acid cycle (or Krebs cycle) occurs in the cytosol, while in eukaryotic organisms (e.g. yeast), it takes place in the mitochondrial matrix.

Overall, two molecules of CO_2 are produced and three molecules of NAD^+ are reduced to $NADH + H^+$ during a turn of the citric acid cycle. Besides, two electrons and two protons released from pyruvic acid are used to reduce a molecule of flavin adenine dinucleotide (FAD) into $FADH_2$, similar to that of NAD^+. Finally, there is a release of chemical energy, which allows the synthesis of an ATP molecule. To summarize, for each molecule of glucose oxidized, four molecules of CO_2, six molecules of $NADH + H^+$, two molecules of $FADH_2$, and two molecules of ATP are generated.

1.2.2.3 Electron Transport Chain and Oxidative Phosphorylation

In a process called *oxidative phosphorylation*, the energy stored in the reduced coenzymes (NADH and $FADH_2$) is released to produce ATP molecules. During this process, each coenzyme is oxidized and gives its two energy-rich electrons to an electron transport chain, or respiratory chain, located in the plasma membrane in prokaryotes and the inner membrane of mitochondria in eukaryotes. The chain consists of complex organic acceptors that transfer electrons through a series of redox reactions (Figure 1.3). As the electrons are transferred in this chain, the energy that they contain is released by stages and is used to activate chemiosmosis. This creates a gradient of H^+ protons whose energy is reinvested, through the enzyme ATP synthetase, in the production of three molecules of ATP per molecule of oxidized NADH and two molecules of ATP per molecule of oxidized $FADH_2$.

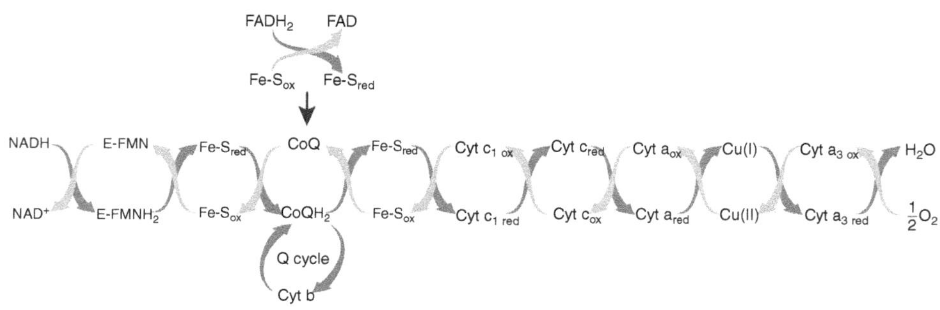

Figure 1.3 The electron transport chain showing the respiratory complexes.

At the end of the transport chain, the electrons exhausted of their energy bond with O_2 and H^+ ions to form a water molecule. To summarize, for each glucose molecule that enters glycolysis, 34 molecules of ATP will be produced by oxidative phosphorylation (i.e. 30 molecules coming from NADH and 4 coming from $FADH_2$). Further reading about the electron transport chain could be found in Campbell (2015).

1.2.3 Anaerobic Respiration

Some microbial species use anaerobic respiration to obtain their energy in the absence of O_2. During this process, the electrons that are removed from organic nutrients such as glucose follow the same pathways as in aerobic respiration, except that the final acceptor is not O_2, but another inorganic molecule (e.g. sulfate, nitrate, etc.). The sulfate ion is generally reduced to hydrogen sulfide, while nitrate ion can be reduced to nitrite, nitrogen oxide, or molecular nitrogen. Some bacteria reduce carbonate to methane. The number of ATP molecules produced by anaerobic respiration varies from one organism to another and from one metabolic pathway to another. This number is generally less than the 38 mol of ATP generated by aerobic respiration, the energy yield is lower, and anaerobic microorganisms usually grow slower than aerobic ones.

1.2.4 Fermentation

Some microbial species can obtain their energy in the absence of O_2 through the catabolic pathway of fermentation. The only difference compared to respiration is in the final electron acceptor (Angelidaki et al. 2011; Dunford 2012; Madigan et al. 2015). In this case, ATP is produced without the Krebs cycle or an electron transport chain involved. This metabolic pathway does not require O_2, because ATP comes exclusively from glycolysis, and the last electron acceptor is an organic molecule such as pyruvic acid (or a derived molecule). Figure 1.4 summarizes the metabolic pathway of fermentation.

As shown in Figure 1.4, glucose is oxidized during glycolysis to form two molecules of pyruvate. The electrons and protons released during this pathway are captured by the NAD^+ to be reduced to NADH + H^+. As shown above, two molecules of ATP are produced during glycolysis. To regenerate the NAD^+, the NADH + H^+ must be reoxidized; otherwise, the oxidation of glucose will stop and glycolysis too. During this oxidation, electrons and protons are directly transferred to pyruvate or one of its derivatives. The reduction of these final electron acceptors results in the formation of many different compounds, which provide a great variety of types of fermentation. At the same time, the NAD^+ is regenerated and can engage in another round of glycolysis. The goal is to provide an uninterrupted supply of NAD^+, which allows uninterrupted oxidation of glucose.

During fermentation, all ATP is produced solely by glycolysis, which implies a much lower energy yield compared to aerobic respiration (2 mol of ATP against 38 in prokaryotes). Considering that glucose oxidation is partial, a large part of the energy originally contained in glucose remains stored in the chemical bonds of the final fermentation product (e.g. ethanol, lactic acid, etc.). Fermentation microorganisms must, therefore, compensate for this shortfall by the oxidation of a larger quantity of substrate.

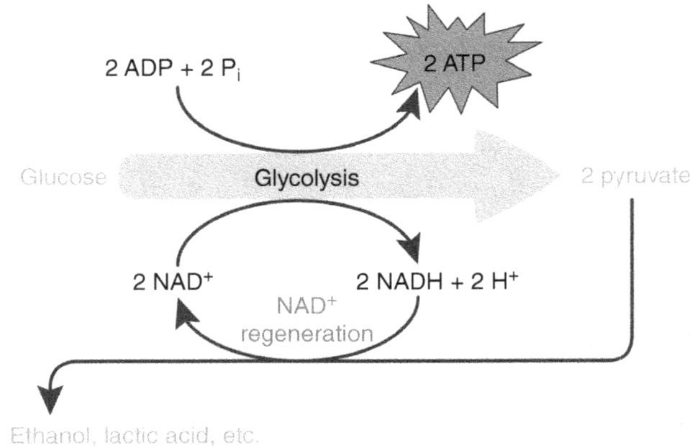

Figure 1.4 Schematic representation of fermentation and energy generation.

Different microorganisms can metabolize organic substrates (e.g. monosaccharides, amino acids, glycerol, etc.) (Angelidaki et al. 2011) to produce organic products such as acids, alcohols, and gases (Wilkins and Atiyeh 2012). This transformation occurs when all essential conditions and factors (e.g. temperature, pH, sugar concentration, culture medium, dissolved O_2, and other micronutrients) required to the growth of microorganisms are provided (Smith 2009). Besides, the microorganism used must be viable, genetically stable, and able to resist several factors including the high concentrations of substrate, salt, and product, and sometimes to the presence of inhibitors, especially when using an industrial by-product, or a lignocellulosic hydrolyzate.

Under anaerobic conditions, pyruvate is fermented to a wide range of fermentation products; most of them are of industrial importance (Figure 1.5). In general, microorganisms can be grouped into two classes according to the number of compounds that will be produced during fermentation: homofermentaries and heterofermentaries. Homofermentative microorganisms use a fermentation pathway that leads to the production of a single compound. For example, some lactic acid bacteria can oxidize glucose to produce only lactic acid. Heterofermentative microorganisms, on the other hand, use a fermentation pathway that generates several compounds after the oxidation of the substrate. For example, the bacterium *E. coli* can produce several types of organic acids during the fermentation of glucose. The following are among the most occurring industrial fermentative products: ethanol, lactic acid, propionic acid, butyric acid, and acetone (Eş et al. 2017; Lin et al. 2014; Navarrete-Bolaños et al. 2013; Ruijschop et al. 2008).

1.3 Microorganisms Used in Fermentation Processes

Fermentation can take place under anaerobic or aerobic conditions with the help of microorganisms (Gänzle 2015; Ghosh et al. 2015; Kutyna et al. 2012). The most used microorganisms in bioprocesses are bacteria and fungi (yeasts and molds).

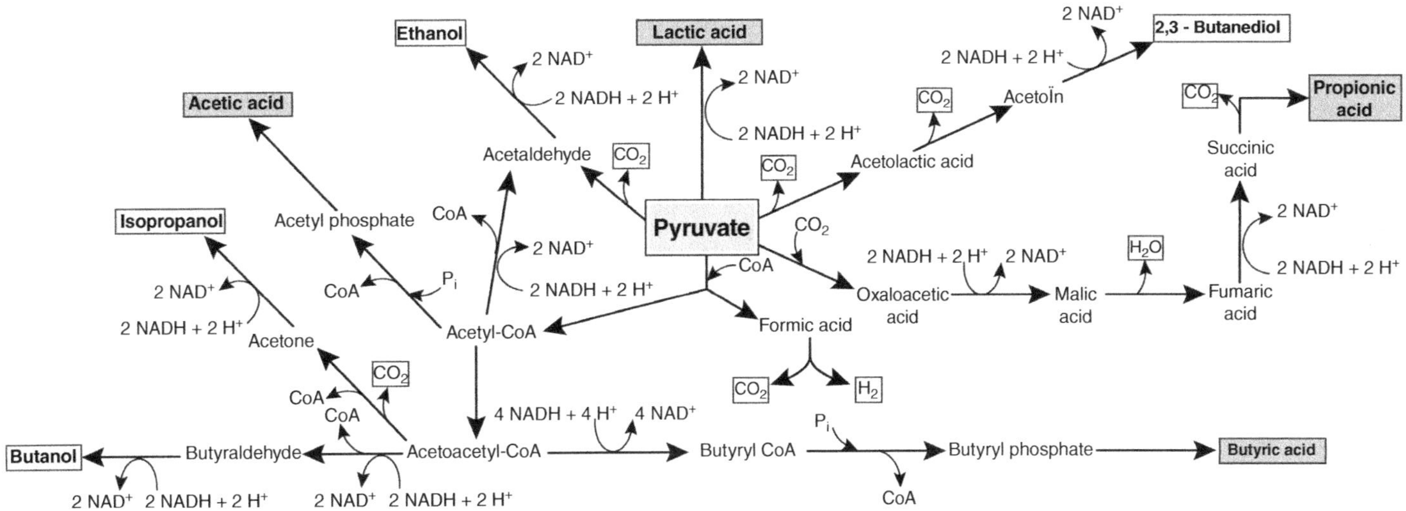

Figure 1.5 Main terminal reactions of catabolic fermentations using pyruvate.

1.3.1 Bacteria

Bacteria are prokaryotes (Figure 1.6a) and include two distinct categories: archaebacteria or archaea ("ancient" bacteria) and eubacteria ("true" bacteria). The group containing almost all the species used in industrial bioprocesses is eubacteria (Waites et al. 2001) and will, therefore, be detailed below.

Eubacteria group could be divided into 12 subgroups: (i) Proteobacteria; (ii) Gram-positive eubacteria; (iii) Cyanobacteria; (iv) Chlamydia; (v) Planctomyces and Pirella; (vi) Bacteroides and Flavobacteria; (vii) Green sulfur bacteria; (viii) Spirochetes and relatives; (ix) Deinococci, radioresistant micrococci, and relatives; (x) Green nonsulfur bacteria and anaerobic phototrophs; (xi) Thermotoga and Thermosulfobacteria; and (xii) Aquifex. Only two of them are of industrial interest: the proteobacteria and the Gram-positive eubacteria.

1.3.1.1 The Proteobacteria

This subgroup constitutes a major kingdom of Gram-negative bacteria that include purple photosynthetic and nonphotosynthetic bacteria such as the Enterobacteriaceae (e.g. *E. coli*), along with *Pseudomonas*, *Hyphomicrobium*, *Thiobacillus*, *Nitrobacter*, and *Vibrio* (Waites et al. 2001).

1.3.1.2 The Gram-Positive Eubacteria

This subgroup could be divided into two major subdivisions. The first one corresponds to the bacteria having low guanine (G) and cytosine (C) base pair content in their DNA and includes *Bacillus*, *Staphylococcus*, *Leuconostoc*, *Streptococcus*, *Clostridium*, *Lactobacillus*,

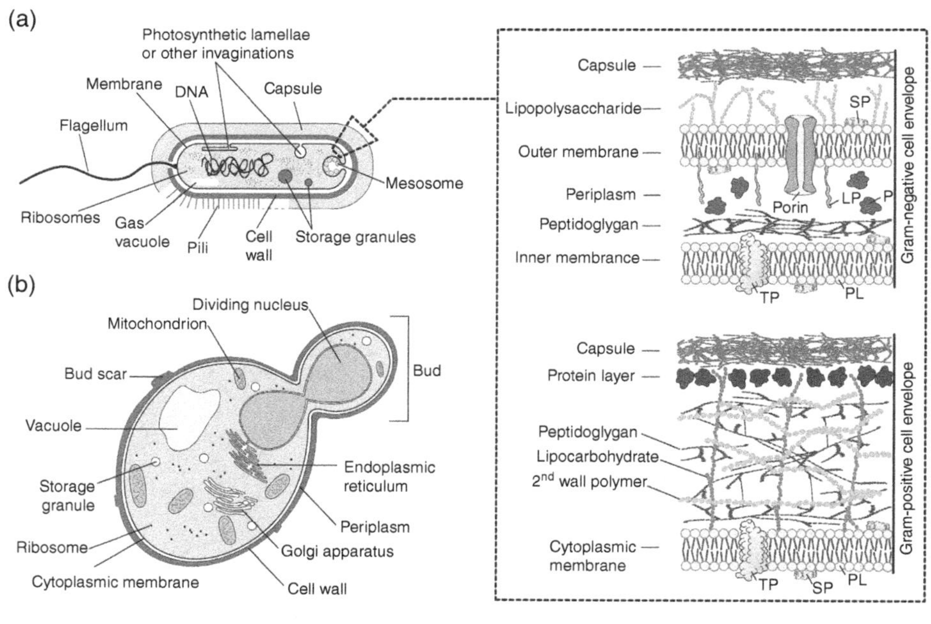

Figure 1.6 Schematic representation of a prokaryotic cell (a) and a budding yeast cell (b). Insert represents a schematic structure of typical Gram-negative and Gram-positive cell envelopes. LPS = lipopolysaccharide; LP = lipoprotein; P = protein; PL = phospholipid; SP = surface protein; TP = transmembrane proteins.

and *Mycoplasma*. The second subdivision corresponds to the bacteria having high guanine (G) and cytosine (C) base pair content in their DNA and includes the actinomycetes (filamentous bacteria, e.g. *Streptomyces*), *Mycobacterium*, *Micrococcus*, and *Corynebacterium*.

The physiological difference between Gram-negative and Gram-positive bacteria is mainly in the composition of their cell envelopes (see insert in Figure 1.6).

1.3.2 Fungi

Fungi are eukaryotic microorganisms that could be divided into filamentous hyphae (also called *molds*) and unicellular fungi (also called yeasts). A relatively low number of filamentous fungi are used at industrial scale in bioprocesses (e.g. *Acremonium*, *Agaricus*, *Aureobasidium*, *Aspergillus*, *Claviceps*, *Coniothyrium*, *Curvularia*, *Cylindrocarpon*, *Fusarium*, *Lentinus*, *Mortierella*, *Mucor*, *Paecilomyces*, *Penicillium*, *Rhizomucor*, *Rhizopus*, *Sclerotium*, *Trametes*, *Trichoderma*, and *Trichosporon*). Filamentous fungi are chemoheterotrophs and nonphotosynthetic. Most of them secrete a wide variety of hydrolytic enzymes (e.g. cellulose, amylase, xylanase, etc.) (de Souza and de Oliveira Magalhães 2010; Payne et al. 2015; Polizeli et al. 2005) that can degrade different polymers (e.g. lignocellulosic materials) into smaller molecules (e.g. monosaccharides, disaccharides, etc.), which are easily absorbed and metabolized. Filamentous fungi germinate from either individual spore or a fragment of hyphae when exposed to suitable environmental conditions (e.g. pH, temperature, etc.). The length of hyphae can increase rapidly at rates reaching several μm/min (Waites et al. 2001). Further reading about the filamentous fungi can be found in Quintanilla et al. (2015).

Yeasts are unicellular fungi (Figure 1.6b) with great industrial importance, notably *S. cerevisiae*, the major yeast used in alcoholic fermentation (see Chapter 2). Yeasts are heterotrophic, and most of them can grow in the presence and absence of O_2 (facultative anaerobes), unlike most fungi. Yeasts are not nutritionally demanding as a relatively simple medium composition (e.g. reduced carbon sources, organic and inorganic nitrogen sources such as urea and ammonium salts, respectively, some minerals, and water) allows them to multiply. Sometimes, vitamins (e.g. biotin) are also supplemented to allow the optimal growth of the yeasts. Yeasts of industrial importance include the genera of *Blakeslea*, *Candida*, *Hansenula*, *Kluyveromyces*, *Pachysolen*, *Phaffia*, *Pichia*, *Rhodotorula*, *Saccharomyces*, *Xanthophyllomyces*, *Yarrowia*, and *Zygosaccharomyces* (Waites et al. 2001). Thousands of examples have been reported in the literature to describe the potential of these yeasts in different industrial sectors (Defavari do Nascimento and Pickering 2017; Drévillon et al. 2018; Koubaa et al. 2020; Peris et al. 2018).

The selection of a microbial species to perform a bioprocess does not only require its ability to synthesize a potentially useful compound or to carry out a particular metabolic pathway. Indeed, most of the industries seek strains that can meet other important criteria, which will allow the optimization of the biological process and maximize profitability. Such criteria include:

1) The ability of the strain to grow quickly on inexpensive organic substrates (e.g. molasses, corn liquor, whey, etc.);
2) The ability of the strain to perform in a simple and fast way the sought-after transformations with high efficiency and a minimum of energy consumption;

3) The strain must be genetically stable (low mutation rate) to maintain its production capacity over time;
4) The strain must be specialized in the synthesis of products that are easy to extract and separate; and
5) The strain should not be pathogenic.

Wild-type strains are usually unable to meet the above-mentioned criteria. Indeed, they often have a limited performance that must be amplified to reach the industrial requirements. In fact, wild-type microorganisms have usually metabolic regulation mechanisms, often of the negative feedback type, allowing them to produce naturally only the quantity of enzymes and metabolites they need to survive in a competitive environment. By short-circuiting these mechanisms with genetic modifications, more productive microorganisms could be generated. Besides, the wild strain may have certain undesirable characteristics that can also be modified, such as sensitivity to a bacteriophage, the trend to generate a lot of foam in a liquid medium, or the synthesis of a by-product that is difficult to eliminate during the downstream purification steps. In all cases, there is an interest in implementing a genetic improvement program for the strain involving the use of mutations or recombinant DNA technology (Figure 1.7). Further reading about microbial transformation techniques can be found in Han (2004). The review of the literature showed hundreds of examples of genetic modifications of microbial strains toward their

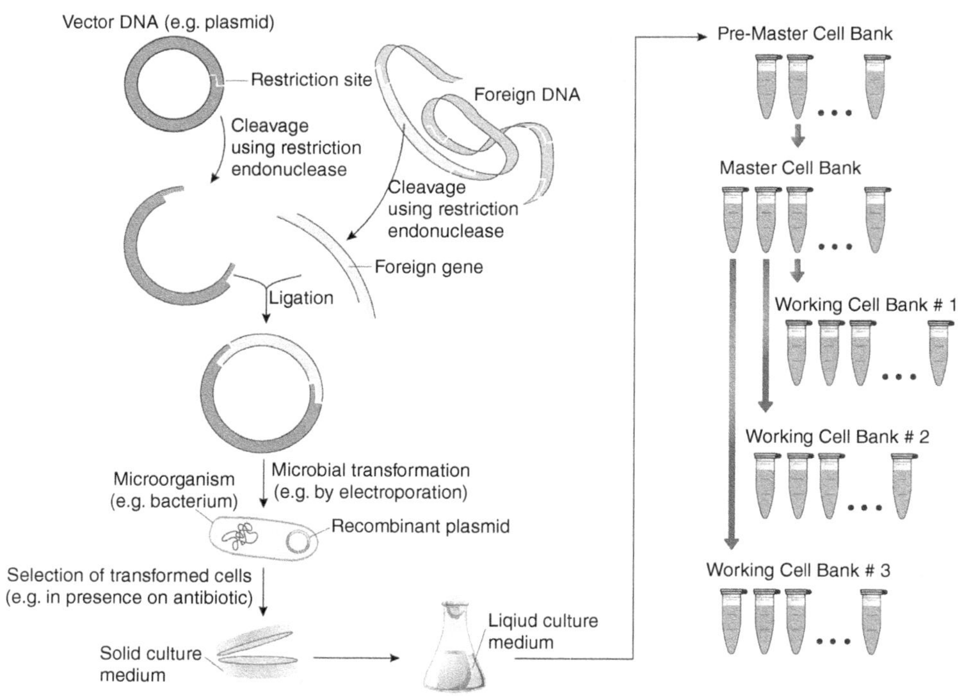

Figure 1.7 Diagram of a typical gene cloning methodology and cell banking system preparation.

use in industrial bioprocesses. The major metabolic engineering achievements in the recent years include microbial production of amino acids (e.g. L-valine, L-threonine, L-lysine, and L-arginine), bulk chemicals (e.g. 1,4-butanediol, 1,4-diaminobutane, 1,5-diaminopentane, 1,3-propanediol, butanol, isobutanol, and succinic acid), and drugs (e.g. artemisinin) (Lee and Kim 2015).

1.4 Fermentation Technology

The main challenge facing bioprocess specialists is the technological transfer from the laboratory to the industrial scale. Indeed, it is usually quite simple to cultivate a microorganism in an Erlenmeyer flask containing a few hundred milliliters of culture medium. However, the same operation is much more complicated in industry, where high culture volumes are processed. Nonetheless, whether in the laboratory or the industry, certain steps must be taken to cultivate a microorganism and make it produce a molecule of commercial interest.

In this regard, the fermentation industry provides a basic model in which a sequence of steps is commonly followed in the majority of bioprocesses. Whatever the strain used and the product sought, there are six major points to take into consideration when developing an industrial fermentation:

1) The formulation of an adequate culture medium that will promote the growth of the strain used and the production of the molecule of interest;
2) Sterilization of the culture medium, the bioreactor, and accessory components to prevent another microorganism developing there and contaminating the fermentation;
3) The development of an inoculum that will be used to produce a pure culture of the selected strain in sufficiently high concentration and volume to be able to adequately inoculate the bioreactor;
4) The growth of the strain within the bioreactor, under optimal and controlled production conditions;
5) Recovery and purification of the product from the culture (i.e. downstream processing steps); and
6) The elimination of effluents and other residues from the process.

The inoculum represents the biomass (e.g. cells, spores, mycelium, etc.) required for seeding the culture in a bioreactor. On one hand, it must provide a sufficient quantity of cells in homogeneous suspension to achieve rapid growth in the bioreactor, while, on the other hand, it must be free from any microbial contamination. To achieve this, an industrial lyophilized or frozen strain from a working cell bank (Figure 1.7) is used and grown in a liquid medium stirred in an Erlenmeyer flask to obtain a significant volume of culture. This volume will be then transferred aseptically to a small bioreactor, commonly called *preculture*, which has the function of developing a very high and metabolically active cellular biomass concentration. These stages may require several days of work and, in the case of fermentation, are carried out in very large volumes; it is often even a real fermentation before the main fermentation. Generally, the volume of the inoculum represents between

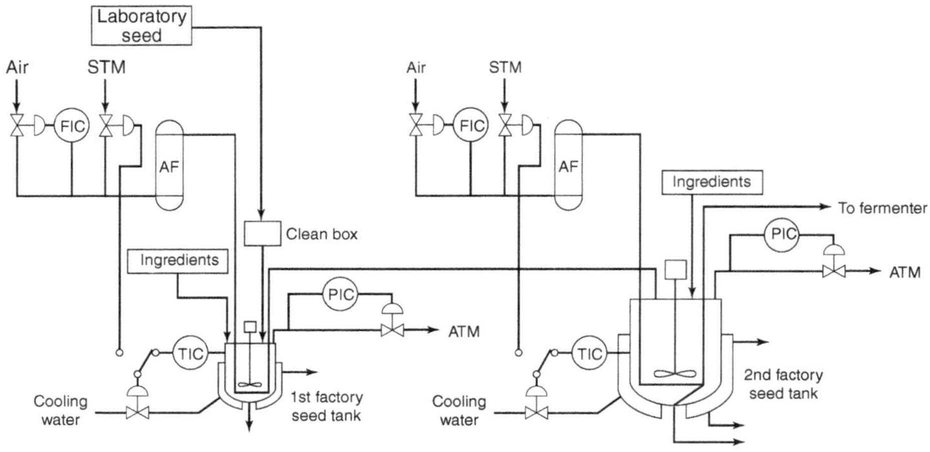

Figure 1.8 Factory seed culture equipment. AF = air filter; FIC = flow indication control; STM = steam; TIC = temperature indication control; PIC = pressure indication control; ATM = atmosphere. *Source:* Oka (1999).

5 and 10% of the volume of the main bioreactor where the fermentation will take place. Seeding steps at an industrial scale are presented in Figure 1.8.

In the field of industrial fermentation bioprocesses, developing a profitable process requires the consideration of some criteria for the culture medium:

1) The environment must be as inexpensive as possible (i.e. the cost of acquiring and storing raw materials must be affordable).
2) Raw materials should be available on an annual basis and ideally in the local market.
3) The quality of the raw materials must be constant, which allows obtaining similar results in terms of yield and productivity between the different batches.
4) The substrate must demonstrate good physicochemical stability during storage.
5) The medium must be easy to sterilize by usual techniques since certain media could be viscous or very heavily loaded with solid matter that affect the sterilization efficiency.
6) The medium must have an acceptable viscosity, since a too viscous medium is unfavorable for the aeration and homogenization of the substrate, in addition to causing a higher energy demand for agitation and a greater risk of formation of unwanted foam.
7) The raw materials must be able to guarantee the quality of the finished product (i.e. be free of toxic substances) and have the lowest possible content of impurities to facilitate the recovery and purification stages at the end of the process.

Any industrial fermentation aims to produce a molecule of interest in the highest quantity, in the shortest time, and at the lowest possible cost. For this, a microorganism must be cultivated under controlled physicochemical conditions within a large volume enclosure specially designed for this purpose: the bioreactor or fermenter. One of the most used bioreactors at the industrial scale is the stirred tank type, schematized in Figure 1.9. Other types of bioreactors exist and are described in Chapter 2.

Mainly, two types of bioreactors exist. The first one allows performing nonaseptic cultures such as brewing and effluent treatment, and the second one requires aseptic

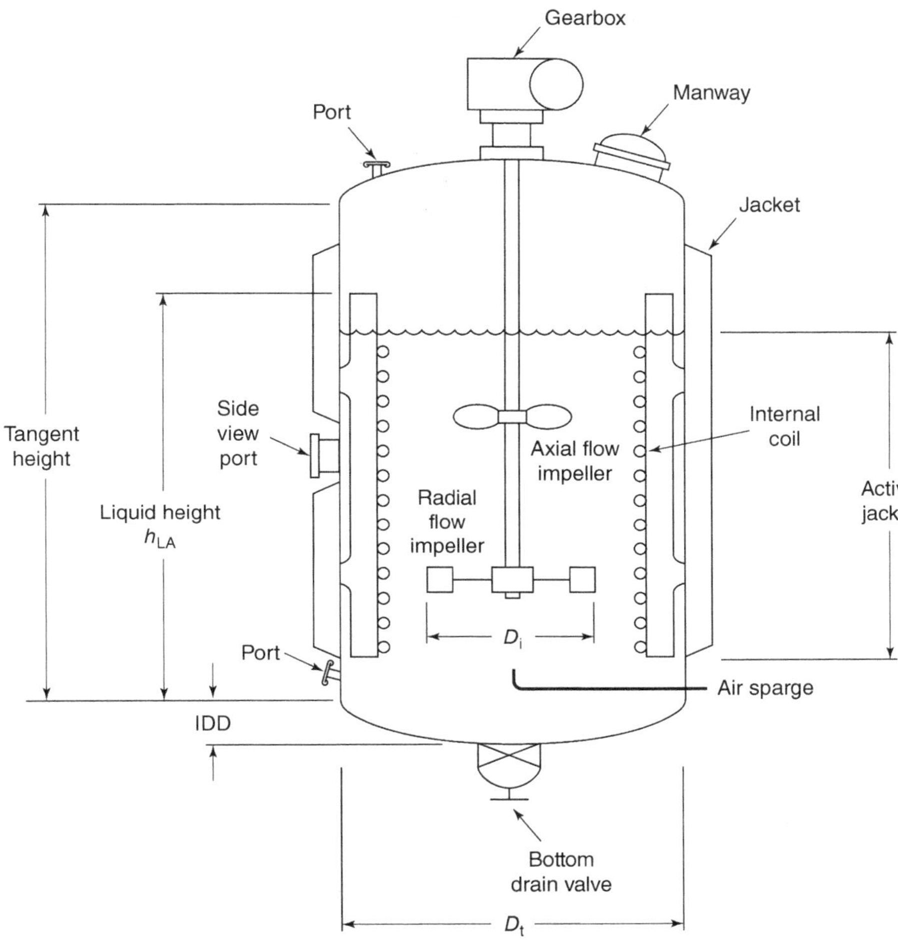

Figure 1.9 Fermenter vessel schematic and terminology. *Source:* Charles (1999).

conditions for successful product formation, such as antibiotics, vitamins, polysaccharides, and recombinant proteins. Aseptic bioreactors can be sterilized repeatedly and can process a large volume of culture. Several peripheral systems are coupled to the tank and allow controlling automatically the parameters affecting the microbial growth such as the temperature, the pH, and the dissolved O_2. Within the bioreactor, the microorganisms are suspended in the aqueous nutrient medium containing the necessary substrates for microbial growth and product synthesis. Changing the culture conditions has a great impact on the bioprocess performances, as discussed in Chapter 3.

In a bioreactor, fermentation occurs in a multiphase state: a gas phase where CO_2, O_2, and N_2 can be exchanged, a liquid phase mainly composed of the aqueous medium, and a solid phase mainly composed of microorganisms and solid substrates. These phases must be perfectly mixed to ensure the most efficient and homogenous heat and mass transfer. To ensure a perfect mixing of the suspension, the shaft of the bioreactor is usually equipped with several impellers. Although a wide range of impeller types exist in the market, the six flat-bladed (Rushton) turbine impellers are used in the majority of bioreactors. Usually, three to five impellers are mounted and spaced at intervals equivalent to one tank diameter along the shaft to avoid a swirling type of liquid movement.

An optimal production system could be achieved by taking a few requirements that include:

1) Avoiding the contamination of the culture, which can be achieved by designing a bioreactor where the inputs and outputs of the fermenter are controlled;
2) No leakage or water evaporation from the medium can occur;
3) Agitation of the medium should be perfect to allow a homogenous distribution of O_2 in the medium and the maintenance of O_2 concentration above the critical value; and
4) The fermentation parameters that include the pH, temperature, etc. must be automatically controlled.

A successful process occurs when all the parameters required for the growth and production of a molecule of interest are brought together. Yield and productivity constitute the two main aspects to be quantified during fermentation. They indicate the conversion efficiency of the substrate into cells and products. The following equations are given for culture in batch mode.

The yield in biomass $Y_{X/S}$ is given by Eq. (1.3) and corresponds to the mass (g) of dry cell biomass produced per gram of substrate consumed ($g\,g^{-1}$).

$$Y_{X/S} = \frac{X - X_0}{S_0 - S} \tag{1.3}$$

where X_0 and X correspond, respectively, to the initial and final biomass concentrations ($g\,l^{-1}$). S_0 and S correspond, respectively, to the initial and final carbon substrate concentrations ($g\,l^{-1}$).

The yield in product $Y_{P/S}$ is given by Eq. (1.4) and corresponds to the weight in grams of product produced per gram of substrate consumed ($g\,g^{-1}$).

$$Y_{P/S} = \frac{P - P_0}{S_0 - S} \tag{1.4}$$

where P_0 and P correspond, respectively, to the initial and final product concentrations ($g\,l^{-1}$). S_0 and S correspond, respectively, to the initial and final carbon substrate concentrations ($g\,l^{-1}$).

The productivity in biomass P_X is given by Eq. (1.5) and corresponds to the weight in grams of biomass produced per liter of culture per hour ($g\,l^{-1}\,h^{-1}$).

$$P_X = \frac{X_t - X_0}{t - t_0} \tag{1.5}$$

where X_0 and X_t correspond, respectively, to the initial biomass concentration and that at time t ($g\,l^{-1}$). t_0 and t correspond, respectively, to the initial and final time (h).

The productivity in product P_p is given by Eq. (1.6) and corresponds to the weight in grams of product produced per liter of culture per hour ($g\,l^{-1}\,h^{-1}$).

$$P_P = \frac{P_t - P_0}{t - t_0} \tag{1.6}$$

where P_0 and P_t correspond, respectively, to the initial product concentration and that at time t (g l^{-1}). t_0 and t correspond, respectively, to the initial and final time (h).

1.5 Conclusions

This chapter summarized the different steps to conduct a conventional fermentation with a special emphasis on energy metabolism, the different microorganisms used at industrial scale, as well as process parameters and performances. Nonetheless, the field of bioprocesses includes as well the culture of plant and animal cells, algae, and environmental processes. After fermentation, depending on the intracellular or extracellular localization of the synthesized product, cell lysis is required before starting the different downstream processing steps. Fermentation optimization could be performed using different approaches such as the use of the design of experiment approach and the implementation of emerging technologies to assist the fermentation.

References

Actis, L.A., Tolmasky, M.E., and Crosa, J.H. (1999). Bacterial plasmids: replication of extrachromosomal genetic elements encoding resistance to antimicrobial compounds. *Front. Biosci.* (4): D43–D62.

Angelidaki, I., Karakashev, D., Batstone, D.J. et al. (2011). Biomethanation and its potential. In: *Methods in Enzymology, Methods in Methane Metabolism, Part A* (eds. A.C. Rosenzweig and S.W. Ragsdale), 327–351. Academic Press.

Bérdy, J. (2005). Bioactive microbial metabolites. *J. Antibiot. (Tokyo)* 58: 1–26.

Campbell, M.K. (2015). *Biochemistry*. Belmont, CA: Cengage Learning Brooks/Cole. ed.

Cao, H., Chen, X., Jassbi, A.R., and Xiao, J. (2015). Microbial biotransformation of bioactive flavonoids. *Biotechnol. Adv.* 33: 214–223.

Carnes, A.E. and Williams, J.A. (2014). Plasmid fermentation process for DNA immunization applications. In: *DNA Vaccines: Methods and Protocols, Methods in Molecular Biology* (eds. M. Rinaldi, D. Fioretti and S. Iurescia), 197–217. New York, NY: Springer New York.

Carnes, A.E., Hodgson, C.P., and Williams, J.A. (2006). Inducible *Escherichia coli* fermentation for increased plasmid DNA production. *Biotechnol. Appl. Biochem.* 45: 155–166.

Charles, M. (1999). Fermenter design. In: *Encyclopedia of Bioprocess Technology: Fermentation, Biocatalysis, and Bioseparation* (eds. M.C. Flickinger and S.W. Drew), 1157–1189. New York / Chichester / Weinheim / Brisbane / Singapore / Toronto: Wiley.

Cooper, G.M. (2000). *The Cell: A Molecular Approach*. Washington, D.C.: ASM Press; Sinauer Associates.

de Souza, P.M. and de Oliveira Magalhães, P. (2010). Application of microbial α-amylase in industry – a review. *Braz. J. Microbiol.* 41: 850–861.

Defavari do Nascimento, D. and Pickering, W. (2017). *Plant-Based Genetic Tools for Biofuels Production*. Bentham Science Publishers.

Drévillon, L., Koubaa, M., and Vorobiev, E. (2018). Lipid extraction from *Yarrowia lipolytica* biomass using high-pressure homogenization. *Biomass Bioenergy* 115: 143–150. https://doi.org/10.1016/j.biombioe.2018.04.014.

Dunford, N.T. (2012). *Food and Industrial Bioproducts and Bioprocessing*. Chichester, West Sussex, UK: Wiley Blackwell.

Eş, I., Khaneghah, A.M., Hashemi, S.M.B., and Koubaa, M. (2017). Current advances in biological production of propionic acid. *Biotechnol. Lett.* 39: 635–645.

Ferrier, D.R. (2017). *Lippincott Illustrated Reviews: Biochemistry*, 7e. Philadelphia Wolters Kluwer.

Gänzle, M.G. (2015). Lactic metabolism revisited: metabolism of lactic acid bacteria in food fermentations and food spoilage. *Curr. Opin. Food Sci.*, Food Microbiology • Functional Foods and Nutrition 2: 106–117.

Garlapati, V.K., Shankar, U., and Budhiraja, A. (2016). Bioconversion technologies of crude glycerol to value added industrial products. *Biotechnol. Rep.* 9: 9–14.

Ghosh, K., Ray, M., Adak, A. et al. (2015). Microbial, saccharifying and antioxidant properties of an Indian rice based fermented beverage. *Food Chem.* 168: 196–202.

Griffiths, A.J.F., Miller, J.H., Suzuki, D.T. et al. (eds.) (2000). *An Introduction to Genetic Analysis*, 7e. New York: W.H. Freeman.

Han, L. (2004). Genetically modified microorganisms. In: *The GMO Handbook: Genetically Modified Animals, Microbes, and Plants in Biotechnology* (ed. S.R. Parekh), 29–51. Totowa, NJ: Humana Press.

Jeandet, P., Vasserot, Y., Chastang, T., and Courot, E. (2013). Engineering microbial cells for the biosynthesis of natural compounds of pharmaceutical significance. *Biomed. Res. Int.* 2013 https://doi.org/10.1155/2013/780145.

Jurtshuk, P. (1996). Bacterial metabolism. In: *Medical Microbiology* (ed. S. Baron). Galveston (TX): University of Texas Medical Branch at Galveston.

Kalaichelvan, P.T. and Pandi, I.A. (2019). *Bioprocess Technology*. MJP Publisher.

Kieliszek, M., Kot, A.M., Bzducha-Wróbel, A. et al. (2017). Biotechnological use of *Candida* yeasts in the food industry: a review. *Fungal Biol. Rev.* 31: 185–198.

Koubaa, M., Imatoukene, N., Drévillon, L., and Vorobiev, E. (2020). Current insights in yeast cell disruption technologies for oil recovery: a review. *Chem. Eng. Process. Process Intensif.* 150: 107868. https://doi.org/10.1016/j.cep.2020.107868.

Koutinas, A.A. (2017). Fermented dairy products. In: *Current Developments in Biotechnology and Bioengineering* (eds. A. Pandey, M.Á. Sanromán, G. Du, et al.), 3–24. Elsevier.

Kutyna, D.R., Varela, C., Stanley, G.A. et al. (2012). Adaptive evolution of *Saccharomyces cerevisiae* to generate strains with enhanced glycerol production. *Appl. Microbiol. Biotechnol.* 93: 1175–1184.

Lee, S.Y. and Kim, H.U. (2015). Systems strategies for developing industrial microbial strains. *Nat. Biotechnol.* 33: 1061–1072.

Lin, X., Fan, J., Wen, Q. et al. (2014). Optimization and validation of a GC–FID method for the determination of acetone-butanol-ethanol fermentation products. *J. Chromatogr. Sci.* 52: 264–270.

Madigan, M., Martinko, J., Bender, K. et al. (2015). *Brock Biology of Microorganisms*, 15e. Boston: Pearson.

Matassa, S., Boon, N., Pikaar, I., and Verstraete, W. (2016). Microbial protein: future sustainable food supply route with low environmental footprint. *Microb. Biotechnol.* 9: 568–575.

Misra, K.C. (2011). *Introduction to Geochemistry: Principles and Applications*. Wiley Blackwell.

Navarrete-Bolaños, J.L., Fato-Aldeco, E., Gutiérrez-Moreno, K. et al. (2013). A strategy to design efficient fermentation processes for traditional beverages production: prickly pear wine. *J. Food Sci.* 78: M1560–M1568.

Oka, T. (1999). Amino acids, production processes. In: *Encyclopedia of Bioprocess Technology: Fermentation, Biocatalysis, and Bioseparation* (eds. M.C. Flickinger and S.W. Drew), 89–100. New York / Chichester / Weinheim / Brisbane / Singapore / Toronto: Wiley.

Otles, S. and Ozyurt, V.H. (2019). Probiotic and prebiotic beverages. In: *Functional and Medicinal Beverages* (eds. A.M. Grumezescu and A.M. Holban), 447–458. Academic Press.

Payne, C.M., Knott, B.C., Mayes, H.B. et al. (2015). Fungal cellulases. *Chem. Rev.* 115: 1308–1448.

Peris, D., Pérez-Torrado, R., Hittinger, C.T. et al. (2018). On the origins and industrial applications of *Saccharomyces cerevisiae* × *Saccharomyces kudriavzevii* hybrids. *Yeast* 35: 51–69. https://doi.org/10.1002/yea.3283.

Polizeli, M.L.T.M., Rizzatti, A.C.S., Monti, R. et al. (2005). Xylanases from fungi: properties and industrial applications. *Appl. Microbiol. Biotechnol.* 67: 577–591.

Quintanilla, D., Hagemann, T., Hansen, K., and Gernaey, K.V. (2015). Fungal morphology in industrial enzyme production – modelling and monitoring. In: *Filaments in Bioprocesses, Advances in Biochemical Engineering/Biotechnology* (eds. R. Krull and T. Bley), 29–54. Cham: Springer International Publishing.

Raveendran, S., Parameswaran, B., Ummalyma, S.B. et al. (2018). Applications of microbial enzymes in food industry. *Food Technol. Biotechnol.* 56: 16–30.

Ruijschop, R.M.A.J., Boelrijk, A.E.M., and te Giffel, M.C. (2008). Satiety effects of a dairy beverage fermented with propionic acid bacteria. *Int. Dairy J.* 18: 945–950.

Sanchez, S. and Demain, A.L. (2009). Microbial primary metabolites: biosynthesis and perspectives. In: *Encyclopedia of Industrial Biotechnology: Bioprocess, Bioseparation, and Cell Technology* (ed. M.C. Flickinger), 1–16. American Cancer Society https://doi.org/10.1002/9780470054581.eib167.

Sindhu, R., Pandey, A., and Binod, P. (2017). Design and types of bioprocesses. In: *Current Developments in Biotechnology and Bioengineering* (eds. C. Larroche, M.Á. Sanromán, G. Du and A. Pandey), 29–43. Elsevier.

Smith, J.E. (2009). *Biotechnology*. Cambridge; New York: Cambridge University Press.

Sousa, F., Passarinha, L., and Queiroz, J.A. (2009). Biomedical application of plasmid DNA in gene therapy: a new challenge for chromatography. *Biotechnol. Genet. Eng. Rev.* 26: 83–116.

Vitorino, L.C. and Bessa, L.A. (2017). Technological microbiology: development and applications. *Front. Microbiol.* 8 https://doi.org/10.3389/fmicb.2017.00827.

Waites, M.J., Morgan, N.L., Rockey, J.S., and Higton, G. (2001). Microbial cell structure and function. In: *Industrial Microbiology: An Introduction*, 7–20. Wiley Blackwell.

Wilkins, M.R. and Atiyeh, H. (2012). Fermentation. In: *Food and Industrial Bioproducts and Bioprocessing* (ed. N.T. Dunford), 185–203. Wiley.

2

Current Developments in Industrial Fermentation Processes

Mehrdad Niakousari[1], *Maryam Razmjooei*[1], *Maryam Nejadmansouri*[1], *Francisco J. Barba*[2], *Krystian Marszałek*[3,5], *and Mohamed Koubaa*[4]

[1] *Department of Food Science and Technology, College of Agriculture, Shiraz University, Shiraz, Iran*
[2] *Faculty of Pharmacy, Preventive Medicine and Public Health, Food Science, Toxicology and Forensic Medicine Department, Universitat de València, València, Spain*
[3] *Department of Fruit and Vegetable Product Technology, Prof. Wacław Dąbrowski Institute of Agricultural and Food Biotechnology, Warsaw, Poland*
[4] *ESCOM, UTC, EA 4297 TIMR, Compiègne, France*
[5] *Department of Food Technology and Human Nutrition, Institute of Food Technology and Nutrition, College of Natural Science, University of Rzeszow, Rzeszow, Poland*

2.1 Introduction

Advances in the fermentation industry are mainly focused on enhancing fermentation processes in order to obtain better yields and higher quality commercialized products (Soccol et al. 2017). Innovative fermentation technologies cover a broad range of products in three industries: organic acids and enzymes in the food industry, antibiotics and recombinant proteins in the pharmaceutical industry, and the production of biofuels in the chemical industry to decrease our future dependence to products and chemicals produced from fossil fuels (Ojha et al. 2017). Therefore, scaling up the fermentation process from laboratory to large-scale bioreactors should be considered in order to transform the optimal operating conditions for producing these products.

Scaling up is great matter of concern in the production of fermenters since two different aspects should be considered. The first one is related to the size shifting from a lab-scale to a bigger pilot plant fermenter which is not a huge deal in terms of engineering and calculation. The second aspect, however, is keeping the stable concentration of the desired products in both small and large sizes of the equipment during fermentation that unfortunately follows a gradient pattern. Pilot- (20–2000 l) and laboratory-scale (0.5–20 l) fermenters have been used in the past to screen for situations with maximum volumetric yield and productivity. This, however, is very expensive, owing to the decrease in production time for screening in a full-scale reactor (Gernaey 2015). Although the results of experiments conducted on laboratory and pilot scale are not comparable to huge industrial bioreactors, extra pilot plant experiments should be done to investigate the effect of extra process

Fermentation Processes: Emerging and Conventional Technologies, First Edition.
Edited by Mohamed Koubaa, Francisco J. Barba, and Shahin Roohinejad.

factors on process performance at large scale. By understanding and applying innovative developments in the fermentation process such as industrial fermentation media, identification of fermentation systems, inhibition in fermentation, fermentation process modeling, and fermentation optimization in large industrial bioreactors, we are able to improve productivity and open new perspectives for this industry.

The purpose of this chapter is to consider the main achievements in industrial fermentation in the food, chemical, and pharmaceutical sectors as well as to investigate the current developments in industrial fermentation from different aspects to improve the productivity and efficiency of the fermentation process.

2.2 Main Achievements in Industrial Fermentation

Various fermentation processes are used for a variety of products. Cultivating microorganisms in bioreactors and fermenters, also known as bioprocesses, requires knowledge in microbiology, biochemistry, and engineering sciences (Najafpour 2015). The molecules produced by fermentation are mainly used in food, agricultural, pharmaceutical, and chemical industries (Table 2.1). Only the bioprocesses used in the food, chemical, and pharmaceutical industries will be discussed in this chapter. The names of some industrial fermentation companies with their fields of activities and website addresses have been represented in Table 2.2 (https://www.ventureradar.com/similar/Ginkgo%20BioWorks/e5fd843c-54a2-4bfb-8590-6255daf8a4b3).

2.2.1 Fermentation Processes in Food Industry

Food fermentation is based on the chemical transformation of organic compounds (i.e. substrates) into others (i.e. products) through the action of microorganisms. This catalysis involves microbial enzymes belonging to different kinds of microorganisms that include bacteria, yeasts, and molds (Corma et al. 2007). The food fermentation industry offers a chance for the consumers to provide better quality, palatable, and safer products, as well as energy-efficient processes (Pereira and Vicente 2010). Therefore, although fermentation was originally developed to preserve perishable agricultural products, since then, this technology has been developed as a tool to create attractive organoleptic attributes to the food products (Di Cagno et al. 2013). In addition, fermentation may also contribute to reducing toxins and antinutritional factors in food materials, as well as enhancing their nutritional value. For instance, the fermentation of soybean into products such as tempeh (fermented dehulled soybean with meat-like texture and flavor), natto (fermented soybean with a strong flavor and slimy texture), and soy sauce (fermented soybean and wheat making a dark brown condiment) leads to decreasing the content of some antinutritional factors such as trypsin inhibitors and phytic acid, as well as to the hydrolysis of complex soy proteins into more bioavailable and digestible amino acids and peptides (Chen et al. 2012; Soni and Dey 2014).

Fermented food products have become a center of interest in the human diet, mainly because of the health benefits brought by the nutritional metabolites and probiotic microorganisms (Figure 2.1). Hundreds of fermented food products are now commercially available, and some can even prevent or treat a wide range of illnesses. For example, fermented

Table 2.1 Products provided by biological processes.

Field	Product
Chemicals	Ethanol, acetone, butanol
	Organic acids, enzymes
	Perfumeries
	Polymers
Pharmaceuticals	Antibiotics
	Enzymes
	Enzyme inhibitors
	Monoclonal antibodies
	Steroids
	Vaccines
Energy	Ethanol (gasohol)
	Methane (biogas)
Food	Dairy products, baker's yeast
	Beverages (beer, wine)
	Food additives
	Amino acids
	Vitamin B
	Proteins (SCP)
Agriculture	Animal feeds (SCP)
	Waste treatment
	Vaccines
	Microbial pesticides
	Mycorrhizal inoculants

and spiced kimchi (napa cabbage) has many beneficial properties: antiobesity, antiaging, anticancer, and anticonstipation (Kim et al. 2011). In another example, Kefir, which is a fermented milk drink similar to a thin yogurt that is made from kefir grains, decreases lactose intolerance symptoms, stimulates the immune system, decreases cholesterol levels, and has anticarcinogenic and antimutagenic properties (Guzel-Seydim et al. 2011). Although the majority of health benefits regarding fermented foods lack well-founded scientific grounds (Marsh et al. 2014), many findings and *in vitro* experiments confirm some of these observations. In 2016, fermented foods were ranked among the top 10 food trends and have continued to grow in popularity over the past few years (Reilly et al. 2014). Furthermore, food factories are willing to invest in this type of technology either by commercializing some traditional fermented foods (e.g. kefir and kombucha) or by promoting new fermented products based on traditional ones (e.g. bionade, which is a fermented beverage based on malt flavored using kombucha, and Rythem, as the starter culture). Many probiotic products based on soy and cereals are also being commercialized due to allergies

Table 2.2 Names of some industrial fermentation companies with their fields of activities and website addresses.

Company	Location	Year	Field of activity	Website
Green Biologics	United Kingdom	2003	Developing and delivering new green alternatives	http://www.greenbiologics.com
Zymergen	USA	2013	A molecular technology company	http://www.zymergen.com
Genomatica	USA	2000	Transformative manufacturing processes	http://www.genomatica.com
Verdezyne	USA	2005	Developing fermentation processes for producing renewable fuels and chemicals	http://www.verdezyne.com
Caribou Biosciences	USA	2011	A developer of cellular engineering based on CRISPR technologies	http://cariboubio.com
Manus Bio	USA	2012	Recreating plant processes in microorganisms to produce natural ingredients through fermentation	http://www.manusbio.com
Evolva	Switzerland	2004	Develop, make, and sell natural ingredients that provide health	http://www.evolva.com
Amyris	USA	2003	Providing sustainable alternatives to a broad range of petroleum-sourced products	http://www.amyris.com
Arzeda	USA	2008	Harnessing the power of computational and synthetic biology to create new enzymes and chemicals	http://arzeda.com
Lygos	USA	2011	Engineering microbes to convert sugars into high-value, industrial chemicals	http://www.lygos.com
Industrial Microbes	USA	2014	Building fermentation technology to convert methane and carbon dioxide into valuable, versatile building-block chemicals	http://www.imicrobes.com
BioAmber	USA	2008	Combining industrial biotechnology, an innovative purification process and chemical catalysis	http://www.bio-amber.com
Calysta	USA	Unknown	Creating innovative industrial products from sustainable sources	http://www.calysta.com

Ingenza	United Kingdom	2002	Applying synthetic biology to the manufacture of industrial products including enhanced biofuels	http://www.ingenza.com
BioSyntia	United Kingdom	2012	Developing microbial technology to produce chemicals and fuels from renewable resources	https://www.biosyntia.com/
CHAIN Biotechnology	United Kingdom	Unknown	Product development company based at the Imperial Incubator on the campus of Imperial College, London	http://www.chainbiotech.com
Deinove	France	2006	Development and commercial exploitation of innovative, environmentally friendly products for production of biofuels	http://www.deinove.com
Novomer	USA	2004	Developing a technology platform to produce performance polymers, plastics, and other chemicals	http://www.novomer.com
Fermentalg	France	2014	Produces molecules of interest and bypasses conventional methods using petrochemicals or even industrial fish farming	https://www.fermentalg.com/
Global Bioenergies	France	2008	Revolutionizing the hydrocarbon industry by converting bio-sourced feedstocks into plastics, rubbers, and fuels	http://www.global-bioenergies.com
Barry Callebaut	Switzerland	1996	Barry Callebaut is among the world's largest cocoa producers and grinders	http://www.barry-callebaut.com
Biotork	USA	2008	Focusing on the development of microbial strains and bioconversion processes	http://biotork.com
Yield10 Bioscience, Inc.	USA	1992	Developing and commercializing bioplastics	http://www.yield10bio.com/
Cereduce	Sweden	2010	The purpose of the company is to put scientific findings related to baker's/brewer's yeast to use in the food and beverage industries	http://www.cereduce.com
Kalion, Inc.	USA	Unknown	Transforming the traditional chemical industry using traditional, green, fermentation technology	http://www.kalioninc.com
Photanol	Netherlands	2008	Converting CO_2 and sunlight into valuable organic compounds	http://www.photanol.com

(Continued)

Table 2.2 (Continued)

Company	Location	Year	Field of activity	Website
Magenta Biolabs	Unknown	Unknown	Producing high-quality hyaluronic acid for the cosmetic industry	https://magenta-biolabs.webflow.io/
Osprey Biotechnics	Unknown	Unknown	Development of microbial products and bioproducts for environmental, industrial, and agricultural applications	http://www.ospreybiotechnics.com
Cysal GMBH	Germany	Unknown	Producing dipeptide arginyl aspartate on an industrial scale	http://www.cysal.de
3f Bio	United Kingdom	2015	Transforming the production economics for meat-free food products	https://www.3fbio.com
Bict	Italy	2002	Production of enzymes, hydrolysates, peptides, and secondary metabolites	http://www.bict.it
Phytowelt GreenTechnologies GMBH	Germany	1998	Develops process, product, and plant for flavors and fragrances, active ingredients, raw materials	http://www.phytowelt.com
Byosis Group	Netherlands	Unknown	Extending knowledge with digestion and fermentation processes	http://www.byosis.com
Eucodis Bioscience	Austria	2007	Application-driven enzyme engineering company in pharmaceutical, biotechnology	http://www.eucodisbioscience.com
Galactic	Belgium	1994	Producing an extensive range of natural products by fermentation	http://www.lactic.com
Microbial Bioinformatics	Unknown	2011	Bioinformatics Centre (NBIC), Top Institute Food and Nutrition (TIFN)	https://www.nbic.nl, https://www.tifn.nl/
Puratos	Belgium	1919	International group with full range of innovative products, raw materials, and application expertise in the bakery, patisserie, and chocolate sectors	http://www.puratos.com
OENON Holdings	Japan	1924	Fermentation technology nurtured for many years through alcoholic beverage	http://www.oenon.jp
Nature BioScience	India	2014	Development, manufacture, and supply of analytical reagents, enzymes, and biochemicals for multiple industries	http://www.naturebioscience.com

Maps Enzymes	India	1975	Pioneer biotechnology company producing enzymes and microorganisms	http://www.mapsenzymes.com
Imenz Bioengineering	Netherlands	1998	Developing processes to improve the sustainable production of bio-based chemicals from renewable raw materials	http://www.imenz.com/
Ennolys	France	1992	Assisting aromaticians, food manufacturers, perfumeries, and nutraceuticals	http://ennolys.fr
Aum Enzymes	India	2007	A metabolic engineering firm specializing in safe microbial–origin industrial technical grade enzymes produced by solid-state fermentation and submerge fermentation	http://aumenzymes.com
Carbolosic	USA	2012	Manufacturing fine specialty chemicals from both fermentation and cellulosic materials	http://www.carbolosicenergy.com
Ecolyse	USA	2007	Providing products and services to remediate ethanol production disruption as well as a variety of other bacterial, fungal, and algae problems in industrial settings	http://www.ecolyse.com
Korea Institute of Industrial Technology	USA	1989	Production of itaconic and fumaric acids via fungal fermentation based on noble industrial biotechnology	www.kitech.re.kr
Algavia	Unknown	Unknown	Using standard industrial fermentation process to grow algae	http://algavia.com
GO2 water	USA	2007	Environmental solutions company with proven carbon-negative technology for reclaiming water, energy, and nutrients from municipal, industrial, and agricultural wastewater	http://www.go2intl.com
Microbiogen	Australia	Unknown	Global leading developer of enhanced yeasts for the production of fuel ethanol and other industrial applications	http://www.microbiogen.com
Microbial Discovery Group	USA	2007	Providing bioaugmentation products and solutions for a multitude of industries	http://www.mdgbio.com
IMBP	Belgium	2002	Production of commercialized microbial products	http://www.imbp.be

(Continued)

Table 2.2 (Continued)

Company	Location	Year	Field of activity	Website
Alco Bio Fuel	Belgium	2005	One of Belgium's main bio-ethanol production units	http://www.alcobiofuel.com
$Green_2$ Chem Consulting	Belgium	Unknown	A technology development company focused on services to industries active in sustainable chemistry and biochemistry	http://www.green2chem.com
Mycologix	United Kingdom	Unknown	Fundamental research into microbial decay mechanisms of wood	http://www.mycologix.co.uk
Avecom	Belgium	1995	Steering and optimizing microbial processes with a focus on customer needs	http://avecom.be
Remat	USA	Unknown	Combining biology, automation, and fermentation	http://www.refactoredmaterials.com
Microbiogen Australia	Australia	Unknown	Using its advanced breeding and screening techniques to develop new strains of yeast (*Saccharomyces cerevisiae*)	http://microbiogen.com.au
Bio-on	Italy	Unknown	Creating natural products/solutions based on renewable resources or agricultural processing waste materials	http://www.bio-on.it
Mycelia	Belgium	1985	With a production capacity of 30.000 l of spawn weekly	http://www.mycelia.be
Citrique Belge	Belgium	1929	Delivering citric acid and/or associated coproducts in fermentation process	http://www.citriquebelge.com
Biogastec	Belgium	2005	Constructing and managing biogas installations at agricultural and industrial scales	http://www.biogastec.com
Bio Base Europe	Belgium	2009	Consisting of a Bioprocess Pilot Plant in Ghent and a Training Facility in Terneuzen	http://www.bbeu.org
Algist Bruggeman	Belgium	1949	Leading manufacturer of baker's yeast for both traditional and industrial bakeries	http://www.algistbruggeman.be
Biotrend – Inovação – Engenharia em Biotecnologia, S.A.	Portugal	Unknown	Using genetically modified microorganisms in which multistep pathways have been cloned	http://www.biotrend.biz

Fermenco aps	Denmark	Unknown	Improvement of industrial microorganisms through genetic engineering	http://www.fermenco.com
Butylfuel	USA	Unknown	A biorefining and industrial biotechnology company focused on producing high-value chemicals, fuels, and energy from renewable carbon resources	http://www.butanol.com
Phocone AG	Switzerland	2011	Developing and producing novel mid-infrared laser modules specifically well suited for simultaneous multigas detection	http://www.phocone.com
LS9	USA	2006	Production of sustainable products for the fuel and chemical industries	http://www.ls9.com
OPX Biotechnologies	USA	2007	Using EDGE™ microbe engineering platform to manufacture bio-based chemicals and fuels	http://www.opxbiotechnologies.com
TerraVia (aka Solazyme)	USA	2003	A renewable oil and bioproducts company that transforms a range of low-cost plant-based sugars into high-value tailored oils	http://terravia.com

Figure 2.1 Novel and traditional fermented foods. *Source:* Sauerkrauft: miss.lemon/Adobe Stock; Kimchi: sasimoto/Adobe Stock; Tempeh: hary/Adobe Stock; Natto: norikko/Adobe Stock; Kefir: expressiovisual/Adobe Stock; Kombucha: LIGHTFIELD STUDIOS/Adobe Stock; Soy Sauce: ddukang/Adobe Stock; Lactic fermented vegetable mixes: Artem Shadrin/Adobe Stock; Beverages fermented by kefir and kombucha starters: MeganBetteridge/Adobe Stock.

related to consumption of dairy proteins, lactose, and gluten intolerances, as well as some lifestyle choices, such as veganism (Gupta and Abu-Ghannam 2012).

2.2.1.1 Alcoholic Beverages

One of the most important products in industrial fermentation is ethanol, which is produced by yeasts (Figure 2.2). *Saccharomyces cerevisiae* is the most popular strain of yeast used for the production of alcoholic beverages (e.g. beer, wine, etc.). Since the major biotechnological process in fermentation involves the conversion of sugars to alcohol, it is important to understand the physiology of yeast so as to enhance the industrial production of alcohol. Rapid approaches of analyzing the yeast physiology and fermentation factors with new instrumentation are reported in Table 2.3 (Walker and Walker 2018).

The history of fermentation started in around with brewing beer. Today, alcoholic beverages are strongly related to the economies of many countries, with over 100 billion liters produced per year around the world. The largest ethanol producer in the world is the United States with an average over a million barrels (159 million liters) per day and 16 billion gallons (60 billion liters) per year. More than 85% of the world's ethanol is produced by the United States and Brazil, which is responsible for beverage, manufacturing baked products; coloring, flavoring and biocontrol agents; probiotics; enzymes; chemical materials; therapeutic proteins; biofuel alcohol production; and biopharmaceuticals (Walker and Walker 2018) (Table 2.4). Although the fermentative abilities of yeasts have been exploited for many years, there are still many aspects of industrial-scale ethanol production that need to be learnt.

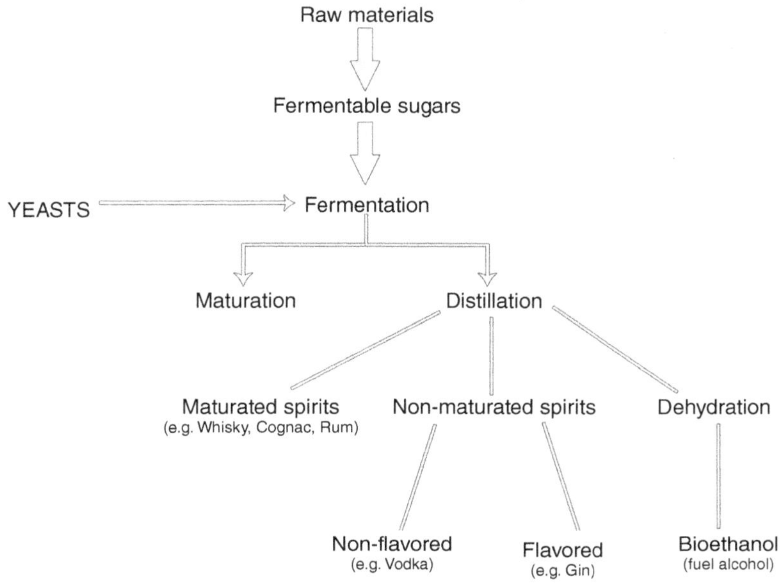

Figure 2.2 General schemes for beverage alcohol and bioethanol production.

Table 2.3 Rapid methods of analyses of yeast physiology and fermentation parameters with modern instrumentation.

Instrument	Website address
Yeast Activity Monitor	www.nalco.com
Yeast Monitor and Countstar	www.aberinstruments.com
Cellometer and Celigo Image Cytometer	www.nexcelom.com
Incyte and Dencytee biomass sensors	http://www.hamiltoncompany.com/products
Ankom Gas Production System	www.ankom.com

Source: Walker and Walker (2018).

2.2.1.2 Enzymes

Industrial fermentation processes make use of submerged and solid-state bioreactors, which are operated in batch, fed-batch, and continuous modes. Submerged fermentation processes are used for the production of alcoholic beverages, dairy beverages, and condiments such as vinegar. In contrast, solid state fermentation (SSF) is used in most traditional food fermentation processes and different types of enzymes (Teixeira and Vicente 2017).

SSF is a process where microorganisms grow in a condition without free water, or with low amount of free water. SSF has several biotechnological advantages over submerged fermentation, such as the higher activity and quality of the extracts, no requirement of organic solvents, lower investment and operating costs, decreased downstream processing, and higher productivity (Martín-Sampedro et al. 2012; Singh and Satyanarayana 2008). In addition, SSF provides an environmentally friendly way with many advantages, such as the hindering of

Table 2.4 Global world ethanol production in recent years (millions of liters).

Geographic location	2016	2017	2018
North and Central America	59752	60910	61380
South America	27104	26696	26950
Asia	5292	5705	6302
Europe	4866	5246	5486
Africa	150	160	175
Total	97359	98927	100513

Information from Licht (2018) World Ethanol and Biofuels Report. https://www.agra-net.com/agra/world-ethanol-and-biofuels-report.
Source: Walker and Walker (2018).

environmental pollution, prevention of agricultural waste, or agglomeration of by-products (Kumar and Kanwar 2012). On the other hand, it has some drawbacks, such as limited application at industrial level, lower heat conductivity of solid substrates, and insufficient free-flowing water, which create an obstacle in the scale-up of the process (Mazaheri and Shojaosadati 2013). These days, the majority of SSF research works are related to enzyme production such as cellulose, xylanase, amylase, laccase, lipase, pectinase, and protease (Table 2.5). Mainly, cellulases and xylanases are at the center of interest. Various industries produce high amount of waste residues which are difficult to dispose of. These wastes can be used as nutrient origin for the production and growth of different metabolites from microbial origins (Soccol and Vandenberghe 2003). In the past few decades, there had been an upward tendency for using the waste of agriculture as carbon source under SSF to generate different ranges of useful secondary metabolites from microorganisms and enzymes (Sodhi et al. 2005).

2.2.2 Fermentation Processes in Chemical Industry

Recently, there has been a growing interest in the commercialization of microbial fermentation processes to generate chemicals, fuels, and polymers from renewable biomass-based feedstock (Procentese et al. 2017). Customers prefer using biodegradable, natural, and environmental friendly products; therefore, these bioproducts show an environmentally maintainable option to their petrochemical analogs and fossil resources (Isikgor and Becer 2015). Furthermore, these days, in different sectors, there is a worldwide concern about residue production. The use of agro-food residues, especially agricultural waste, is at the center of interest from economic and environmental aspects (Mondala 2015). According to the Food and Agricultural Organization, global food waste amount is estimated to 1.3 billion tons (Gustavsson et al. 2011). In industrialized countries, such as in Europe and North America, there is a great deal of wastage at the consumption stage, whereas in less developed countries, large amounts of food waste are produced at the primary and middle stages of the supply chain (Gustavsson et al. 2011). Recently, many efforts have been made to use food waste as a bioresource due to the functionalized molecules collected in them to generate valuable products, such as organic acids, biofuels, and triacylglycerols (TAG), which are

Table 2.5 Different types of enzymes produced by solid-state fermentation.

Product (enzyme)	Microorganism	Substrate/support	Productivity	Bioreactor	References
Amylase	*Aspergillus oryzae*	Starch and soybean meal	$22118\,U\cdot g^{-1}$, 12 days	Flasks	Sahnoun et al. (2015)
Amylase	*Bacillus licheniformis*	Date waste	$209\,U\cdot g^{-1}$, 7 days	Flasks	Afrisham et al. (2016)
Amylase	*Bacillus* sp.	Wheat straw	$6900\,U\cdot g^{-1}$, 5 days	Flasks	Qureshi et al. (2016)
Amylase	*Thermomyces* sp.	Soy and bread waste	$39900\,U\cdot g^{-1}$, 4 days	Cylindrical reactor (4.5 l)	Cerda et al. (2016)
Amylase	*Aspergillus flavus*	Wheat bran	$733\,U\cdot g^{-1}$	Flasks	Elkhouri and Bakri (2016)
Amylase	*Aspergillus niger*	Bermuda grass, corncob	$2176\,U\cdot g^{-1}$	Flasks	Amaro-Reyes et al. (2016)
Amylase	*Aspergillus terreus*	Pearl millet	$19.2\,U\cdot g^{-1}$	Flasks	Sethi et al. (2016)
Amylase	*Pleurotus ostreatus*	Pineapple waste, rice straw	$93.7\,U\cdot g^{-1}$	Flasks	Rashad et al. (2016)
Amylase	*Monascus sanguineus*	Beetroot	$0.029\,U\cdot ml^{-1}$	Flasks	Tallapragada et al. (2017)
Amylase	*Aspergillus niger*	Rice husk, bagasse, corncob	$81.86\,U\cdot ml^{-1}$	Flasks	Aliyah et al. (2017)
Cellulase	*Rhizopus oryzae*	Wheat bran	$437\,U\cdot g^{-1}$, 5 days	Flasks	Pandey et al. (2016)
Cellulase	*Trichoderma reesei*	Sorghum stover	$959.53\,IU\cdot g^{-1}$	Pilot scale	Idris et al. (2017)
Cellulase	*Myceliophthora thermophila*	Sugarcane bagasse, wheat bran	$12\,FPU\cdot gds^{-1}$	Flasks	Kilikian et al. (2014)
Cellulase	*Pseudoxanthomonas taiwanensis*	Coffee husk	$10\,FPU\cdot g^{-1}$	Bioreactors (4.5 l)	Cerda et al. (2016)
Laccase	*Pleurotus ostreatus*	Wheat bran	$32450\,U\cdot g^{-1}$, 7 days	Flasks	El-Batal et al. (2015)
Laccase	*Pyrenophora phaeocomes*	Rice straw	$10859\,U\cdot g^{-1}$, 4 days	Flasks	Rastogi et al. (2016)
Laccase	*Pleurotus ostreatus*	Polyurethane foam	$18030\,U.l^{-1}$	Flasks	Álvarez-Cervantes et al. (2016)

(Continued)

Table 2.5 (Continued)

Product (enzyme)	Microorganism	Substrate/support	Productivity	Bioreactor	References
Laccase	*Pleurotus nebrodensis*	Wheat straw	$172\,U\cdot ml^{-1}$	Flasks	Aslam et al. (2016)
Laccase	*Coriolopsis gallica*	Saw dust	$4880\,U.l^{-1}$	Flasks	Daâssi et al. (2016)
Laccase	*Tricholoma giganteum*	Wheat straw	$2.69\times10^5\,U.\,g^{-1}$	Flasks	Patel and Gupte (2016)
Laccase	*Ganoderma lucidum*	Rice husks and straw	$10\,927\,U.kg^{-1}$, 10 days	Pilot scale	Postemsky et al. (2017)
Laccase	*Tinea versicolor*	Cornstalk	$2600\,U\cdot g^{-1}$	Flasks	Adekunle et al. (2017)
Laccase	*Marasmiellus palmivorus*	Pineapple leaf	$627.7\,IU\cdot ml^{-1}$	Flasks	Aiswarya et al. (2018)
Lipase	*Burkholderia cenocepacia*	Sugarcane bagasse, sunflower seed	$72.3\,U\cdot g^{-1}$, 96 h	Flasks	Liu et al. (2016)
Lipase	*Aspergillus niger*	Rice bran	$38.67\,U\cdot g^{-1}$, 5 days	Flasks	Utami et al. (2017)
Lipase	*Trichoderma*	Oil palm empty fruit bunch	$0.390\,U\cdot g^{-1}$	Flasks	Musa et al. (2017)
Lipase	*Rhizopus microspores*	Wheat bran	$113\,U\cdot g^{-1}$, 20 h	Pilot packed-bed bioreactor	Pitol et al. (2017)
Lipase	*Aspergillus niger*	Canola meal	—	—	Mulinari et al. (2017)
Lipase	*Aspergillus oryzae*	Castor de-oiled cake	$25\,U\cdot g^{-1}$, 6 days	Flasks	Jain and Naik (2018)
Pectinase	*Aspergillus oryzae*	Citrus pulp and sugarcane bagasse	$40\,U.g^{-1}$, 18–24 h	Pilot-scale packed-bed	Biz et al. (2016)
Pectinase	*Rhizopus delemar*	Apple pomace	$61.50\,U\cdot g^{-1}$	Flasks	Pathania et al. (2018)
Pectinase	*Aspergillus niger*	Wheat bran	$22\,U\cdot g^{-1}$, 20 h	Pilot-scale packed-bed	Finkler et al. (2017)
Pectinase	*Aspergillus tamari*	Wheat bran	$101.05\,U\cdot ml^{-1}$	Flasks	Shanmugavel et al. (2018)
Protease	*Aspergillus terreus*	Chickling vetch peels	$5266.8\,U\cdot ml^{-1}$, 96 h	Static flasks	Sethi et al. (2016)
Protease	*Aspergillus niger*	Wheat bran, soybean meal	$262.78\,U\cdot g^{-1}$, 48 h	Static flasks	de Castro et al. (2015)
Protease	*Aspergillus oryzae*	Canola cake	1,388.25	Laboratory-scale system	Freitas et al. (2015)

Protease	*Consortium*	Anaerobically digested sludge	52 230 $U \cdot g^{-1}$, 14 days	Reactors (10 l)	Yazid et al. (2016)
Protease	*Aspergillus spp.*	Wheat bran	450 $U \cdot g^{-1}$	Flasks	Ortiz et al. (2016)
Protease	*Bacillus nealsonii*	Locust bean	70 $U \cdot g^{-1}$	Flasks	David et al. (2018)
Protease	*Aspergillus oryzae*	Wheat bran	2331 ± 39 $U \cdot g^{-1}$	Flasks	Zhao et al. (2019)
Xylanase	*Thermoascus aurantiacus*	Sugarcane bagasse, rice bran	176 $U \cdot g^{-1}$, 196 h	Static flasks	Chanwicha et al. (2015)
Xylanase	*Aspergillus tubingensis*	Sorghum straw	5177.23 $U \cdot g^{-1}$, 120 h	Static flasks	Adhyaru et al. (2016)
Xylanase	*Rhizopus oryzae*	Wheat bran	273.83 $U \cdot g^{-1}$, 120 h	Static flasks	Pandey et al. (2016)
Xylanase	*Aspergillus awamori*	Cocoa meal	66.5 $U \cdot g^{-1}$, 48 h	Flasks	Amorim et al. (2017)
Xylanase	*Aspergillus niger*	Oil palm empty fruit bunches	3246 $IU \cdot g^{-1}$	Flasks	Ajijolakewu et al. (2017)
Xylanase	*Pseudoxanthomonas taiwanensis*	Coffee husk	48 $U \cdot g^{-1}$	Bioreactors (4.5 l)	Cerda et al. (2017)
Xylanase	*Phomopsis stipata*	Sugarcane bagasse	694 33 $U \cdot g^{-1}$	Flasks	Marques et al. (2018)
Xylanase	*Penicillium roqueforti*	Sugarcane bagasse	19.95 $IU \cdot g^{-1}$	Flasks	Souza et al. (2018)
Xylanase	*Talaromyces stipitatus*	*Parthenium hysterophorous*	582 $IU \cdot g^{-1}$	Flasks	Bharti et al. (2018)
Xylanase	*Aspergillus niger*	Wheat bran	2919 ± 174 $U \cdot g^{-1}$, 48 h	Flasks	Khanahmadi et al. (2018)

achieved at low prices by using raw materials, containing depleted sugar beet cossettes, rice hulls, wheat bran, wheat straw, corn stover, etc. (Díaz et al. 2017), which do not involve competition with the food chain (Bellasio et al. 2015).

2.2.2.1 Biofuels

Transportation is the second largest energy-consuming sector in the world; with more than 95% of the energy is assigned to fossil fuels (Kumar et al. 2018). Although fossil fuels such as gasoline, crude oil, and natural gas cause serious environmental problems, they still dominate the global energy market (Diaz et al. 2018). The cost of producing fossil fuels can be decreased by utilizing adequate lignocellulosic biomass as a raw material. Between them, liquid fuels are more interesting because of their suitability with the current ignition engines (Roy et al. 2013). Bioethanol is widely used worldwide as a renewable transportation fuel, and it is predicted that it will supply 7% of the total worldwide energy as a transportation fuel in the near future (Rathore et al. 2016). It has many advantages, including high yields and less greenhouse gas emissions (Wang et al. 2012). The biggest bioethanol producer in the world is the United States, and it devotes a production of 15.25 billion gallons (58% of the total worldwide production). The majority of bioethanol in the United States is produced from corn using dry grind processing (Somavat et al. 2016). Conventional dry grind processing consists of the sequential steps of grinding, liquefaction, saccharification, fermentation, distillation, and coproduct recovery (Figure 2.3). In the past 10 years, the wholesale ethanol cost has changed extensively between \$1.18 and \$3.51 per gallon (an average of \$1.96 gal^{-1}). Agro-food by-products have also been used for the production of butanol and acetone as well as ethanol (Table 2.6). Butanol should, however, be eliminated from the broth during the fermentation in order to avoid the microbial growth inhibition. Many technologies can be used to achieve this goal: adsorption, pervaporation, distillation, gas stripping, reverse osmosis, liquid–liquid extraction, etc. Furthermore, by enhancing the strains, we are able to increase their tolerance to butanol, as well as their yield and productivity (Li et al. 2014). The only commercial bio-butanol manufacturer in the world is Lignicell Refining Biotechnologies Ltd. in China (Jiang et al. 2015). Hydrogen (H_2) is also produced from nonrenewable raw substances (Table 2.6). For industrial hydrogen production and achieving high yield, cell immobilization technology can be utilized in a continuous manner.

2.2.2.2 Organic Acids

Generating organic acids through the fermentation of agro-food by-products/residues plays a significant role in the food and chemical industries. Organic acids such as butyric, succinic, itaconic, lactic, fumaric, and malic acids are used extensively in cosmetics and pharmaceutical products as an additive, textile, plasticizers, surfactants, detergents and in the food industries as flavoring agents (Table 2.7) (Diaz et al. 2018). It is predicted that 140 000 metric tons of renewable succinic acid will be attainable worldwide in the next three years. An interesting application of succinic acid is as an acidulate, pH regulator in fermentation and antimicrobial agent. The American company Charles Pfizer Co. was the only producer of the itaconic acid until 1979. Fermentation technology is used to produce itaconic acid; however, the purification and fermentation processes are still financially demanding and ineffective (Mondala 2015). The American company Myriant is working

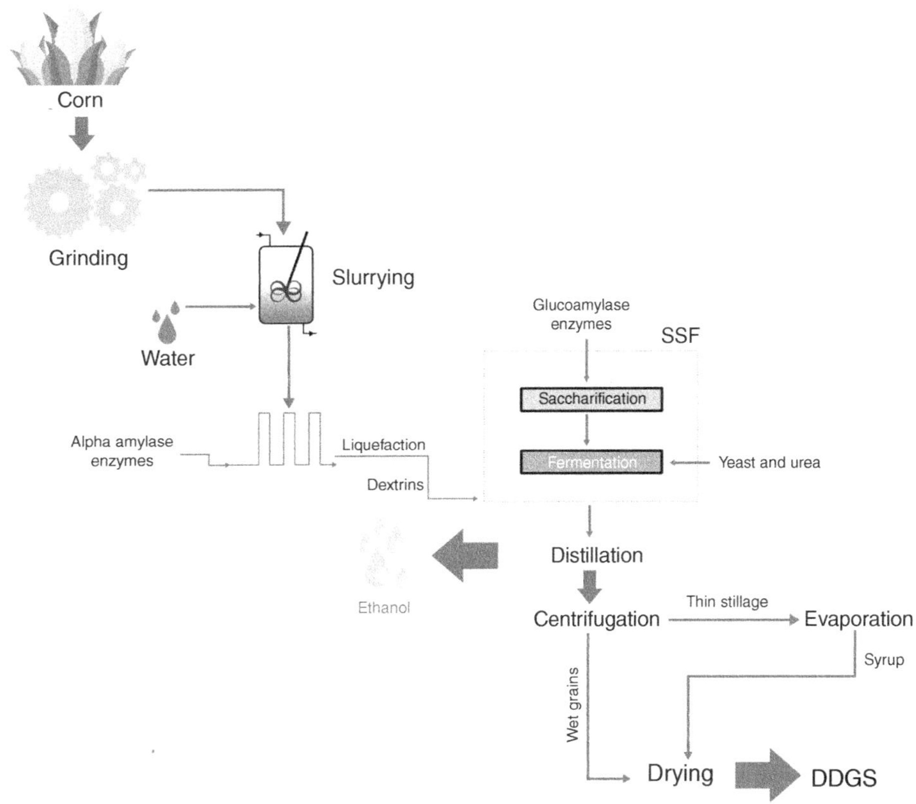

Figure 2.3 Schematic of dry grind process for ethanol production from corn (SSF – Solid State Fermentation; DDGS – Dried Distillers Grains with Solubles).

Table 2.6 Productions of biofuels by sugar fermentation of agro-food residues.

Biomass	Sugars	Strategy	Microorganism	Yield/concentration	Product	References
Rice straw	Glucan/hemicellulose/lignin	SSF	*Saccharomyces cerevisiae*	$0.08\,g{\cdot}g^{-1}$ ub	Ethanol	Jung et al. (2015)
Corn stover	Glucan/hemicellulose/lignin	SSF	*Cryptococcus curvatus*	$0.16\,g{\cdot}g^{-1}$ ub	Ethanol	Kim et al. (2015)
Corn stover	pb/glucan/xylan	SSF	*Saccharomyces cerevisiae*	$71.40\,g{\cdot}l^{-1}$	Ethanol	Qureshi et al. (2015)
Rice straw	pb/glucan/xylan	SSCF	*Saccharomyces cerevisiae*	$0.38\,g{\cdot}g^{-1}$ pb	Ethanol	Sasaki et al. (2015)
Wheat straw	Glucan/hemicellulose/lignin	SSF	*Kluyveromyces marxianus*	$0.23\,g{\cdot}g^{-1}$ pb	Ethanol	Moreno et al. (2013)
Oil palm	Glucan/hemicellulose/lignin	SSF	*Saccharomyces cerevisiae*	$0.19\,g{\cdot}g^{-1}$ ub	Ethanol	Jung et al. (2013)
Wheat straw	Glucan/hemicellulose/lignin	SSF	*Saccharomyces cerevisiae*	$0.02–0.17\,g{\cdot}g^{-1}$ ub	Ethanol	Theuretzbacher et al. (2015)
Rice straw	Glucan/hemicellulose/lignin	CBP	*Fusarium oxysporum*	$0.12\,g{\cdot}g^{-1}$ ub	Ethanol	Xu et al. (2015)
Lignocellulosic	Hexose/pentose	AF	*Clostridium acetobutylicum*	—	Butanol/Ethanol	Tracy et al. (2012)
Corn stover	Glucan/hemicellulose/lignin	SHF	*Clostridium acetobutylicum*	$0.11\,g{\cdot}g^{-1}$ ub	Butanol	Wang and Chen (2011)
Sugarcane	Glucan/hemicellulose/lignin	SSF	*Clostridium beijerinckii*	$6.86\,g{\cdot}l^{-1}$	Butanol	Su et al. (2015)
Rice straw	—	CBP	*Clostridium thermocellum*	$6.9\,g{\cdot}l^{-1}$	Butanol	Kiyoshi et al. (2015)
Corn stalk	—	SSF	*Thermoanaerobacterium*	$0.007\,g{\cdot}g^{-1}$ ub	Hydrogen	Zhao et al. (2013)
Wheat straw	—	SSF	Digested sewage sludge	$0.005\,g{\cdot}g^{-1}$ ub	Hydrogen	Reilly et al. (2014)

ub = untreated biomass; pb = pretreated biomass; SSF = simultaneous saccharification and fermentation; SSCF = simultaneous saccharification and cofermentation; CBP = consolidated bioprocessing; SHF = separate hydrolysis and fermentation; CBP = consolidated bioprocessing; AF = anaerobic fermentation.

Table 2.7 Productions of organic acids by sugar fermentation of agro-food residues.

Biomass	Sugars	Strategy	Microorganism	Yield/ concentration	Product	References
Corn fiber	—	SHF	*Clostridium tyrobutyricum*	$6.45\,g\cdot l^{-1}$	Butyric acid	Liu et al. (2013)
Rice hulls	—	SHF	*Clostridium tyrobutyricum*	$1.70\,g\cdot l^{-1}$	Butyric acid	Liu et al. (2013)
Corn stover	—	SHF	*Clostridium tyrobutyricum*	$4.24\,g\cdot l^{-1}$	Butyric acid	Liu et al. (2013)
Wheat straw	—	SHF	*Clostridium tyrobutyricum*	$8.06\,g\cdot l^{-1}$	Butyric acid	Liu et al. (2013)
Sugarcane bagasse	Glucan/hemicellulose/ lignin	SHF	*Clostridium tyrobutyricum*	$0.50\,g\cdot g^{-1}$ hs	Butyric acid	Wei et al. (2013)
Fruit and vegetable waste	—	Solid state	*Actinobacillus succinogenes*	—	Succinic acid	Dessie et al. (2018)
Sugarcane bagasse	—	SHF	*Escherichia coli*	$0.79\,g\cdot g^{-1}$ ub	Succinic acid	Xi et al. (2013)
Corn stover	—	SSF	*Actinobacillus succinogenes*	$0.39\,g\cdot g^{-1}$ pb	Succinic acid	Zheng et al. (2010)
Corn stover	—	SHF	*Actinobacillus succinogenes*	$0.74\,g\cdot g^{-1}$ hs	Succinic acid	Salvachúa et al. (2016)
Corn stalk	—	SHF	*Escherichia coli*	$0.39\,g\cdot g^{-1}$ ub	Succinic acid	Wu et al. (2013)
Citrus peels	—	SHF	*Actinobacillus succinogenes*	$0.77\,g\cdot g^{-1}$ hs	Succinic acid	Patsalou et al. (2017)
Rice husks	Glucose	Submerge	*Aspergillus terreus*	$49\,mg\cdot g^{-1}$	Itaconic acid	Pedroso et al. (2017)
Lignocellulosic	Hexoses/pentoses	Fermenter	*Aspergillus terreus*	—	Itaconic acid	Saha et al. (2017)
Lignocellulosic	—	Fermenter	*Candida lignohabitans*	—	Itaconic acid	Bellasio et al. (2015)

(Continued)

Table 2.7 (Continued)

Biomass	Sugars	Strategy	Microorganism	Yield/ concentration	Product	References
Wheat bran	Glucan/hemicellulose/ lignin	SHF	*Lactobacillus pentosus*	$0.73\,g{\cdot}g^{-1}$ hs	Lactic acid	Tirpanalan et al. (2015)
Wheat straw	—	SSCF	*Bacillus coagulans*	$0.46\,g{\cdot}g^{-1}$ ub	Lactic acid	Zhang et al. (2014)
Corn straw	Xylose/glucose	Fermenter	*Rhizopus oryzae*	$0.35\,g{\cdot}g^{-1}$ hs	Fumaric acid	Xu et al. (2010)
—	Xylose/glucose	Fermenter	*Rhizopus arrhizus*	—	Fumaric acid	Liu et al. (2017)

SHF = separate hydrolysis and fermentation; SSF = simultaneous saccharification and fermentation; SSCF = simultaneous saccharification, ub = untreated biomass; pb = pretreated biomass; hs = hydrolyzed sugars.
Source: Diaz et al. (2018).

on the development of fumaric acid as a bio-based chemical, which can replace maleic anhydride in the production of unsaturated polyester resins. Fumaric acid is produced commercially via the catalytic isomerization of petroleum-derived maleic acid, with yields of up to 90% (Zhang et al. 2013).

2.2.2.3 Triacylglycerols and Polyhydroxyalkanoates

Among different lipids, TAG are exploited to generate lipid-based biofuels such as hydrocarbon fuels and biodiesel and also serve as a significant carbon storage molecule in animals, algae, plants, and microorganisms (Kurosawa et al. 2015). Oleaginous microorganisms are needed to generate TAG from agro-food by-products/residues such as corn stover, corncob, wheat straw, rice straw, groundnut shells, etc. because they generate more than 20% of their weight in the form of lipids (Deeba et al. 2017). Oleaginous hydrocarbon-degrading bacteria and oleaginous yeasts require nitrogen-restricting conditions and the existence of an extra carbon source to accumulate lipids (Pirozzi et al. 2015) (Table 2.8). Polyhydroxyalkanoates (PHAs) are a family of biodegradable and natural polymers that can be lowered to water and carbon dioxide (CO_2) by microorganisms in the environment (Du et al. 2001). Polyhydroxybutyrate (PHB) is the most widely used polymer to create coated paper, films, bags, compost, etc. PHAs made from renewable sources can substitute those produced from petrochemical ones, but they are expensive. Increasing the PHA production is induced by the reduction of nitrogen source in the media (Munoz and Riley 2008). Regarding PHB, it has been generated from many agro-food residues including rice bran, wheat straw, rice straw, wheat bran, etc. (Ong et al. 2018).

2.2.2.4 Syngas Fermentation

Today, renewable energy is necessary due to the constantly increasing global population, intensification of the industrial activities, reduction of the fossil fuel resources, as well as the greenhouse effect and the associated global warming (Roy et al. 2015). As a result, over the past few decades, the use of syngas fermentation to produce biofuels for renewable energy has gained more attention (Asimakopoulos et al. 2018; Daniell et al. 2012). Syngas or synthesis gas is a combination of gases (mainly CO_2, carbon monoxide, and H_2) that can be generated from the gasification of biomass. The proportion of each gas depends on the kind of biomass used as feedstock, the form of the gasifier, and the functional parameters of the gasification process (Molino et al. 2016). Gasification technology consists of fluidized-bed gasifiers and a fixed bed. Furthermore, syngas is important from an environmental perspective because it is a significant by-product of some industrial processes such as petroleum refining, steel milling, steam reforming, and others, including ignition and partial oxidation. The European Union has set a long-term goal (until 2050) to reduce its greenhouse gas emissions by 80–95% compared to 1990. Combining fermentation units and syngas on a commercial level has already been tried for the production of bioethanol (Dürre and Eikmanns 2015).

The production of volatile fatty acids (VFAs) using mesophilic alkaline fermentation of waste-activated sludge (WAS) with syngas conversion has attracted much attention. Because it can be utilized as merchandise chemicals, higher value materials such as long-chain fatty acids, alcohols, and PHAs, and also, and it can also be an easily available alternative for the common organic sources of carbon (Cagnetta et al. 2016; Rao et al. 2018).

Table 2.8 Productions of triacylglycerols by sugar fermentation of agro-food residues.

Biomass	Sugars	Strategy	Microorganism	Yield/concentration	Product	References
Corn stover	Glucan/hemicellulose/lignin	SSF	*Cryptococcus curvatus*	$0.16\,g{\cdot}g^{-1}$ pb	Triacylglycerols	Gong et al. (2014)
Corn stover	—	SHF	*Cryptococcus curvatus*	$1.04\,g{\cdot}l^{-1}$	Triacylglycerols	Kim et al. (2015)
Corn stover	—	SHF	*Rhodococcusopacus*	$0.14\,g{\cdot}g^{-1}$ hs	Triacylglycerols	Kurosawa et al. (2014)
Corn stover	—	SHF	*Rhodosporidium toruloides*	$0.07\,g{\cdot}g^{-1}$ ub	Triacylglycerols	Xie et al. (2012)
Wheat straw	—	SHF	*Cryptococcus curvatus*	$0.05\,g{\cdot}g^{-1}$ ub	Triacylglycerols	Yu et al. (2011)

SHF = separate hydrolysis and fermentation; SSF = simultaneous saccharification and fermentation; ub = untreated biomass; pb = pretreated biomass; hs = hydrolyzed sugars.
Source: Walker and Walker (2018).

VFAs have a higher value (revenue of more than five times) compared to biogas, which will raise the total economy to WAS bioconversion process. WAS bioconversion is very challenging, and biodegradation of WAS is difficult, and therefore, the hydrolysis recognition for anaerobic conversion is done at the rate-limiting step (Neumann et al. 2016). Furthermore, measures for prohibiting methanogens are needed for VFA or other products except for the biogas. So, it is critical to show the impact of the addition of syngas on the reduction of WAS and the alterations in the compositions of the microbial community. The addition of syngas to the anaerobic fermentation of WAS can raise VFA concentration, which could influence the conversion productivity of syngas because of inhibition from the products.

The company Ineos Bio was founded in 2008 but started its bioethanol project in 2001. In late 2012, the process was industrialized, leading to the production of ethanol (eight million gallons per year), which was indulgent to most impurities during the gasification production process and electricity (6 MW). The produced ethanol had 99.7% purity; therefore, it could be combined with gasoline. In 2017, Alliance Bio-Products Inc. received the approval of the United States Department of Agriculture (USDA) to buy the Ineos Bio plant with the company planning to enhance its cellulose to sugar technology. Lanzatech is nearby commercializing a process transforming steel mill off-gas to ethanol with two plants in China and Belgium, with an ethanol capacity of 16 and 21 million gallons per year, respectively. The company utilizes the properties of microbial strains for the bioconversion of hybrid separation systems and synthesis gas with recycling water to purify the product and gather the coproducts. Lanzatech will attempt to develop platforms for converting syngas to various products, for instance, butadiene, jet fuel, isopropanol, 2,3-butanediol, and isobutylene (Liew et al. 2016). Another company engaged in syngas fermentation was Coskata which promoted a semicommercial process for the fermentation of syngas to fuel chemicals. The company went out of business in 2015, and the technology was purchased by Synata Bio (Lane 2016).

2.2.3 Fermentation Processes in the Pharmaceutical Industry

Today, the fermentation industry is encountering rapid growth in the products' diversity, location, and scale-up of the processes, accompanied by huge investments for the relocation of large-scale bioprocesses, successful decoding of the human genome with its many innovative therapeutic targets, and the considerable advances in the construction of progressive fermentation systems for making new antibodies and proteins, considering receptor/drug interactions, action of membrane proteins in illness, health, and combating serious diseases such as cancers, Alzheimer's, and asthma (McNeil and Harvey 2008). Although innovative fermentation-derived medicines, including biopharmaceuticals, therapeutic proteins, and antibodies, are more complex and expensive, they have the prospect of enhanced specificity of care and reduced side effects. Therefore, it is an innovative period in clinical medicine as new drugs produced using fermentation processes are entering the market.

2.2.3.1 Drugs

A broad range of industrial applications of biosensor, food, and chemotherapeutic drugs is covered by L-asparaginase. Various strains of yeast and bacteria can produce L-asparaginase, but filamentous fungi are favored for SSF because they provide enough and natural environment for the microorganism, thereby leading to functional product development. Researchers reported that L-asparaginase was produced using rotary and tray bioreactors and optimized tri-substrate combination by using *Aspergillus tubingensis* IBBL1. In tray bioreactor, parameters such as the frequency of mixing and bed height were optimized, which had an influence on mixing methods, substrate loading, and the particle size of the substrate. In rotary bioreactor, baffles are designed for optimized conditions. Greatest L-asparaginase activity of 20.58 $U \cdot gds^{-1}$ (units per gram dry substrate) was observed by utilizing a tray bioreactor with a bed height of 1 cm; in contrast, in the in-house designed rotary bioreactor, L-asparaginase activity of 19.96 $U \cdot gds^{-1}$ was achieved with 1.5 kg substrate loading. These data demonstrate that the in-house designed rotary bioreactor was convenient to manage. Uncomplicated empirical models such as linear, logistic, and exponential models were used to depict growth kinetics (Doriya and Kumar 2018). In addition to L-asparaginase, monacolin K (lovastatin) is used as a drug for hypercholesterolemia therapy and decreasing low-density lipoprotein cholesterol (Di Cagno et al. 2013; Pereira and Vicente 2010; Soccol et al. 2017). Naturally, monacolin K can be generated by various fungal species such as *Aspergillus terreus*, *Monascus spp.*, *Paecilomyces viridis*, and *Penicillium citrinum* (Marsh et al. 2014; Pereira and Vicente 2010). Compared to other fungi, the use of *Monascus spp.*, for example, in the monacolin K production process, has various benefits including the nonpathogenicity and its use as traditional fermentation food in Asian countries. Different grains such as millet, rice, corn, wheat, and barley were used as raw substances for monacolin K production by SSF of *Monascus ruber*. Among these substrates, millet was known to be the best choice for monacolin K production, by which the productivity arrived to 7.12 $mg \cdot g^{-1}$. For increased monacolin K production, the impacts of fermentation time, inoculum volume, and primary moisture content were systematically examined in SSF of *M. ruber*. Results illustrated that the highest production of monacolin K (19.81 $mg \cdot g^{-1}$) could be achieved at the optimal situations (Zhang et al. 2018). In addition, nowadays, rice has been used as substrate for the production of monacolin K by SSF of *Monascus spp.*, by which the productivity of monacolin K could be up to 16.45 $mg \cdot g^{-1}$ (Feng et al. 2016). The use of other grain, agricultural residues, nongrain food, and neutral carrier as substrate has also been reported for producing monacolin K by SSF of *Monascus spp.* For instance, Lee and colleagues in 2006 utilized potato, dioscorea, and cassava as substrates to generate monacolin K by *Monascus purpureus NTU 301*. The results demonstrate that the best substrate for producing monacolin K at 2.584 $mg \cdot g^{-1}$ was dioscorea (Lee et al. 2006).

2.2.3.2 Recombinant Proteins

The industrial microbiologist is able to modify the wild-type organism and supply cultural situations to enhance the yield. This has been gained successfully for several years, by the choice of induced mutants, and the control of the process surroundings of the producing organism, as well as the use of recombinant DNA technology. New industrial production of recombinant proteins in mammalian hosts is influenced by the market changes and technology. In 1986, the first biological drug produced in recombinant mammalian cells was commercialized (Ecker et al. 2015). An evolution is in the change of traditional batch stages into continuous and mixed manufacturing. The health authorities such as the Food and Drug Administration (FDA) support this change. The major groups of recombinant proteins advanced as therapeutic agents are mentioned in Table 2.9 (Walsh 2014). The most ordinary cultivation ways used in biomanufacturing are perfusion and fed-batch. The use of each one depends on various factors connected to the host or to the protein (Bielser et al. 2018). Cells are cultivated, maintained in suspension, or bound to carriers. The batch bioreactor is the easiest mode for operation. After inoculation, cells grow until facing a restriction due to the media consumption and/or the accumulation of a toxic substance. The second ordinary process is fed-batch, where nutrient restrictions are hindered by the addition of highly concentrated feeds at different time sections during the cultivation. Therefore, in fed-batch mode, the culture duration is longer than in batch mode, and the final yield is higher. A simple approach to have a continuous process is called chemostat process. Actually, media is added at a constant flow rate, and the bioreactor substance is taken off at the same flow rate, without any cell retention (Henry et al. 2008). This method is more common for microbial fermentation due to the high division rate of bacteria. Finally, in perfusion, there is a constant *in* and *out* flow, but the cells are kept inside the bioreactor. From the past until now, perfusion is used for cell culture, and various commercial molecules are produced in this way, such as interferon β-1a (Rebif©), recombinant follicle-stimulating hormone (Gonal-f©), and factor-VIII (Kogenate-FS©) (Pollock et al. 2013). The majority of these molecules are changeable, and their indication range is

Table 2.9 The major groups of recombinant proteins developed as therapeutic agents.

Recombinant protein	Therapeutic group	Production system	Treatment
Factor VIII	Blood factors	Mammalian cells	Anemia
Tissue plasminogen activator	Thromobolytics	Mammalian cells, *Escherichia coli*	Clot lysis
Hirudin	Anticoagulants	*Saccharomyces cerevisiae*	Anticoagulant
Insulin	Hormones	*Escherichia coli*, *Saccharomyces cerevisiae*	Diabetes
Human growth hormone	Hormones	*Escherichia coli*, *Saccharomyces cerevisiae*	Hypopituitary dwarfism
Follicle-stimulating hormone	Hormones	Mammalian cells	Infertility
Glucagon	Hormones	*Escherichia coli*, *Saccharomyces cerevisiae*	Type 2 diabetes
Erythropoietin	Growth factors	Mammalian cells	Anemia
Granulocyte macrophage colony	Growth factors	*Escherichia coli*	Bone marrow transplantation
Granulocyte colony	Growth factors	*Escherichia coli*, mammalian cells	Cancer
Interferon-alpha	Cytokines	*Escherichia coli*	Cancers, hepatitis B, leukemia
Interferon-beta	Cytokines	*Escherichia coli*	Cancers, amyotrophic lateral sclerosis, genital warts

Source: Based on Walsh (2014).

expressed in micrograms. Fixed bed, spin filters, and gravity settlers are utilized for cell maintenance. These technologies provide an efficient split because the cells are bounded on a carrier. Physical parameters utilized for cell maintenance are density and particle size, which are larger for the carrier than for the single cell in suspension (Pollock et al. 2013). In another study, improving bacterial production of a recombinant cetuximab-Fab by biased humanization and its use for drug union was considered. Cetuximab is a murine–human illusory monoclonal antibody that is utilized in the treatment of epidermal growth factor receptor of positive cancers. As antibody fragment engineering has appeared as an economic optional to monoclonal antibody drugs via bacterial production, researchers searched for useful mutations for increasing productivity. Amino acid substitutions were chosen to be similar to the humanized sequence to prevent unpredictably elevating immunogenic problems by the mutations. So, FM329 showed high productivity in a fed-batch operation, arriving approximately 3.5 times productivity in comparison to FM318. Additionally, antibody–drug conjugates showed enlarged anticancer activity with retention of epidermal growth factor receptor-binding affinity (Sim et al. 2018).

2.3 Current Developments in Industrial Fermentation

The fermentation development strategy involves the knowledge in identifying the microbial diversity, industrial fermentation media, and variety of fermentation systems and recognition of inhibitors in fermentation, fermentation process modeling, fermentation optimization, etc. in order to achieve high productivity in large-scale fermenters and provision of better conditions for reducing costs.

2.3.1 Microorganisms

The striking biochemical variety of microorganisms has been exploited by the fermentation industry by separating strains from the natural environment to be able to generate products of commercial value. Isolation is the initial step in the screening of microorganisms of potential industrial application. A selection of the major culture collections is given in Table 2.10 (Stanbury et al. 2018). In some cases, it is suitable to design the isolation protocol in ways that the growth of potential producers is stimulated; in contrast, in other cases, the microorganisms must be isolated and the producers identified at a next stage. The important point is that the isolation must be economically viable, and the choice of the culture to be utilized is a compromise between the yield of the organism and the economic restriction of the process.

Today, there are many choices of yeast starter cultures for alcoholic fermentation. Industrial producers of alcohol in novel large-scale fermentation performances do not trust on instinctive fermentations with native yeasts, which are generally used in small-scale craft processes, such as those adopted by winemakers, brewers, or distillers (e.g. Brazilian cachaça production and Mexican agave-based spirits). In industrial semi-continuous fermentation processes, recycled yeasts are used. In some cases, freshly propagated yeasts are added stepwise in seed culture volume, and frequently, yeast cells are kept in lyophilized or cryopreserved vials for subsequent uses. The choice of yeast strain for

Table 2.10 Major culture collections and their corresponding website address.

Culture collection	Website
National Collection of Type Cultures (NCTC)	http://www.phe-culturecollections.org.uk
National Collection of Industrial Food and Marine Bacteria (NCIMB Ltd)	http://www.ncimb.com
National Collection of Yeast Cultures (NCYC)	www.ncyc.co.uk
UK National Collection of Fungus Cultures	http://www.cabi.org
The British Antarctic Survey Culture Collection	http://www.cabi.org
National Collection of Plant Pathogenic Bacteria	http://ncppb.fera.defra.gov.uk
American Type Culture Collection (ATCC)	http://www.lgcstandards-atcc.org/?geo-country=gb
Persian Type Culture Collection (PTCC)	https://irost.org/ptcc/micro-oraganism-databases
Deutsche Sammlung von Mikroorganismen und Zellkulturen (DSMZ)	https://www.dsmz.de
Centraalbureau voor Schimmelcultures (CBS)	http://www.cbs.knaw.nl
Czech Collection of Microorganisms (CCM)	http://www.sci.muni.cz/ccm/index.html
Collection Nationale de Cultures de Microorganismes (CNCM)	https://www.pasteur.fr/
Japan Collection of Microorganisms (JCM)	http://jcm.brc.riken.jp/en

Source: Stanbury et al. (2018).

alcoholic fermentations is important in terms of the economic benefits and product quality. It is noticeable to select an initial strain with the most suitable cell physiological background due to valuable industrial characteristics (Le Borgne 2012). Modern strain development processes are complicated, as shown in Figure 2.4. The purpose of strain development is engineering new yeast strains, which have stress endurance, better fermentation performances, enhanced processes, and high quality of the product. Genetic manipulation strategies are used for improving the traits of yeasts containing substrate usage, ethanol production, and stress endurance in fermentation. If genetically modified (GM) yeasts are used in biofuel alcohol processes and alcoholic beverages, the destination of coproducts should also be considered. Once prototype yeast strains have been checked in laboratory fermentations, they should be examined at pilot plant scale before commercialization (Basso et al. 2008). This technique includes culturing yeast in environmental stress or particular pressure, either in serial subculturing in batches or chemostat continuous cultures, to select a multitude of spontaneous mutant cells for growing in this condition (Caspeta et al. 2016). Various methods including rare mating, hybridization, mutagenesis, and protoplast fusion have been applied for enhancing the yeast traits for fermentation. Mostly, such methods have verified more successful than genetic engineering ways to enhance stress endurance in bioethanol yeast strains (González-Ramos et al. 2016). Nonetheless, they present some disadvantages, for example, the improvement of strains needs high-throughput screening methods to recognize the desired traits.

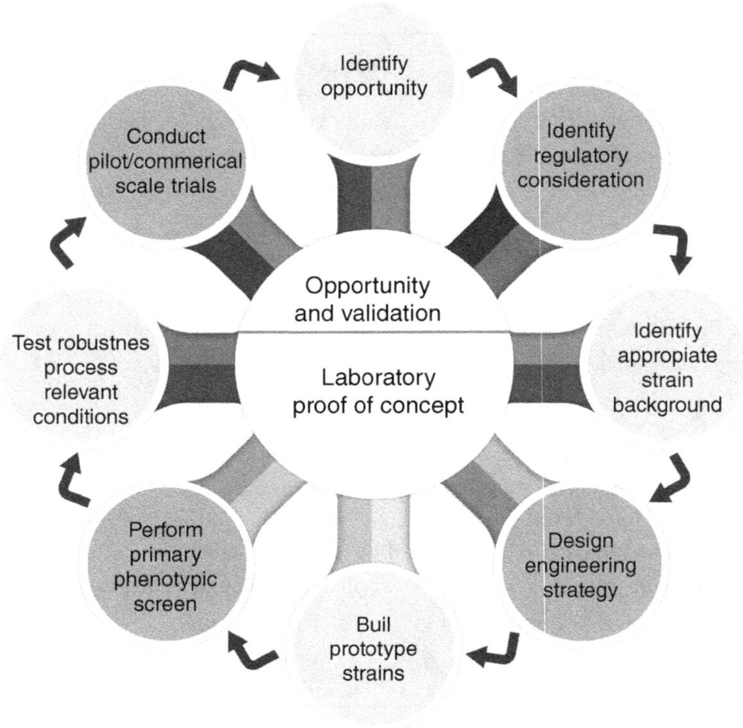

Figure 2.4 Yeast strain developmental cycle for alcohol fermentations.

Furthermore, the methods are not generally suitable for industrial strains with complicated genetic backgrounds. Nowadays, several molecular genetic techniques are available to enhance yeast strains for industrial alcohol fermentations, including gene editing, synthetic biology, and genetic engineering. There are two main kinds of genetic engineering in yeasts (recombinant DNA technology) for optimizing alcohol fermentation, which include transgenic and cisgenic methods. Self-cloning includes conveying genes from one yeast species to the identical species. In comparison to recombinant DNA or transgenic strains, regulatory confirmation of self-cloned yeasts is relatively easier (Walker and Walker 2018). For confirming genetic stability of industrial polyploidy or aneuploid strains of yeasts, homologous recombination of all engineered chromosome copies is necessary. Earlier successes contain modifications of alcohol-producing yeasts to decrease glycerol production. For example, by the introduction of GM yeasts in Brazil, the percentage of glycerol decreased between 20 and 30% in sugarcane ethanol plants, which led to high profitability (Walker and Walker 2018). An especially favorable area is strain engineering for cellulosic ethanol production. The successful operation of great-scale cellulosic ethanol biorefineries is more dependent on novel yeast strain evolvement. For instance, pretreatment and hydrolysis of lignocellulosic substances release a mixture of pentose and hexose sugars, which are not fermentable by natural yeast strains. Generally, GM yeasts with novel metabolic pathways and enzymes can supply the fermentation industries with considerable added value, enhanced product and by-product quality, and the ability to make new better-value by-products. Mainly, for bioethanol fermentations from corn mashes in the United States, different commercial GM strains of *S. cerevisiae* are accessible for use. Some of these strains are outlined in Table 2.11. The industrial widespread of GM microorganisms for the fuel alcohol subdivisions has begun since 2012, and it is anticipated to rise sharply in the future. GM yeasts are vital for second-generation ethanol production processes as well as for nonethanol chemicals and fuels. Apart from ethanol, the other biofuel alcohol generated by yeast fermentation is isopropanol, and GM strains of *S. cerevisiae* are now being used commercially for its production (Walker 2011). Generally, fuel alcohol processes depend on the genetically engineered yeasts, as they are able to enhance ethanol productivity, increase stress endurance, decrease dependency to nutrients or exogenous enzymes, decrease levels of secondary fermentation metabolites (such as glycerol), and alter normally nonfermentable carbon sources. Genetically improved yeasts also have a huge potential in beverage alcohol fermentations. The primary GM yeast used in fermentation was a brewing strain of *S. cerevisiae* advanced by the Brewing Research International in the United Kingdom. This yeast enables to perfect the fermentation of remaining maltodextrins in brewer's wort. The consequent beer (termed Nuffield Lyte) was the world's first beer made utilizing GM yeast and a low-sugar end product. The benefits of GM yeasts in beverage fermentations are considerable. One of them is flavor modification with winemaking yeasts and GM brewing. For instance, adjustment of the genes encoding sulfite reductase in brewer's yeast led to enhance beer taste stability by increasing the levels of sulfite during fermentation (Chen et al. 2012). Other ways of increasing brewing yeast strain fermentation performance consist of genetic modification of sugar carrier genes to raise sugar uptake. An extra vital point correlated to public opinions about GM technology for beverages and foods is mostly negative. Gene-revising techniques are an optional approach to recombinant DNA techniques in yeast, which led to enhanced productivity compared to traditional yeast genetic engineering methods. In particular, the introduction of technology based on Clustered Regularly Interspaced Palindromic Repeats (CRISPR)-associated protein-9 nuclease and CRISPR has proved to be specifically advantageous (Jamal 2018). For yeast gene altering, this technique allows the modification of all gene target sites despite the manipulation of the strain's genetic background. These modern gene-editing techniques can reduce stress endurance in yeast, having tolerance to high temperature, high ethanol, and acetic acid (Yang et al. 2013). In the field of synthetic biology, the conversion of lignocellulose to fermentable substances is one of the most promising areas to produce ethanol. Synthetic biology could be utilized in combination with genetic engineering to recognize the heterologous metabolic pathways of different strains, as well as for the optimization of codon to enhance translational efficiency. For example, an engineered strain of *S. cerevisiae* has recently been modified to express four cellulose-lowering enzymes on the cell surface in order to increase the productivity of bioethanol (Liu et al. 2016). It has attempted to use synthetic biology not only for biofuel industry but also for producing new taste compounds, such as "hoppy" flavors in beer or raspberry ketone in wine by yeast engineering. Synthetic biology is a fast-growing area. It is expected to design and construct synthetic yeast genomes with specific functions in the near future. By explaining the function and sequence of genetic material, the new possibilities of yeast synthetic biology occur in the world.

Table 2.11 Genetically modified *Saccharomyces cerevisiae* yeast strains for bioethanol production.

Commercial name	Substrate	Characteristics	Company (website)
ER-Xpress	Starch	Glucoamylase expression	Leaf Technologies (www.lesaffreadvancedfermentations.com)
Rich Yeast with GA	Starch	Glucoamylase expression	Richmond Chemicals (https://www.richmanchemical.com/
Innova Drive	Starch	Glucoamylase expression	Novozymes (www.novozymes.com)
TransFerm	Starch	Glucoamylase expression	Lallemand (www.lallemandbds.com)
SucraMax	Sucrose	Reduced glycerol	Lallemand (www.lallemandbds.com)
FT 858L	Sucrose	Stress tolerant	Leaf Technologies (www.lesaffreadvancedfermentations.com)
CelluX	Cellulose	Xylose isomerase expression	Leaf Technologies (www.lesaffreadvancedfermentations.com)
Excellulor	Cellulose	Stress tolerant	GlobalYeast (www.globalyeast.com)
CV-110	Cellulose	Xylose isomerase and 1-epimerase expression	Terranol (www.terranol.com)
C5 Fuel	Cellulose	Xylose fermentation and inhibitor tolerance	Lallemand (www.lbds.com)
XyloFerm	Cellulose	Xylose fermentation and inhibitor tolerance	Lallemand/Taurus Energy AB (www.lbds.com)

2.3.2 Fermentation Media

All microorganisms need energy, water, nitrogen, carbon, vitamins, mineral elements, and oxygen (O_2) for their survival and multiplication. In a tiny scale, it is easy to design a medium including pure compounds, but this task is more difficult in a wide-scale process. Some criteria must be considered to create a medium for large-scale production. Actually, it should provide the maximum productivity and a high rate formation and concentration of biomass or product per each gram of substrate consumed. Furthermore, the minimum yield of undesired products should be produced. In addition, product should have a firm quality and should be readily accessible during the year. Finally, it should cause no difficulties during sterilization, media making, and also during agitation, aeration, purification, extraction, and waste treatment (Chapter 1) (Stanbury et al. 1995).

2.3.2.1 Types of Media Sources

The use of cheap substrates such as cereal grains, sugar beet and cane molasses, ammonium salts, sugar carbon sources, nitrates, urea, soya bean meal, corn steep liquor, slaughterhouse waste, and fermentation residues/by-products, such as nitrogen sources, have the tendency to meet most of the parameters required for production media. In contrast, costly pure substances can be selected if the overall price of the whole process can be decreased. Table 2.12a shows the absolute minimum amounts of S, N, P, K Mg, etc. to include in a primary medium recipe. Absolutely, selecting media for large-scale fermentations has been based on accessibility and cost more than on microbial physiology. Table 2.12b shows animal- and plant-based complex media for the industrial cultivation of fungi, bacteria, and yeasts, which are employed in the food industry (Walker 2014).

2.3.2.1.1 Carbon Sources

Most of the food microorganisms are chemoorganotrophs, which acquire their energy and carbon by metabolizing organic substances such as sugars, carbon biopolymers, polyols, alcohols, fatty acids, organic acids, peptides, amino acids, and polypeptides. Sugars constitute the principal fermentable carbon sources for food microorganisms. At laboratory scale, among sugars, glucose is the commonly used sole carbon source for the growth of food microorganisms; in contrast, for industrial fermentation, it is used with other carbon sources such as fructose, sucrose, xylose, maltose, and lactose. Corn steep liquor, molasses, and sulfite waste liquor are complex carbon sources providing mineral and nitrogenous nutrients. Molasses composition varies depending on the geographic location and production process. Corn steep liquor, which results from the steeping of maize, is the water extract that is concentrated to 50% solids. It contains high levels of lactic acid. Sulfite waste liquor is the exhausted liquor, which is obtained from the paper pulping industry, and is remained after digestion of wood to cellulose pulp by calcium bisulfite under pressure and heat (Walker 2014). In industrial yeast alcohol fermentations, the most important raw materials are sugar-rich plants and cereals. Two commercial enzymes – glucoamylase for saccharification and α-amylase for starch liquefaction – are needed for bioethanol processes, which lead to high ethanol concentrations at the end of fermentation, mostly over 20% v/v. Other non-starch polysaccharides could be used as fermentation raw materials. Fermentations of sugar-rich plants are easier to compare with others. For example, crushing of sugarcane leads to the release of sucrose, glucose, and fructose. Yeasts with great invertase activity,

Table 2.12 Minimum quantities of elements to include in an initial medium (a), selected fermentation media for food microorganisms (b), and commonly employed nitrogen sources in food fermentation media (c).

A.

Element	Bacteria	Yeasts	Fungi
Carbon	50–53	45–50	40–63
Hydrogen	7	7	—
Nitrogen	12–15	7.5–11	7–10
Phosphorus	2.0–3.0	0.8–2.6	0.4–4.5
Sulfur	0.2–1.0	0.01–0.24	0.1–0.5
Potassium	1.0–4.5	1.0–4.0	0.2–2.5
Sodium	0.5–1.0	0.01–0.1	0.02–0.5
Calcium	0.01–1.1	0.1–0.3	0.1–1.4
Magnesium	0.1–0.5	0.1–0.5	0.1–0.5
Chloride	0.5	—	—
Iron	0.02–0.2	0.01–0.5	0.1–0.2

B.

Media	Microorganisms
Barley malt wort	Yeasts (*Saccharomyces* spp.)
Cereal wort	Yeasts (*Saccharomyces* spp.)
Ethanol	*Acetobacter spp.*
Rice hydrolysate	*Aspergillus oryzae*, yeasts
Sugarcane and sugar beet molasses	Yeasts, fungi, bacteria
Wine must, fruit juices, honey	Yeasts, lactic acid bacteria
Milk, cheese whey	Lactic acid bacteria, yeasts, fungi
Starch hydrolysates, glucose syrups	Fungi, yeasts, bacteria
Water, CO_2, sunlight	Chlorella, Scenedesmus, Spirulina
Solid substrate media	
Soy, wheat	*Aspergillus oryzae*, yeasts, lactic acid bacteria
Wheat flour	Yeasts, lactic acid bacteria
Peanut press cake	*Neurospora sitophila*
Meat, fish	Lactic acid bacteria, fungi
Plants, vegetables	Lactic acid and other bacteria
Straw, manure, sawdust	*Pleurotus spp. Agaricus bisporus*, *Volvariella volvacea*
Oak wood	*Lentinula edodes*
Milk, curd	*Penicillium spp. Propionibacterium spp.*
Wheat bran	Fungi (e.g. *Aspergillus niger*)
Tea leaves	*Acetobacter xylinum*, *Schizosaccharomyces pombe*

Table 2.12 (Continued)

c.

Organic N sources	Inorganic N sources
Corn steep liquor	$(NH_4)_2SO_4$
Casein hydrolysate	NH_4Cl
Soybean meal	NH_3
Yeast extract	$(NH_4)_2HPO_4$
Barley malt	$(NH_4)_2PO_4$
Dried distillers grains with solubles	NH_4NO_3
Pharmamedia (cottonseed flour)	NH_4OH
Corn gluten meal	Urea
Linseed meal	
Rice and wheat meal	

splitting sucrose into fructose and glucose, are used for sucrose fermentations. Nowadays, commercial microalgae such as the green chlorophyta algae are used as substances for ethanol production (Doan et al. 2012). A wide range of sugars is associated with the osmoregulatory function and cellular structure across various species of microalgae. For example, between chlorophytes, the cell wall of some *Chlorella* spp. consists of hemicelluloses and cellulose; in contrast, the one of *Chlamydomonas* spp. has hydroxyproline glycosides including galactose, glucose, and arabinose (Miller et al. 1972). The carbohydrate productivity from biomass depends on the cultivation conditions and the selected microalgae. Some researchers report the total carbohydrate content of microalgae up to 0.5–0.6 dry weight, which contains monosaccharides and starch (Choi et al. 2010). Among the structural, storage, and physiological carbohydrates present in microalgae, the kinds and amounts of basic monosaccharide ingredients are potentially suitable for conversion into ethanol.

2.3.2.1.2 Nitrogen Sources

Inorganic nitrogen sources such as ammonium salts, gaseous ammonia, diammonium hydrogen phosphate, and ammonium sulfate are extensively used in fermentation processes. Inorganic and organic nitrogen sources in food fermentation media are shown in Table 2.12c. Urea and nitrate can be used as nitrogen sources. Urea, in particular, can be used as a low-priced nitrogen source in some industrial fermentation media, such as molasses. However, it is not suitable for the production of beverages by yeast fermentation due to the formation of carcinogenic ethyl carbamate during the distillation process. Complex organic forms of nitrogen are established in different kinds of hydrolyzed plant protein material, for instance, casein hydrolysate, corn steep liquor, barley malt, soybean meal, and yeast extract supply combinations of amino acids and peptides that inevitably provide higher rates of fermentation and growth than those obtained utilizing inorganic nitrogen sources. Peptones are protein hydrolysates, which are proportionately costly sources of nitrogen for industrial applications (Walker 2014). Alcohol producers should be attentive of available minerals and nitrogen because nitrogen deficiency (typically about $150\,mg{\cdot}l^{-1}$ in cereal mashes) can lead to a decrease in the fermentation yields.

2.3.3 Fermentation Systems

Nowadays, different types of fermenters are used, including batch, fed-batch, semi-continuous, and continuous. However, fed-batch and batch fermentations are the most popular processes used. A new generation of fermentation systems is being developed to increase productivity and yield, which will be discussed below.

2.3.3.1 Solid-State Fermentation Bioreactors

Various kinds of bioreactors are used for SSF processes such as rotating disc reactors, tray bioreactors, column-tray bioreactors, packed-bed bioreactors, rotating horizontal drum bioreactors, air-pressure pulsation solid-state bioreactors, stirred-drum bioreactors, air-lift bioreactors, fluidized-bed bioreactors, and immersion bioreactors (Rodríguez-Couto 2018). Choosing the appropriate bioreactor to carry out an SSF process requires knowledge of the morphology of the microorganism used and also consideration of the process parameters and conditions (e.g. pH, temperature, support type, aeration, moisture, and water activity). The benefits and drawbacks of the four principal kinds of SSF bioreactors are shown in Table 2.13 (Ge et al. 2017). Actually, different categories of SSF bioreactors are commercially available; the first one is based on the mixing system used, i.e. stirred tank bioreactors (e.g. drum stirred, horizontal drum) and static bioreactors (e.g. perforated trays, fixed bed). The second one is in accordance with the kind of aeration (i.e. accompanied or unaccompanied by forced aeration) (Durand 2003). These bioreactors such as packed-bed, tray, fluidized-bed, and horizontal drum bioreactors, were promoted to improve new bioreactors with greater design (Figure 2.5). The design of the bioreactors takes into account the diversity of the substances used as growth media and their characteristics such as size, composition, porosity, strength, and water holding capacity, as well as fungus morphology (the presence or absence of septate hyphae), which affects the selection of the method of agitation (i.e. stirring from time to time, continuously, without stirring). In addition, aeration by forced or diffusion method should be considered.

2.3.3.1.1 Static Bioreactors

Various kinds of static bioreactors are used for SSF, from pilot scale to industrial scale, and are detailed below.

Fixed Bed Bioreactors Packed-bed bioreactors or Raimbault columns are filled with the solid substance or supports (Raimbault and Germon 1976). Column bioreactors are closed systems with compulsory aeration and a controlled temperature. Forced aeration influences the growth of the microorganism by estimating the respiration kinetics (i.e. CO_2 produced and O_2 consumed). These bioreactors are used at laboratory scale. They present some benefits including the restricted bacterial dissemination due to the use of closed system, the fact that the fermentation and the extraction processes of the final product are done in the same bioreactor (Spier et al. 2011), the low temperature gradients because of the convection caused by the air injection in the reactor, the removal of the released CO_2 by the metabolic reactions, and the well process control. The most ordinary design of packed-bed bioreactors contains of a vertical jacketed column packed with the

Table 2.13 Main types of solid-state fermentation bioreactors along with their advantages and disadvantages.

Type of SSF bioreactor	Strategy	Advantages	Disadvantages
Tray	Unforced aeration, no mixing	Less energy consumption	Gas concentration gradients
		Simple operation	Temperature gradients
			Large operating area
			Labor intensive
Packed-bed	Forced aeration, no mixing	Cost, space, and labor-effective	Problem of heat removal on a large scale
		Easy process control	Pressure drop through the bed
			Improper aeration inside the bed
Rotary-drum	Unforced aeration, continuous or intermittent mixing	Gentle and uniform mixing	Particle agglomeration over time
			Continuous rotation may cause shear stress
			Difficulties to control heat and mass transfer
Fluidized-bed	Forced aeration, continuous mixing	Uniform particle mixing	Difficult fluidization for particles of different sizes
		Good aeration	Sticky particles may form agglomerates
		Uniform temperature gradient	Continuous fluidization may cause shear stress
		Channeling and clogging are avoided	

Source: Modified from Ge et al. (2017).

microorganism and the support, which is ventilated from the bottom with the presence of internal heat transfer plates. Therefore, the fundamental design of packed-bed bioreactors is the introduction of air by way of a sieve that upholds the substrate. For enhancing the function of packed-bed bioreactor, a prepilot scale is designed which has some characteristics such as the check and optimizing the airflow rate, air-entrance temperature, and the addition of agitation and water during an SSF process. The geometry of the stirred bed reactor is comparable to that utilized in fixed bed but contains a mixing system which is used for higher scales (Spier et al. 2011).

Perforated Trays Bioreactors Tray bioreactors are the simplest SSF bioreactors. They comprise a thermostatic chamber with flat pierced trays. Humidified air is moved over and under the bed, and water is dispersed to keep the surroundings saturated (Planinić et al. 2016). The function of tray bioreactors is laying out the substrate (from 5 to 15 cm)

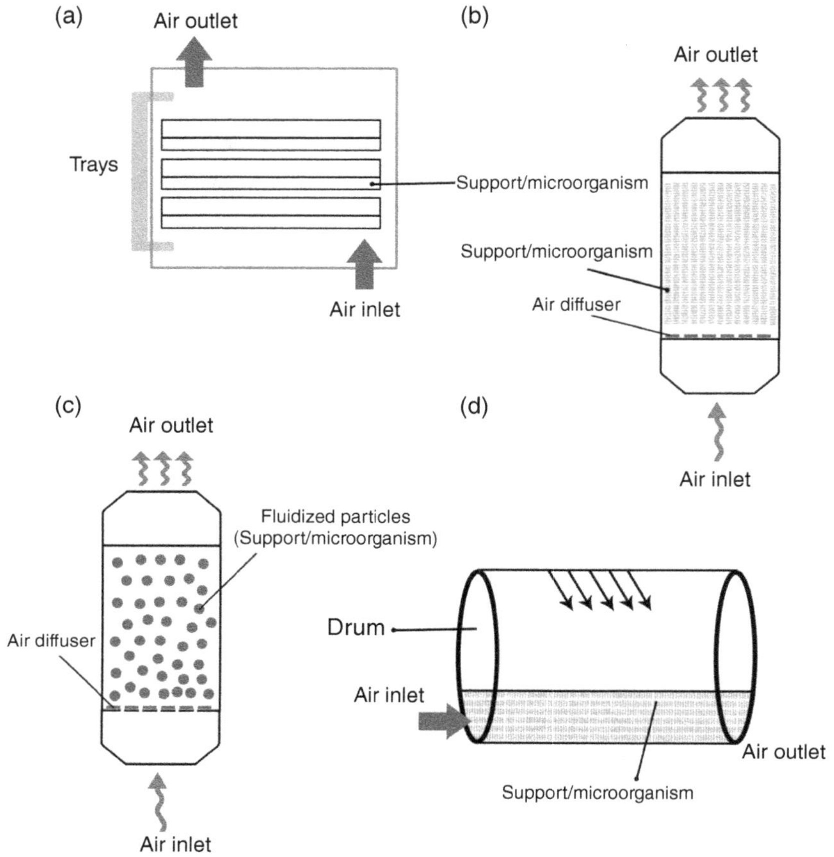

Figure 2.5 Main types of bioreactors: (a) tray, (b) packed-bed, (c) fluidized-bed, and (d) rotary-drum.

on trays, which are made by stainless steel or wood with controlled temperature and humidity, and which are pierced to make easier air convection. The benefit of this system is convenient scale-up, but the disadvantages are that it requires a large amount of space to operate, there is a problem with contamination, it is labor-intensive, and it is difficult to transfer O_2 due to the formation of mycelium, which alters the efficient diffusivity and porosity, especially for higher substrate layers.

Fluidized-Bed Bioreactors Fluidized-bed bioreactors are upright columns including substrate particles settled by the microorganism fluidized regularly by air blown through a drilled base plate at the bottom.

2.3.3.1.2 Stirred Bioreactors

Continuous stirring of the solid medium was done for SSF by a horizontal paddle mixer or spinning drum with or without a water jacket (Durand 2003). The merits of this system are high uniformity of the solid medium and better O_2 convey to the microorganism.

***Koji* Bioreactors** One type of static nonsterile bioreactor is a rotary-type automatic *Koji*, which is extensively used in Asian countries and marketed by Fujiwara in Japan. The treated substance is piled up on a rotary disk based on the working volumes (maximum thickness of 50 cm) and disk diameter. A microcomputer is used for controlling the airflow rate, temperature of the air-entrance, and agitation periods. The disadvantage of this system is the introduction of the substrate before filling the reactor in another equipment (Spier et al. 2011).

Horizontal Drum Horizontal drum bioreactors consist of baffles for medium stirring. The agitation can be irregular or continuous, which leads to damage the mycelium structure (Durand 2003).

Mixing Bioreactor This reactor is completely automatic for process control during sterilization of the medium, fermentation, and recording of information.

Rotating Drum Rotating drum bioreactors comprise a flat cylinder with baffles in the inner wall for uniform and gentle mixing, partly stuffed with the microorganism and support. Rotating the whole bioreactor around its main axis carries out successive mixing. The engineering principle of rotating drum bioreactors consists of generating biofuels using cellulosic materials (Spier et al. 2011). According to Durand (2003), the biggest reactor was a 200-l stainless steel rotating drum, which used 10 kg of steamed wheat bran as substrate for kinetic studies of *Rhizopus* (Durand 2003).

2.3.3.1.3 Recent Patents in Solid-State Fermentation Bioreactors

SSF became at the center of study from the years 1950 and was first patented in 1962 by Riker Laboratories, Inc. Today, a subsidiary of 3M Company is producing pyrogen-free penicillinase, which is used in injection remedy for tackling allergic sensitivity to penicillin (Soccol et al. 2017). Figure 2.6a presents the distribution of SSF-related patent applications among the countries. Above 80% of beverages and fermented foods produced are made via SSF in China (Liu et al. 2004). The majority of recent patent applications respecting bioreactor engineering have been emphasizing for enhanced stirring device and adjustment for continuous cultures, containing link with downstream processes, and better controlling systems. One of the patents engineered by Lang and Zhang (2010) for SSF was an automatic chain-belt-type bioreactor. The novelty of this apparatus is that the inoculating tank is being linked with the fermentation reactor through a spiral conveyer. The transmission and departure of materials are done in an automatic manner; therefore, the input to the second fermenter gets the output from the first one (inoculating tank). At least, a patent receives one International Patent Classification (IPC), which relates to its act and application. A patent may get higher than one IPC containing the invention's discovery, creativeness, and industrial applications. The SSF patent applications in different classes were reported according to their received IPC in Figure 2.6b. The greatest one belonged to beverage and food technologies, which comprised fermented alcoholic and nonalcoholic beverages. In recent times, attention has been devoted to the technologies generating beverages and foods with medicinal characteristics, such as reducing blood fat, antioxidant activity, assisting digestion, and enhancing greater blood distribution.

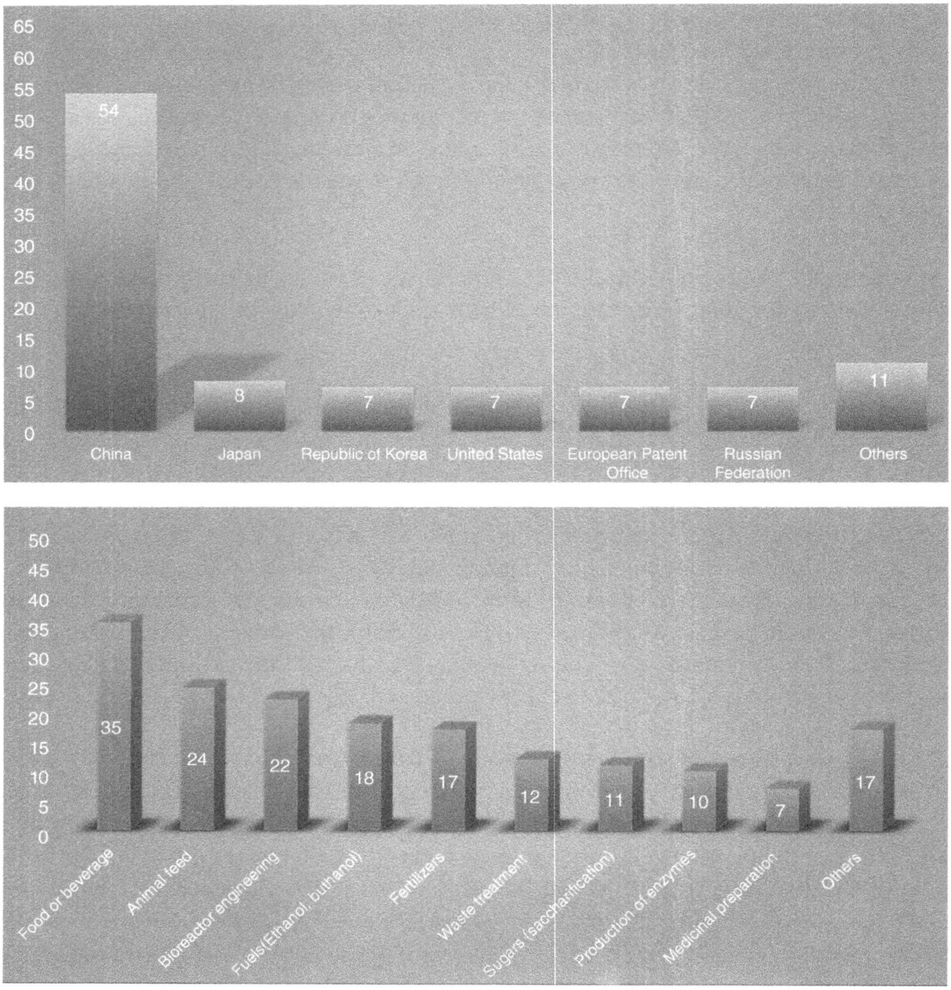

Figure 2.6 Distribution of solid-state fermentation–related patent applications between the countries that contribute more to this technology development (a) and patent applications of solid-state fermentation classified according to their function (b). *Source:* (a) Soccol et al. (2017).

2.3.3.2 Ultrasonic Fermentation Process

The food fermentation industry needs new methods to enhance the quality of the fermented products and the yield along with the novel analytical devices to observe and study the complicated fermentation processes. A new monitoring and processing technology consisting of ultrasound has been considered to improve the process efficiency and yield of food fermentation (Ojha et al. 2017). One of the innovations in fermentation processes is the use of ultrasound for monitoring the advance of fermentation or influencing its progress and controlling the activity of microorganisms. Applications of low- and high-frequency ultrasound have been investigated for several years in the fermentation industry. High-frequency ultrasound (>2 MHz) is used as a nondestructive analytical technique for observing fermentation processes (Novoa-Díaz et al. 2014). It has been widely used for measuring the alterations in chemical composition during fermentation by preparing real-time data on reaction advances. Whereas, low-frequency ultrasound (20–50 kHz) has been employed for enhancing the fermentation rates, mass transfer, and cell penetrability which lead to enhanced efficiency, pasteurization, and other unique processing applications involving wine aging and maturation and deaeration or degassing of alcoholic beverages. Ultrasound is also used for the elimination of microorganisms inhibiting the process. One of the most provoking new fields is the ultrasonic impact on living cells by utilizing lower intensity ultrasound at a sublethal level. The application of ultrasound in biotechnology, especially in the field of enzymes or cells that are activated by ultrasonic waves, is of great interest. Some researchers showed that the cellular metabolism could be modified by the application of low-intensity ultrasonic waves. These waves could also enhance the mass transfer of products and substrates through the border layer or the cellular membrane and wall. In the case of enzymes, the rise in the mass transfer rate of the reagents to the active area appears to be the most significant factor. The use of ultrasound to improve microbial yield in sonobioreactors was reviewed by Chisti (2003). The main key consists of modifying the acoustic energy delivered to a cell suspension and reducing the impact of cavitation by changing the acoustic power and frequency of ultrasound, as well as the exposure time. The influence of low-intensity ultrasound is the result of acoustic and microstreaming. It leads to low cavitation damage, which has beneficial effects such as enhancement of microbial reactions (e.g. fermentation), raised transfer of materials across the cell membrane (e.g. gene transfer [sonoporation]), and activation of enzymes in enzyme-modulated reactions.

2.3.3.2.1 High-Frequency Ultrasound in Fermentation

The traditional procedure of observing fermentation processes and classical chemical analyses takes a long time, and they are out of control. In contrast, ultrasonic waves can be used to either multiphase or homogenous systems and provide useful data about fermentation processes in real time. Measurement systems based on acoustic wave propagation are sanitary, fast, low priced, nonintrusive, and appropriate for automation. For monitoring concentrations in food composition, solutions, food structure, physical state, and molecular properties, online measurement can be utilized. High-intensity ultrasound application in food fermentation is shown in Table 2.14. Various ultrasonic parameters involving power reduction and time of application, pulse-echo, and single pass "through transmission" mode have been employed for the assessment of fluid density in different systems (Kuo et al. 2008). For example, Novoa-Díaz et al. (2014) reported that an alteration in ultrasonic speed is strongly related to the concentration (0–8 $g \cdot l^{-1}$) of lactic and malic acids during red wine fermentation.

Another application of ultrasound is the measurement of the concentration of yeast cells in liquid suspensions by higher frequency ultrasound (>15 MHz). For achieving this goal, a backscattering ultrasound method is used with an ultrasonic receiver or emitter broadband-focused transducer centered at 75 MHz. The technique represented enhanced sensitivity detection of yeasts at low concentration of 10^4 cells$\cdot ml^{-1}$ (Elvira et al. 2016). After backscattering, the pulse spectrum arriving at the transducer by a yeast cell depends on the reduction of the medium, nature and size of the cell, and the traits of the sound wave.

Table 2.14 Application of high- and low-frequency ultrasound in food fermentation.

Product	Ultrasound frequency	Applications	References
Alcoholic fermentation	2 MHz	Density and ultrasonic velocity in the ternary mixture (water–ethanol–sucrose)	Resa et al. (2005)
Wine fermentation	1 MHz	Monitoring of malolactic fermentation process	Novoa-Díaz et al. (2014)
Yogurt fermentation	3.7 MHz	Monitoring of phase change from liquid to gel	Ogasawara et al. (2006)
Malolactic fermentation in wines	1 MHz	Predict the end point of the malolactic fermentation process, malic and lactic acid concentrations	Amer et al. (2015)
Beer fermentation	2 MHz	Ternary system water–maltose–ethanol with respect to density, speed of sound, and temperature	Hoche et al. (2014)
Sweet whey fermentation	Low frequency	To improve fermentation process of sweet whey	Barukčić et al. (2015)
Probiotic fermented milk	20 kHz	Milk fermentation	Nguyen et al. (2012)
Rice wine fermentation	20 kHz	To accelerate the aging of different wines	Chang and Chen (2002)
Red wine fermentation	26 kHz	As an alternative means to control microbial wine spoilage	Luo et al. (2012)
Model wine with ethanol solution	25 kHz	Enhancement of release phenolics from oak chips into the wine model	Tao et al. (2014)
Model wine with water–alcohol	40 kHz	Yeast lysis of light lees in model wine	García Martín et al. (2013)
Red wine and model with water/ethanol	50 kHz	Accelerate aging on lees of red wines and its repercussion in sensorial parameters	Kulkarni et al. (2015)
Red wine and the model wine	100 kHz	Investigate the formation of free radicals generated by ultrasound	Zhang et al. (2016)

2.3.3.2.2 Low-Frequency Ultrasound

Numerous applications of low-frequency ultrasound in food processing have been reported in food fermentation processes such as foam destruction, enhancing microorganism and enzyme performances, emulsification, and enhancing end product safety and quality. The application of low-frequency ultrasound in food fermentation is outlined in Table 2.14.

2.3.3.3 Electrofermentation

Industrial fermentation technologies have several limitations in different aspects application, environmental sustainability, and economic possibilities. Fermentations are redox-imbalanced, so for combating this problem, some traditional strategies should be applied such as the addition of cosubstrates, gas sparging (O_2, H_2, N_2, air), and pH control. Also, checking the chemical conditions of the medium was an important option for consideration (e.g. chelating agents, base or acid, antifoam agents, buffers). In addition, the cost of downstream processing for product purification and separation might be responsible for 80% of the total cost. Therefore, by announcing electrodes, the possibility of controlling and optimizing the fermentation environments for achieving higher purity products can be obtained. A new method which was recently used for achieving higher purity products and stimulation of microbial cell growth is electrofermentation (EF) (Cate and Ball 2016). EF is the combination of conventional industrial fermentation with electrochemistry. A required electrical field causes effects on the microbial metabolism and fermentation environment in an oxidative or reductive mode. The other advantages of this method consist of generating target biochemicals with enhanced selectivity, restriction in the use of supplements for redox balance or pH control, and rise of carbon efficiency. Many researches have used the electrical potentials to fermentative microbial populations (Figure 2.7). The principles of EF consist of the microbial transformation of organic molecules improved by mediated electron transfer. As a carbon source, microorganisms utilizing inorganic mixture as electron acceptors or givers are named lithoautotrophic or lithotrophic for organic compounds or CO_2, respectively. This conduction can be obtained via direct transmembrane transport and extracellular electron transfer, by utilizing cytochromes, nanowires, oxidoreductase, NADH–ubiquinone, hydrogenase, and other membrane-bound enzymes. Actually, chemoorganoheterotrophic organisms that usually find electron acceptors or givers in the shape of organic molecules can be electrochemically active (Kracke et al. 2015; Schievano et al. 2016). VFAs and alcohols are fascinating target products in fermentation because they can be compared by hydrophobicity and volatility, although as a result of strong toxicity to bacteria, they are problematic in production. An electrochemical extraction can extract them from a fermentation broth, and this can hinder product restraint (Gildemyn et al. 2015). The extraction yield greatly depends on the product concentration, although it must be below the inhibition thresholds. Because of toxicity, a particular balance is needed for the production and recovery of VFAs.

2.3.4 Fermentation Optimization

The optimized processes of fermentation are the commercial key performances at industrial scale. The optimization of the fermentation process contains the strategic analysis of the stability of a project. Optimization takes place the possibility of the production in the selected organism. This shows that an expression system has been constructed, and the theoretical system should be regarded as the optimum. The purpose of fermentation process optimization is reaching the volumetric production rates as high as possible and achieving high yield, which decrease the price of the downstream processing steps. The fermentation development strategy involves the knowledge of the strain behavior and the recombinant proteins expressed. The optimization of a heterologous protein production begins with the construction of a well-designed recombinant strain (Thiry and Cingolani 2002). Researchers considered high productivity of prolyl aminopeptidase from recombinant *Bacillus subtilis* under batch fermentation based on examination of fermentation kinetics. Prolyl aminopeptidase in conjunction with other enzymes demonstrated well characteristics, which were suitable for reduction of rice protein, and the hydrolysate of rice protein had α-glucosidase suppressive activity. Therefore, prolyl aminopeptidase has

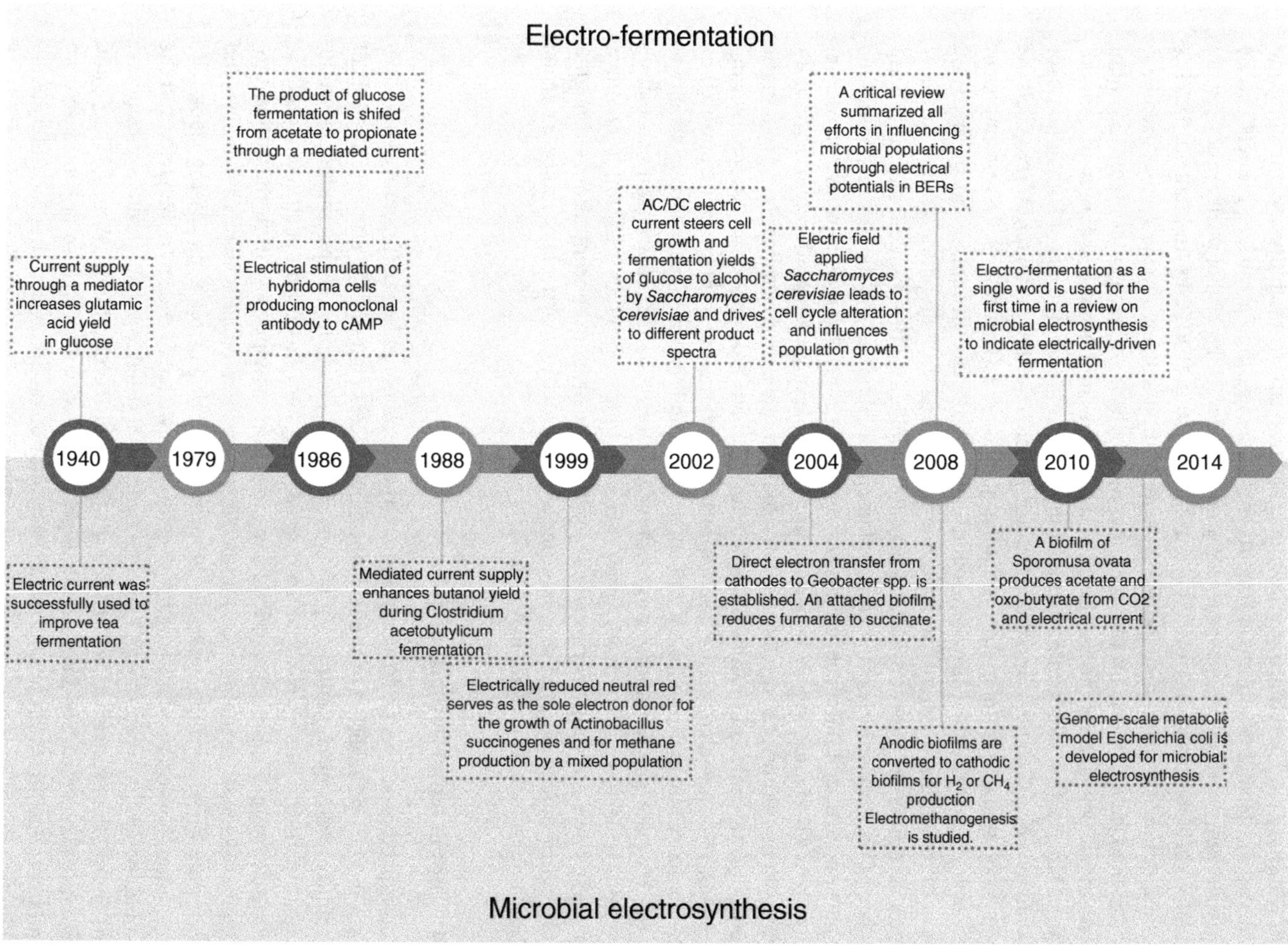

Figure 2.7 Historical overview of efforts to influence fermentations by electrical current.

good ability in the arrangement of bioactivity substances and in food processing. In spite of that it is expensive, the yield of prolyl aminopeptidase was increased by batch fermentation (Wang et al. 2018). Various factors affect the quality of fermented products such as the cultivation methods, the climatic conditions, the geographical surroundings (Bokulich et al. 2014), as well as the conditions of yeast and fermentation process. For example, researchers optimized the fermentation conditions and flavor compounds of greengage wine by utilizing principal compound design experiments and response surface methodology (RSM), involving fermentation temperature, sugar component, and ratio of material to liquid as factors. The results illustrated that the optimized conditions such as material-to-liquid ratio, sugar content, and fermentation temperature for fermented greengage wine were 1:20, $180\,g{\cdot}l^{-1}$, and 20°C, respectively (Tian et al. 2018). Yang et al. (2018) considered optimization of ectopic fermentation system (EFS). EFSs are composed of complicated microbial communities, which contain useful thermophilic microbes combined with litter, which are used to overwhelm the restrictions of existing wastewater cares (Guo et al. 2013; Yang et al. 2018). EFS operates in a dynamic manner on the basis of the inoculation of complicated microbial agents and continuous addition of assembled cow wastewater. In comparison to common static fermentation, this method is more advanced due to the proportionate stability and lower impact of the outer environment. Researchers appraised the impacts of padding materials, various bacterial species, and addition of cow fertilizer on EFS action in a small-scale fermentation bed. EFSs including generally maize straw with 20% cow fertilizer and injected with six thermophilic bacterial strains presented great fermentation action. The survival of external source of thermophilic bacteria, mainly *Bacillus*, after fermentation showed that the inoculated bacteria perform a significant function in advancement of EFS performance. Although the optimization is time-consuming and expensive, the influence of the strain stability is recognized, at least for a number of generations required for large-scale fermentation and cell banking, including the preparation of the precultures. Expression systems on the basis of plasmids are occasionally unstable, and the stability depends on the host strain, size of the added DNA, duplicate number of plasmids, and culture conditions (e.g. temperature, media composition, growth rate, and phase). The choice of the producing strain and the expression vector depend on the constitutive or inducible expression into the cytoplasmic compartment or the secretion into the outer medium. The optimization of the codon usage of the gene of interest for facilitating the expression in the chosen microorganism is necessary. Using a shake flask for immediate expression screening to find a clone producing the recombinant protein at a high level of expression is another point of view.

On the other hand, different factors are associated for scaling up a fermentation process such as the medium sterilization, number of generations, the alteration probability, the optimum temperature and pH controlling, agitation, pressure, and aeration. The great approach for scaling up a procedure is firstly to scale-down to the pilot scale of the culture situations, which will be utilized at the ultimate scale of production (Mandenius and Brundin 2008). Then, by increasing the scale, the broth will become more heterogeneous. The fermentation protocol can be improved at the ultimate scale of production, when the validation of all equipment for excellent manufacturing practices is approved, which contain the suitability of the installation, sterilizing in place (SIP), acceptable of cleaning in place (CIP), repairs, and calibration scheme (Singh and Satyanarayana 2008).

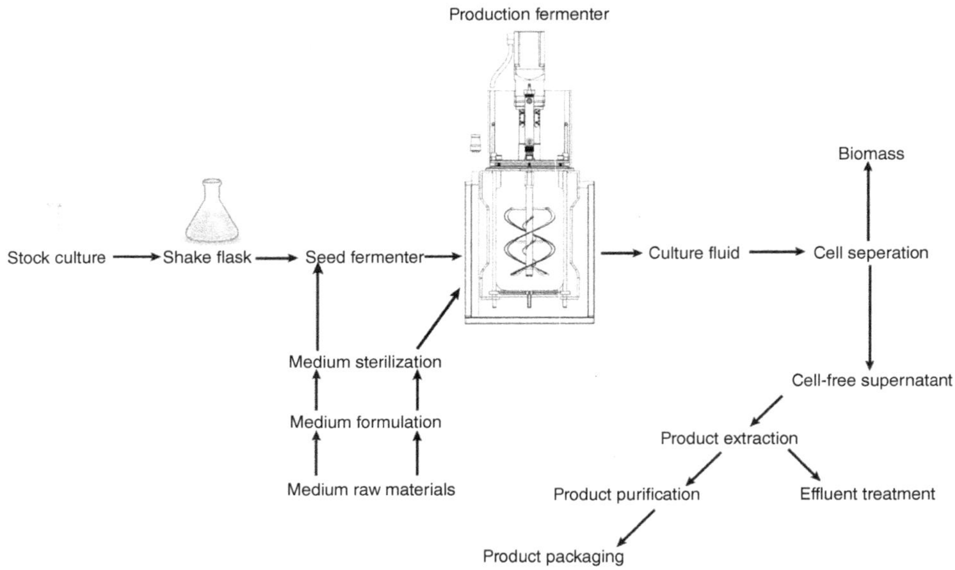

Figure 2.8 Schematic of fermentation process.

The development of fermentation process considers the feedback from the downstream processing development, the final scale of production, and the good manufacturing practice requirements. The main goal of optimization is increasing the production, at the primary stage; this process is done at a laboratory-scale purification process, and some quality control tools are available to quantify and assess the quality of the product (Figure 2.8). The fermentation protocol influences the impurity profile and the efficacy of the downstream processing. Also, the solubility or insolubility of the protein of interest can be determined by considering the fermentation conditions, which affect the downstream processing and consequently the quality and yield of the purified product. The fermentation must be improved concomitantly with the refinement because they will influence each other. The relationship between the development of fermentation and the downstream processing approves the success of the process scaling up. Fermentation process optimization can be done by altering one factor at a time or by changing various factors at the same time and seeking mutual actions by utilizing statistical examination. This examination provides an opportunity for imagining the interactions between empirical variables, which are influencing the yield or the characteristic of the product (Formenti et al. 2014).

However, fermentation can be performed in continuous mode; continuous culture is not usual in the industry because of the high contamination risk. Although batch processes are uncomplicated, the just approach to achieve a high cell density is the fed-batch operation with high complexity but with controlling the metabolism of the strain. For example, in *S. cerevisiae*, the high concentration of glucose induces the Crabtree effect by entering the yeast in a fermentative metabolism. Feed limiting the glucose concentration in the broth allows the cells to be grown to high density in a respiratory metabolism for *S. cerevisiae* (Gernaey et al. 2012) and prevents the production of acetate by *E. coli*. For fed-batch processes, exponential nourishment permits the constant growth rate of the cells, which is good for expression of recombinant proteins. At high growth rate, the soluble O_2 is a restricting factor. For methylotrophic yeast, it is essential that the cells be kept in respiratory metabolism to prevent an excessive gathering of methanol in the broth, which is rapidly toxic. It is useful to supplement air with O_2 gas. Pressure can be increased in the medium for dissolving more O_2 and increases the concentration of dissolved CO_2, which is not suitable. A limitation of dissolved O_2 is not always harmful to good expression; for example, for an enzyme cloned into *E. coli* in an L-arabinose-concluded vector. The moment for induction of recombinant protein expression is optimum when the O_2 uptake rate is increased to the greatest amount but the dissolved O_2 concentration is zero (Lapin et al. 2010).

Optimization of the pH medium is vital, especially for yeast-discharging protein because they can grow in a broad extend of pH values. The selection of pH depends on the stability of the recombinant protein expressed. Working with *Pichia pastoris* at low pH permits protease reduction to be prevented. This yeast is not able to grow at pH value under pH 2.2. Also, temperature has an important role in the solubility of the expressed protein and on the yield. Reducing the temperature from 30 to 25°C during the methanol initiation phase results in a four times rise in yield production of galactose oxidase cloned in *P. pastoris* (Sipiczki 2011).

In order to optimize the foam separation, biosurfactants have several advantages compared to chemically produced surfactants. Although production of biosurfactants is expensive (Sekhon Randhawa and Rahman 2014), there are various methods for decreasing the cost of their production. This includes the adaptation of the producing microbial strains by genetic engineering, the use of industrial by-products as a growth substance, and the optimization of the fermentation conditions. The downstream processing of emulsifiers from fermentation broth devotes about 70% of biosurfactant production costs (Anic et al. 2017). One of the problems in the fermentation processes is foam generation that is enriched in biosurfactants. Elimination of foam is done by three methods consisting of the addition of antifoam agents in a chemical manner, the use of constructed apparatus in a mechanical way, and in thermal form. All of these methods show an extra process step and raise the process intricacy and cost. Different approaches are used for the optimization of the biosurfactants' separation, which contain solvent extraction, adsorption, sedimentation, selective crystallization, and flotation. Nowadays, adsorption or desorption or chromatographic techniques are utilized for biosurfactant segregation from collapsed foam. Adsorption is done with wood-based activated carbon, hydrophobic polymeric Amberlite XAD resins, and silica gels with hydrophobic surface modification. Among them, polystyrene resins show long adsorption and elution times. The factors which restrict the feasibility of these processes are low adsorption capacity and huge amount of solvents needed for time-intensive product extraction and reformation with the greater number of adsorbents, which lead to operational prices and high investment (Kosaric and Vardar-Sukan 2015). The adsorption happens on the surface of the particles and in the swelling gel form of the adsorbent due to the porous structure of gel-type polystyrene resins. On the contrary, adsorbents based on silica keep adsorption positions just on the pore channels of particles and surface of particles.

Different ways for design and improvement of efficiency of a fermentation process are used. The determination of the nutritional needs of the organism, and the design of a medium based on the optimal concentration of each nutrient, is one of these approaches. Nonetheless, it has two disadvantages: (i) it takes much time to investigate each nutrient and decide its optimum level and (ii) the use of commercial substrates (e.g. soy bean meal, corn steep liquor, etc.), which have a complex composition of numerous nutrients is difficult (Geiger 2014). The problem of planning and optimizing fermentation processes can be managed rapidly using some statistical methods. A statistical technique for determining the important variables from enormous sets of variables through a set of designed experiments is known as Design of Experiment, which is divided to Evolutionary Optimization (EVOP) and RSM. Actually, these statistical methods are organized protocols that assure reaching the optimum conditions.

In EVOP, the process slowly evolves toward the optimum. It is based on the methods improved by Spendley et al. (1962). The method is a repetitive process in which a simple figure is created by operating one more test than the number of variables to improve efficiency. A simple procedure is designed to discover the optimum by increasing the reaction surface in the direction of the extreme slope, i.e. path with maximal rise in productivity. The protocol begins by the production of an uncomplicated figure. The simple figure is a triangle when two variables are optimized, a triangular pyramid when three variables improve the efficiency, expanding to an $n+1$ polyhedron, where n is the number of variables to be optimized. The empirical emphasize with the deficient response is removed and a recent point created by reflection of the removable point through the center of the simple figure. This procedure is followed until a maximum is obtained. EVOP has some limitations due to its repetitive nature; it is a slow process, which needs numerous steps. Also, it supplies just restricted data about the impacts of the variables and their interactions. Furthermore, EVOP reports the nearest optimum; thus, it is unfamiliar if this optimum is a local optimum or the absolute optimum for the whole process. Although EVOP has some limitations, it is a highly practical optimization method. EVOP is robust, can manage numerous variables at the same time, and constantly leads to an optimum. In addition, little information is required to know about the system before starting the process. Theoretically, the process enhances at each stage of the optimization plan, making it perfect for a production circumstance (Spendley et al. 1962).

RSM is the best way for fermentation optimization. On the one hand, it determines the optimum conditions, and on the other hand, it gives the useful data to plan a process. RSM is a method of improvement efficiency utilizing generally statistical methods based on the particular factorial designs of "Box and Wilson" and "Box and Behnken." It is a systematic way to determine the best situations joining particular experimental plans with Taylor first- and second-order equations. The RSM process settles the surface of the Taylor enlargement curve, which depicts responses such as productivity and impurity level. The Taylor equation is utilized in mathematics for approaching an extensive diversity of continuous works. Some criteria should be used for RSM. Firstly, the factors that are important for the process are familiar. RSM programs are restricted in the number of variables (4–5) that they are planned to manage. Fortunately, for scaling up of most fermentations, the diversity of variables for optimization is restricted, which consist of an aeration rate, phosphate level, temperature, back pressure, carbon and nitrogen sources, pH, power input, stirring rate, carbon-to-nitrogen (C/N) ratio, magnesium and sulfur levels, and soluble O_2 level. Secondly, the factors must be different without stopping over the empirical range checked.

For instance, the variables of pH and stirring and aeration rates are continuous and can be utilized in an RSM model. In contrast, variables such as carbon and nitrogen sources are interrupted and not able to be optimized by RSM. Thirdly, there is a mathematical work, which correlates the response to the factors. Many calculations have prevented the use of RSM because they are difficult and time-consuming. In a successful manner, many computer programs are attainable for doing these calculations, such as SASt, which is expensive and sophisticated and PC-based programs, SPSS-Xt, E-Chipt, and X STATt, which are inexpensive (Adinarayana and Ellaiah 2002). RSM has several advantages, such as huge amount of data from tests, obliges you to design, distinguish the duration time of the project, donates data about the mutual actions among variables, numerous replies at the same time, and gives data for plan and optimization of a procedure. On the other hand, it has some disadvantages such as not reporting the reason for what has happened and not predicting outside the range of study.

2.3.5 Fermentation Process Modeling

Because of the complicated nature of the biological systems, some circumstances are not understood, but models can be used for their determination, and empirical relations are frequently proposed (Table 2.15). The mechanistic fermentation patterns are the union of basic first-doctrine patterns of the physical procedures and empirical patterns for growth kinetics and metabolic rates. The main purpose of fermentation process modeling is the use of process systems engineering methods in the frame of fermentation schemes, and this desire is to be achieved by using mechanistic and computational fluid dynamics (CFD) of models which are attainable for learning fermentation processes (Mears et al. 2017).

2.3.5.1 Mechanistic Models

Mechanistic fermentation process patterns are on the basis of heat, mass, and momentum leveling, completed with a proper mathematical formulation of the fundamental mechanisms such as the kinetic manifestations to express process dynamics. However, developing these patterns needs resources, time, and important process perception, such as biomass growth rates (Gernaey et al. 2010). The kinetic manifestations are empirical and provide a less complicated and perfected sight of a complicated biological mechanism such as microbial growth kinetics recognized by Monod expression. Mechanistic patterns are practical for depicting the process and supporting the development of process. Mechanistic patterns are believed to be the most appropriate for offline fermentation process evolvement (Figure 2.9). Although numerous models are appropriate for depicting the input–output connections of a system, it is just with a mechanistic pattern that we can check new conditions or equipment. Controlling operating situations (e.g. pH, headspace force, agitation speed, aeration rate, and temperature) can lead to the fermentation process development from laboratory scale to industrial scale. To obtain this scale-up protocol, it is influential to comprehend equipment restrictions at various scales of performance (Chopda et al. 2016). Mechanistic fermentation patterns are appropriate at this level to estimate process sensitivity to alterations in the process situations. The application of mechanistic models consists of disconnected batch scheming and process evolution, as well as online process scanning of key process parameters such as substrate, product, and biomass concentrations (Chéruy 1997).

Table 2.15 Summary of the physical and biological model terms and their typical model type.

Modeled term	Considered factor	Model types	Purpose
Physical model	Volume or mass balance	Mechanistic	A volume or mass balance can be described mechanistically based on knowledge of the system inputs and outputs, such as substrate addition, acid and base addition, antifoam addition, evaporation, and sample volume removal.
Physical model	Viscosity	Empirical	Empirical models have been developed to describe the viscosity, and this is often proportional to the biomass concentration and power input.
Physical model	Oxygen mass transfer coefficient	Empirical (CFD)	Simple empirical correlations are typically used to describe the oxygen mass transfer coefficient in fermentation processes, assuming a constant factor across the full volume.
Physical model	Mixing/fluid flow	Empirical mechanistic CFD	The mixing may be described simply in terms of the power input into the liquid volume. This assumes a homogenous fluid and an equal power input to the whole liquid volume.
Physical model	Temperature	Mechanistic	Heat generation due to metabolic activity and mixing may be modeled, in addition to the rate of heat loss due to a cooling jacket or coil.
Biological model	Biomass growth	Empirical mechanistic	A commonly applied, simple empirical model to describe growth kinetics is the Monod equation, which is used to describe the growth rate of the organism as a function of the maximum growth rate and substrate concentration. Extended models are also applied to include product or substrate inhibition.
Biological model	Biomass concentration	Empirical mechanistic	The solving of the biomass and product concentration is mechanistic; however, it relies on empirically derived fermentation parameters for growth rate and yield coefficients.
Biological model	Oxygen uptake rate	Empirical mechanistic	The OUR is often modeled as the rate of oxygen uptake required for growth and production in addition to a constant term for maintenance, which is proportional to the biomass concentration.
Biological model	Dissolved oxygen concentration	Empirical mechanistic	The DO is described mechanistically as the balance between the OUR due to metabolism and the oxygen transfer rate (OTR) from the gas phase.

CFD = computational fluid dynamic; OUR = oxygen uptake rate; DO = dissolved oxygen concentration.

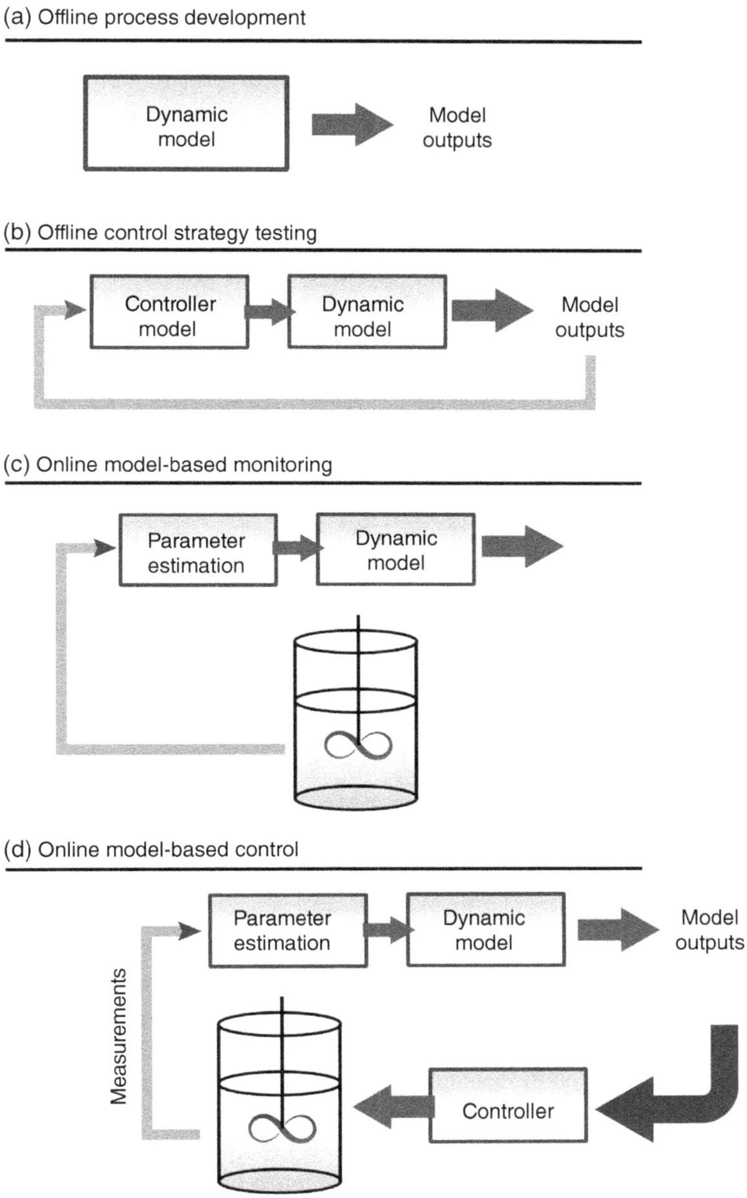

Figure 2.9 Suitable model structures for applying mechanistic process models for (a) offline process development, (b) offline control strategy testing, (c) online model-based monitoring, and (d) online model-based control.

A mechanistic model is based on the process back force, aeration assessment, and agitation speed. Mechanistic patterns are used for process development and optimization. Fermentation procedures are mainly operated in fed-batch mode, so plan of the primary situations for the batch, on the basis of the physical limitations, should be considered

(Gernaey et al. 2010). For this reason, mechanistic patterns are advantageous for considering the situations outside of the data set utilized to improve them. For instance, the application of mechanistic patterns to laboratory-scale filamentous fungal fermentation systems was considered. Joined model acted as a state assessor, which supplies chances to refer more progressive online scanning and control methods. Furthermore, the application of patterns to the process evolution phase allowed optimization of the beginning batch conditions. For prediction of the biomass and product concentration, a state assessor pattern is advanced based on a stoichiometric leveling for online applications to predict the unfamiliar states in real time. This is in the basis of online gas and flow rate assessment. Thus, it is predicted that, the state assessor will be a costly device for online process control and optimization. There is restricted application of smooth sensors in the fermentation trade in spite of the benefits of real-time process comprehending (Luttmann et al. 2012; Mears et al. 2016).

2.3.5.2 Computational fluid dynamics

Stirring up, aeration plan and scale-up have been arranged on the basis of empirical relations and engineering knowledge. Scale-up variables consist of fixed power-to-volume ratio, tip accelerate, volumetric airflow rate, and constant Reynolds number. Also, other parameters such as container geometry, blending intensities, activation conditions, feed position, physical properties, etc. can be considered (Leng et al. 2009). CFD is a great apparatus that can be utilized to encourage the scaling up and down of fermenters and for learning blending and the possible incident of gradients in a tank as well, but one of the challenges of CFD is access to laboratory- and pilot-scale data.

2.3.6 Inhibition of Fermentation Processes

The accumulation of inhibition factors throughout fermentation goes along with a progressive reduction in the value of product. Therefore, the recognition of the different inhibition factors helps to reduce their negative impacts on productivity (Elbeshbishy et al. 2017).

2.3.6.1 Substrate Inhibition

Lower amount of substrate is significant for activating growth and to hinder resporulation of spore-creating bacteria (Van Ginkel and Logan 2005). Increasing the concentrations of substrate in an optimum extent leads to high production of H_2 in dark fermentation. The activity of microbes which produce H_2 can be restrained by various approaches such as the presence of VFAs, lower intracellular pH, high H_2 fractional pressure, and high substrate concentrations (Ciranna et al. 2014). Thus, several approaches are used to prevent substrate inhibition including the optimization of the substrate concentration, the organic burden rate, and the ratio of nutrients to microorganism. Table 2.16 illustrates some examples of suppressive substance concentrations reported in the literature. Actually, having high concentration of biomass in the absence of light for a bio-H_2 fermentation system can reduce the substance restraint impacts and allow the performance of fermentation systems at extravagant organic loading rate. Thus, the biomass recycling from the fermentation effluent flow by microfiltration and the use of immobilized cells with circulating media can be alternative ways for substrate inhibition control (Van Ginkel and Logan 2005). Based on these strategies, efficient bioreactors for biohydrogen production are mentioned in Table 2.17.

Table 2.16 Summary of substrate, pH, carbon-to-nitrogen ratio, carbon-to-phosphorous ratio, metal ions, and temperature inhibition reported in literature.

Substrates	Seed sludge	Substrate concentration (g COD.l^{-1})			Temperature	Mode	H_2 yield	References
		Range	**Optimum**	**Inhibition**				
Glucose	*Caloramator celer*	1.2–58	9.63	>40	67	Batch	2.95 mol $H_2 \cdot mol^{-1}$ glucose	Guo et al. (2010)
Glucose	Agricultural soil	11–43	11–32	43	30	Continuous	2.2 mol $H_2 \cdot mol^{-1}$ glucose	Lee et al. (2011)
Sucrose	Anaerobic digester sludge	11–56	45	56	35	Continuous	12.121 $H_2 \cdot l^{-1}$ reactor.day^{-1}	Kyazze et al. (2006)
Sucrose	*Clostridium butyricum*	5–30	20	30	37	Batch	2.78 mol $H_2 \cdot mol^{-1}$ sucrose	Chen et al. (2005)
Starch	Anaerobic sludge	5–60	20	> 20	35	Batch	2.2 mol $H_2 \cdot mol^{-1}$ hexose	Lin et al. (2008)
Xylose	*Clostridium butyricum*	5–40	20	40	37	Batch	0.73 mol $H_2 \cdot mol^{-1}$ xylose	Lo et al. (2008)
Xylose	Anaerobic digester sludge	10–100	20	> 20	35	Continuous	2.25 mol $H_2 \cdot mol^{-1}$ xylose	Lin and Cheng (2006)
Whey powder	Anaerobic sludge	6–31	21	> 21	55	Batch	1.03 mol $H_2 \cdot mol^{-1}$ glucose	Kargi et al. (2012)
Winery wastewater	Sewage sludge	14–36	14	36	35	Continuous	1.89 mol $H_2 \cdot mol^{-1}$ hexose	Yu et al. (2002)
		pH						
		Range	Optimum					
Glucose	*Clostridium* sp. YM1	5–7	6.5		37	Batch	1.7 mol $H_2 \cdot mol^{-1}$ glucose	Lee et al. (2008)
Glucose	Anaerobic granular sludge	3.88–8.12	7.5		37	Batch	1.46 mol $H_2 \cdot mol^{-1}$ glucose	Guo et al. (2010)
Glucose	Anaerobic digester sludge	5.9–6.3	6.2		15–34	Continuous	359 mmol $H_2 \cdot l^{-1}$ reactor	Lin and Lay (2004)
Glucose	Anaerobic digester sludge	5–6	5.5		55–64	Continuous	1.21 mol $H_2 \cdot mol^{-1}$ glucose	Oh et al. (2004)
Sucrose	*Clostridium butyricum*	5.5–6.5	5.5		37	Batch	2.78–2.91 mol $H_2 \cdot mol^{-1}$ sucrose	Chen et al. (2005)

(Continued)

Table 2.16 (Continued)

Substrates	Seed sludge	Substrate concentration (g COD.l^{-1})			Temperature	Mode	H_2 yield	References
		Range	Optimum	Inhibition				
Starch	Anaerobic sludge	5–7	5–5.5		35	Batch	0.8–1.1 mol H_2·mol^{-1} hexose	Lin et al. (2008)
Xylose	Sewage sludge	5–8	6.5		35	Batch	1.3 mol H_2·mol^{-1} xylose	Lin and Cheng (2006)
Cellulose	Cow dung sludge	5.5–9	7.5		55	Batch	2.8 mmol H_2·g^{-1} cellulose	Lin et al. (2008)
Cornstalk waste	Cow dung compost	4–9	7		36	Batch	141.56 ml H_2·g^{-1} TVS	Zhang et al. (2007)
Winery wastewater	Acclimated municipal sewage	4.5–6	5.5		35	Continuous	4.6 l H_2·g^{-1} VSS day	Yu et al. (2002)
Wheat bran	Anaerobic digester sludge	4.5	5.2		35	Continuous	176 ml H_2·g^{-1} hexose	Noike (2002)
Municipal waste	Anaerobic digester sludge	4.5	5.2		35	Continuous	48 ml H_2·g^{-1} hexose	Noike (2002)
		C/N ratio						
		Range	Optimum					
Cellulose	Sludge compost	5–34	5		60	Continuous	272 ml H_2·g^{-1} hexose	Ueno et al. (2001)
Glucose	*Clostridium acetobutylicum*	36–238	70		30	Batch	350 ml H_2·g^{-1} glucose	Kalil et al. (2008)
Sucrose	Anaerobic sewage sludge	40–130	47		35	Batch	4.8 mol H_2·mol^{-1} sucrose	O-Thong et al. (2008)
Sucrose	Anaerobic digester sludge	40–190	137		25	Continuous	3.5 mol H_2·mol^{-1} sucrose	Anzola-Rojas et al. (2015)
		C/P ratio						
		Range	Optimum					
Wheat powder	Anaerobic sludge	50–1000	1000		37	Batch	281 ml H_2·g^{-1} Starch	Argun et al. (2008)

Palm oil effluent	Anaerobic sludge	450–650	559		60	Batch	6.33 l $H_2 \cdot l^{-1}$ substrate	O-Thong et al. (2008)
Sucrose	Anaerobic sewage sludge	8.7–800	120		35	Continuous	186.8 mmol $H_2 \cdot l^{-1}.day^{-1}$	Lin and Lay (2004)
		Metal ion						
		Range	Optimum	Metal ion				
Glucose	Anaerobic digestion sludge	0–21.9	10.9	Fe	35	Membrane reactor	1.6 mol $H_2 \cdot mol^{-1}$ glucose	Lee et al. (2009)
Glucose	*Clostridium*	0–100	25	Fe	35	Batch	2.2 mol $H_2 \cdot mol^{-1}$ glucose	Alalayah et al. (2009)
Sucrose	Anaerobic digestion sludge	0–12 800	200	Fe	35	Batch	1.32 mol $H_2 \cdot mol^{-1}$ sucrose	Zhang and Shen (2006)
Organic waste	Waste-activated sludge	0–14 000	1000	Fe	35	Batch	1.12 mol $H_2 \cdot mol^{-1}$	Lanza et al. (2012)
Food waste	Waste-activated sludge	5030–28 700	9830	Na	37	Batch	1.88 mol $H_2 \cdot mol^{-1}$	Cao and Zhao (2009)
Sucrose	Anaerobic digestion sludge	0–16 000	2000	Na	37	Batch	2.88 mol $H_2 \cdot mol^{-1}$ sucrose	Hao et al. (2006)
Glucose	*Clostridium butyricum*	0–20 000	0	Na	37	Batch	2.7 mol $H_2 \cdot mol^{-1}$ glucose	Junghare et al. (2012)
Organic waste	Waste-activated sludge	0–100	0	Ni	35	Batch	3.3 mol $H_2 \cdot mol^{-1}$	Lanza et al. (2012)
Glucose	Anaerobic digestion sludge	0–50	0.1	Ni	35	Batch	2.38 mol $H_2 \cdot mol^{-1}$ glucose	Wang et al. (2012)
Glucose	*Clostridium acetobutylicum*	0–1000	0	Mg	30	Batch	2.84 mol $H_2 \cdot mol^{-1}$ glucose	Alshiyab et al. (2008)
Glucose	*Clostridium acetobutylicum*	0–500	0	Ca	30	Batch	2.84 mol $H_2 \cdot mol^{-1}$ glucose	Alshiyab et al. (2008)

(Continued)

Table 2.16 (Continued)

Substrates	Seed sludge	Substrate concentration (g COD.l^{-1})			Temperature	Mode	H_2 yield	References
		Range	**Optimum**	**Inhibition**				
Sucrose	Anaerobic digestion sludge	0–300	150	Ca	35	Anaerobic blanket	3.6 mol $H_2 \cdot mol^{-1}$ sucrose	Chang and Lin (2006)
Cassava	Anaerobic granular sludge	0.02–0.19	0.12	Mo	37	Batch	1.36 mol $H_2 \cdot mol^{-1}$	Dong-Jie et al. (2011)
		pH		Temperature				
				Range	Optimum			
Glucose	*Clostridium* sp. YM1	6.5		25–39	37	Batch	1.7 mol $H_2 \cdot mol^{-1}$ glucose	Lee et al. (2008)
Glucose	*Ethanoligenens harbinense*	5		20–44	37	Batch	1.34 mol $H_2 \cdot mol^{-1}$ glucose	Xing et al. (2008)
Glucose	Anaerobic digester sludge	6.2		15–34	30–34	Continuous	359 mmol $H_2 \cdot l^{-1}$ reactor. day^{-1}	Lin and Lay (2004)
Glucose	Anaerobic digester sludge	5.5		55–64	55	Continuous	1.21 mol $H_2 \cdot mol^{-1}$ glucose	Oh et al. (2004)
Sucrose	*Thermoanaerobacterium*	6.25		45–70	60	Batch	2.53 mol $H_2 \cdot mol^{-1}$ hexose	O-Thong et al. (2008)
Sucrose	Sewage sludge	6.7		30–45	40	Continuous	3.88 mol $H_2 \cdot mol^{-1}$ sucrose	Lee et al. (2006)
Xylose	Waste-activated sludge	7		30–55	40	Batch	1.3 mol $H_2 \cdot mol^{-1}$ xylose	Lin et al. (2008)
Starch	Municipal sewage sludge	8.5		37–55	55	Batch	1.44 mmol $H_2 \cdot g^{-1}$ starch	Lee et al. (2008)
Food waste	*Clostridium beijerinckii*	5.5		30–45	40	Batch	128 ml $H_2 \cdot g^{-1}$ COD	Kim et al. (2008)
Organic waste	Anaerobic digester sludge	7		37–55	55	Semi-continuous	360 ml $H_2 \cdot g^{-1}$ VS	Valdez-Vazquez et al. (2005)

Table 2.17 Summary of different advanced bioreactor configurations for biohydrogen production.

Reactor configuration	Substrate	Optimum OLR (g Substrate/l-d)	Temperature (°C)	H_2 yield	References
AnMBR	Glucose	22	23	0.2 mol H_2.d-l^{-1}	Shen et al. (2010)
AnFBR using activated carbon as media	Sucrose	48	40	3.76 mol $H_2 \cdot mol^{-1}$	Wu et al. (2012)
UAFBR using recycled polyethylene	Sucrose	42	25	3.22 mol $H_2 \cdot mol^{-1}$	Fontes Lima et al. (2013)
UAFBR using recycled polyethylene	Glucose	42	25	1.51 mol $H_2 \cdot mol^{-1}$	Fontes Lima et al. (2013)
Mixed immobilized sludge reactor	Food waste	40	55	85.6 ml $H_2 \cdot g^{-1}$	Han et al. (2015)
UASB reactor	Galactose	180	37	2.25 mol $H_2 \cdot mol^{-1}$	Sivagurunathan et al. (2016)
UASB reactor	Glucose	18	39	1.61 mol $H_2 \cdot mol^{-1}$	Zhao and Yu (2008)

AnFBR = anaerobic fluidized-bed reactor; UASB = upflow anaerobic sludge blanket; UAFBR = upflow anaerobic fixed-bed reactor; AnMBR = anaerobic membrane reactor; OLR = organic loading rate.

2.3.6.2 pH Inhibition

The responsive operating variables influencing dark fermentation is pH, which is able to alter microbial multitude structures, by-products spectrum, and intracellular metabolic roles (Guo et al. 2010). In acidic pH (lower than 4), low amounts of H_2 are produced due to the alcoholic and acidic metabolites creation. However, neutral and alkaline pH (upper than 7) has been established to make easier propionate gathering more than H_2 production. As a result, keeping the pH in a maximum extend is vital to improve H_2 production and avoid inhibition (Table 2.16).

2.3.6.3 Inhibition by Undissociated Acids

Restraint by undetached acids is related to pH alterations and substance inhibition. The products of dark fermentation are H_2, butyrate and acetate as the principal soluble metabolites and submerged electron. However, the important collection of these dissolved metabolites because of the rise in ionic strength and restraint by undetached acids is able to prevent the metabolic action of H_2 manufacturer (Ciranna et al. 2014). The pKa assess for butyrate and acetate are 4.8 and 4.75, respectively. So, unsegregated acids enter the microbial cell wall and liberate the protons within the cell at greater intracellular pH, which influences on the microbial growth and metabolism. The gathering of anions in the cell intensifies the osmolarity of the cytoplasm and expands the cell swelling force, whereas the collection of protons reduces the pH in the cell. The acidic cellular pH can stimulate the fundamental enzymes for solvent production. In addition, proton can be gone along with a reduction of the adenosine triphosphate (ATP) in a cell and cause repression of the use of substrate and microbial growth.

2.3.6.4 Temperature Inhibition

Temperature influences the action of enzymes for producing H_2, the substrate bioreduction speed, and metabolism of H_2 manufacturers (Guo et al. 2010). By-products spectrum and change in the metabolic routes of various bacteria can be checked by temperature because various bacteria become superior at different temperatures. It is prominent to improve efficiency of the temperature to move the metabolic routes in the direction of butyrate and acetate production without alcohol and solvent production (Table 2.16). Actually, great operating temperature (upper than 60°C) can inhibit H_2 manufacturer by way of disabling of fundamental enzymes for growth and denaturation of cell proteins.

2.3.6.5 Nitrogen Inhibition

Although great concentration of ammonia nitrogen and the ratio of C/N can prevent the dark fermentation process, nitrogen is consumed by fermentative microbes as a nutrient (Argun et al. 2008). Ammonia can be manufactured by bioreduction of proteins, urea, and amino acids. Great ammonia nitrogen can inhibit the action of H_2 producers because of impeding in pH of cell, enhancement of maintenance energy needed of the microbial cells, and repression the actions of enzymes dependable for H_2 production. The principal inhibitory form of nitrogen that can pierce the microbial cell wall without limits is NH_3. Acidic pH is suitable for dark fermentation, where the greatest ammonia should keep in the ionized form, which is less toxic. The consumption of carbon by anaerobic fermentative microorganisms is 25–30 times faster than nitrogen (Kalil et al. 2008), sustaining an appropriate C/N ratio in the feedstock is influential to prevent nitrogen repression (Table 2.16).

2.3.6.6 Inhibition by Phosphate

One of the regulating factors of pH for microbial metabolism is phosphate, which has been established to enhance the production of H_2 in dark fermentation, by shortening the microbial lag-phase time (Zhou et al. 2013). The carbon-to-phosphorous (C/P) ratio inhibiting the dark fermentation process is reported in Table 2.16.

2.3.6.7 Inhibition by Sulfide

Sulfate is found in organic squander and wastewater and is used as a substance for dark fermentation (Hwang et al. 2011). Decrease in sulfate can occur under anaerobic conditions throughout fermentation. Also, production of sulfide can be done by bioreduction of proteins including sulfur such as methionine and cysteine. Great sulfide concentrations have been recognized to be toxic or suppressive to H_2-producing microbes in dark fermentation and diminish the bioavailability of metals. H_2S is able to pierce the microbial outer cell wall and alter natural properties of cell proteins by creating cross-links among polypeptide chains.

2.3.6.8 Inhibition by Lactic Acid Bacteria

A particular way for restraining dark fermentation is using lactic acid bacteria (LAB), which is known as an antimicrobial agent and prevents dark fermentation in several ways: for instance, acidification by the production of lactic acid, synthesis of polypeptide antibiotics such as bacteriocins, production of antimicrobial agents, and production of hydrogen peroxide. It has been found that disabling the activity of antibiotics produced by LAB is very impressive for increasing of H_2 produces. However, it is not economical and cannot be used for practical application (Gomes et al. 2015).

2.3.6.9 Inhibition by Metals

Low concentrations of various metal ions invigorate the growth of cell, action of enzyme cofactor, and other metabolic pathways involved in dark fermentation. In whatever way, high concentrations of them prevent the fermentation procedures. The restraint by metal ions consists of various mechanisms including the inhibition of the conveying fundamental ions and nutrients, collection of metals in a cell, and inhibition of the membrane action (Lemire et al. 2013). In addition, fermentation can be affected by the chemical characteristics of the metal ions such as the concentration of organic matters, the pH, and the chelating agents.

2.3.6.10 Inhibition by Phenolic and Furanic Mixtures

Nowadays, there is an increasing interest in using lignocellulosic biomass, which consists of cellulose, hemicellulose, and lignin as raw materials for dark fermentation. These carbohydrates are encapsulated in a complicated opaque arrangement and not obtainable to fermentative microbes for putting in practical use. Thus, for making them approachable to microbes, several pretreatment methods, for example, steam detonation and alkaline or weaker acid treatment are generally applied (Kumar et al. 2015).

2.4 Conclusions

The great number of fermentation applications has been displayed under regulated laboratory circumstances. Industrial acceptation of this technology is restricted because of the meaningful challenges faced in industrial scale-up. This chapter discussed important considerations of the main achievements of fermentation in food, chemical, and pharmaceutical industry. The growing interest of this subject by people has been observed in everything to promote their health, longevity, and reduce their energy sources. Also, many improvements and developments in industrial fermentation from different aspects such as genetic engineering of microorganisms, development of bioreactor configurations for fermentation, application of new technologies such as ultrasound and EF, optimization of the operational fermentation parameters, consideration of different media sources, fermentation process modeling, and recognition of inhibition in fermentation were reported, which led to high yields and cost reduction. In whatever way, for these attempts to be affluent, close cooperation among academia and industry will be demanded, and it will permit immense comprehension of developments in industrial fermentation that would help in procedure scale-up and industry acceptation.

References

Adekunle, A.E., Zhang, C., Guo, C., and Liu, C.-Z. (2017). Laccase production from *Trametes versicolor* in solid-state fermentation of steam-exploded pretreated cornstalk. *Waste and Biomass Valorization* 8: 153–159.

Adhyaru, D.N., Bhatt, N.S., Modi, H.A., and Divecha, J. (2016). Insight on xylanase from *Aspergillus tubingensis* FDHN1: production, high yielding recovery optimization through statistical approach and application. *Biocatalysis and Agricultural Biotechnology* 6: 51–57.

Adinarayana, K. and Ellaiah, P. (2002). Response surface optimization of the critical medium components for the production of alkaline protease by a newly isolated *Bacillus* sp. *Journal of Pharmacy & Pharmaceutical Sciences* 5: 272–278.

Afrisham, S., Badoei-Dalfard, A., Namaki-Shoushtari, A., and Karami, Z. (2016). Characterization of a thermostable, $CaCl_2$-activated and raw-starch hydrolyzing alpha-amylase from *Bacillus licheniformis* AT70: production under solid state fermentation by utilizing agricultural wastes. *Journal of Molecular Catalysis B: Enzymatic* 132: 98–106.

Aiswarya, C., Nayana, P., and Nambisan, P. (2018). Data of optimization of laccase production by *Marasmiellus palmivorus* LA1 under solid state fermentation using one factor at a time method. *Data in Brief* 17: 1276–1282.

Ajijolakewu, A.K., Leh, C.P., Wan Abdullah, W.N., and Lee, C.k. (2017). Optimization of production conditions for xylanase production by newly isolated strain *Aspergillus niger* through solid state fermentation of oil palm empty fruit bunches. *Biocatalysis and Agricultural Biotechnology* 11: 239–247.

Alalayah, W.M., Kalil, M.S., Kadhum, A.A.H. et al. (2009). Effect of environmental parameters on hydrogen production using *Clostridium saccharoperbutylacetonicum* N1-4 (ATCC 13564). *American Journal of Environmental Sciences* 5: 80–86.

Aliyah, A., Alamsyah, G., Ramadhani, R., and Hermansyah, H. (2017). Production of α-Amylase and β-Glucosidase from *Aspergillus niger* by solid state fermentation method on biomass waste substrates from rice husk, bagasse and corn cob. *Energy Procedia* 136: 418–423.

Alshiyab, H., Kalil, M.S., Hamid, A.A., and Yusoff, W.M.W. (2008). Trace metal effect on hydrogen production using *C. acetobutylicum*. *Journal of Biological Sciences* 8: 1–9.

Álvarez-Cervantes, J., Sánchez, C., Díaz, R., and Díaz-Godínez, G. (2016). Characterization of production of laccases, cellulases and xylanases of *Pleurotus ostreaus* grown on solid-state fermentation using an inert support. *Revista Mexicana de Ingenieria Quimica* 15: 323–331.

Amaro-Reyes, A., Gracida, J., Huizache-Peña, N. et al. (2016). On-site hydrolytic enzymes production from fungal co-cultivation of Bermuda grass and corn cob. *Bioresource Technology* 212: 334–337.

Amer, M.A., Novoa-Díaz, D., Puig-Pujol, A. et al. (2015). Ultrasonic velocity of water–ethanol–malic acid–lactic acid mixtures during the malolactic fermentation process. *Journal of Food Engineering* 149: 61–69.

Amorim, G.M., Oliveira, A.C., Gutarra, M.L.E. et al. (2017). Solid-state fermentation as a tool for methylxanthine reduction and simultaneous xylanase production in cocoa meal. *Biocatalysis and Agricultural Biotechnology* 11: 34–41.

Anic, I., Nath, A., Franco, P., and Wichmann, R. (2017). Foam adsorption as an ex situ capture step for surfactants produced by fermentation. *Journal of Biotechnology* 258: 181–189.

Anzola-Rojas, M.d.P., Gonçalves da Fonseca, S., Canedo da Silva, C. et al. (2015). The use of the carbon/nitrogen ratio and specific organic loading rate as tools for improving biohydrogen production in fixed-bed reactors. *Biotechnology Reports* 5: 46–54.

Argun, H., Kargi, F., Kapdan, I.K., and Oztekin, R. (2008). Batch dark fermentation of powdered wheat starch to hydrogen gas: effects of the initial substrate and biomass concentrations. *International Journal of Hydrogen Energy* 33: 6109–6115.

Asimakopoulos, K., Gavala, H.N., and Skiadas, I.V. (2018). Reactor systems for syngas fermentation processes: a review. *Chemical Engineering Journal* 348: 732–744.

Aslam, S., Asgher, M., Hussain, F., and Khan, N.A. (2016). Exploration of optimum operating conditions for enhanced laccase enzyme production by *Pleurotus nebrodensis* WC 850 through response surface methodology. *The Journal of Animal & Plant Sciences* 26: 794–804.

Barukčić, I., Lisak Jakopović, K., Herceg, Z. et al. (2015). Influence of high intensity ultrasound on microbial reduction, physico-chemical characteristics and fermentation of sweet whey. *Innovative Food Science & Emerging Technologies* 27: 94–101.

Basso, L.C., de Amorim, H.V., de Oliveira, A.J., and Lopes, M.L. (2008). Yeast selection for fuel ethanol production in Brazil. *FEMS Yeast Research* 8: 1155–1163.

Bellasio, M., Mattanovich, D., Sauer, M., and Marx, H. (2015). Organic acids from lignocellulose: *Candida lignohabitans* as a new microbial cell factory. *Journal of Industrial Microbiology & Biotechnology* 42: 681–691.

Bharti, A.K., Kumar, A., Kumar, A., and Dutt, D. (2018). Exploitation of *Parthenium hysterophorous* biomass as low-cost substrate for cellulase and xylanase production under solid-state fermentation using *Talaromyces stipitatus* MTCC 12687. *Journal of Radiation Research and Applied Science* 11: 271–280.

Bielser, J.-M., Wolf, M., Souquet, J. et al. (2018). Perfusion mammalian cell culture for recombinant protein manufacturing – a critical review. *Biotechnology Advances* 36: 1328–1340.

Biz, A., Finkler, A.T.J., Pitol, L.O. et al. (2016). Production of pectinases by solid-state fermentation of a mixture of citrus waste and sugarcane bagasse in a pilot-scale packed-bed bioreactor. *Biochemical Engineering Journal* 111: 54–62.

Bokulich, N.A., Thorngate, J.H., Richardson, P.M., and Mills, D.A. (2014). Microbial terroir for wine grapes is conditioned by cultivar, vintage, and climate. *Proceedings of the National Academy of Sciences of the United States of America* 111: E139–E148.

Cagnetta, C., Coma, M., Vlaeminck, S.E., and Rabaey, K. (2016). Production of carboxylates from high rate activated sludge through fermentation. *Bioresource Technology* 217: 165–172.

Cao, X. and Zhao, Y. (2009). The influence of sodium on biohydrogen production from food waste by anaerobic fermentation. *Journal of Material Cycles and Waste Management* 11: 244–250.

Caspeta, L., Chen, Y., and Nielsen, J. (2016). Thermotolerant yeasts selected by adaptive evolution express heat stress response at 30 °C. *Scientific Reports* 6: 1–9. https://doi.org/10.1038/srep27003.

de Castro, R.J.S., Ohara, A., Nishide, T.G. et al. (2015). A versatile system based on substrate formulation using agroindustrial wastes for protease production by *Aspergillus niger* under solid state fermentation. *Biocatalysis and Agricultural Biotechnology* 4: 678–684.

Cate, J.H.D. and Ball, A.S. (2016). Editorial overview: energy biotechnology. *Current Opinion in Biotechnology* 38: v–vii.

Cerda, A., El-Bakry, M., Gea, T., and Sánchez, A. (2016). Long term enhanced solid-state fermentation: inoculation strategies for amylase production from soy and bread wastes by *Thermomyces* sp. in a sequential batch operation. *Journal of Environmental Chemical Engineering* 4: 2394–2401.

Cerda, A., Gea, T., Vargas-García, M.C., and Sánchez, A. (2017). Towards a competitive solid state fermentation: cellulases production from coffee husk by sequential batch operation and role of microbial diversity. *Science of the Total Environment* 589: 56–65.

Chang, A.C. and Chen, F.C. (2002). The application of 20 kHz ultrasonic waves to accelerate the aging of different wines. *Food Chemistry* 79: 501–506.

Chang, F.-Y. and Lin, C.-Y. (2006). Calcium effect on fermentative hydrogen production in an anaerobic up-flow sludge blanket system. *Water Science and Technology* 54: 105–112.

Chanwicha, N., Katekaew, S., Aimi, T., and Boonlue, S. (2015). Purification and characterization of alkaline xylanase from *Thermoascus aurantiacus* var. *levisporus* KKU-PN-I2-1 cultivated by solid-state fermentation. *Mycoscience* 56: 309–318.

Chen, W.-M., Tseng, Z.-J., Lee, K.-S., and Chang, J.-S. (2005). Fermentative hydrogen production with *Clostridium butyricum* CGS5 isolated from anaerobic sewage sludge. *International Journal of Hydrogen Energy* 30: 1063–1070.

Chen, K.-J., Lee, C.-F., Rick, J. et al. (2012). Fabrication and application of amperometric glucose biosensor based on a novel PtPd bimetallic nanoparticle decorated multi-walled carbon nanotube catalyst. *Biosensors and Bioelectronics* 33: 75–81.

Chéruy, A. (1997). Software sensors in bioprocess engineering. *Journal of Biotechnology, Bioinformatics I* 52: 193–199.

Chisti, Y. (2003). Sonobioreactors: using ultrasound for enhanced microbial productivity. *Trends in Biotechnology* 21: 89–93.

Choi, S.P., Nguyen, M.T., and Sim, S.J. (2010). Enzymatic pretreatment of *Chlamydomonas reinhardtii* biomass for ethanol production. *Bioresource Technology* 101: 5330–5336.

Chopda, V.R., Gomes, J., and Rathore, A.S. (2016). Bridging the gap between PAT concepts and implementation: an integrated software platform for fermentation. *Biotechnology Journal* 11: 164–171.

Ciranna, A., Ferrari, R., Santala, V., and Karp, M. (2014). Inhibitory effects of substrate and soluble end products on biohydrogen production of the alkalithermophile *Caloramator celer*: kinetic, metabolic and transcription analyses. *International Journal of Hydrogen Energy* 39: 6391–6401.

Corma, A., Iborra, S., and Velty, A. (2007). Chemical routes for the transformation of biomass into chemicals. *Chemical Reviews* 107: 2411–2502.

Daâssi, D., Zouari-Mechichi, H., Frikha, F. et al. (2016). Sawdust waste as a low-cost support-substrate for laccases production and adsorbent for azo dyes decolorization. *Journal of Environmental Health Science and Engineering* 14 (1) https://doi.org/10.1186/s40201-016-0244-0.

Daniell, J., Köpke, M., and Simpson, S.D. (2012). Commercial biomass syngas fermentation. *Energies* 5: 5372–5417.

David, A., Singh Chauhan, P., Kumar, A. et al. (2018). Coproduction of protease and mannanase from *Bacillus nealsonii* PN-11 in solid state fermentation and their combined application as detergent additives. *International Journal of Biological Macromolecules* 108: 1176–1184.

Deeba, F., Pruthi, V., and Negi, Y.S. (2017). Fostering triacylglycerol accumulation in novel oleaginous yeast *Cryptococcus psychrotolerans* IITRFD utilizing groundnut shell for improved biodiesel production. *Bioresource Technology* 242: 113–120.

Dessie, W., Zhang, W., Xin, F. et al. (2018). Succinic acid production from fruit and vegetable wastes hydrolyzed by on-site enzyme mixtures through solid state fermentation. *Bioresource Technology* 247: 1177–1180.

Di Cagno, R., Coda, R., De Angelis, M., and Gobbetti, M. (2013). Exploitation of vegetables and fruits through lactic acid fermentation. *Food Microbiology* 33: 1–10.

Díaz, A.B., Marzo, C., Caro, I. et al. (2017). Valorization of exhausted sugar beet cossettes by successive hydrolysis and two fermentations for the production of bio-products. *Bioresource Technology* 225: 225–233.

Diaz, A.B., Blandino, A., and Caro, I. (2018). Value added products from fermentation of sugars derived from agro-food residues. *Trends in Food Science & Technology* 71: 52–64.

Doan, Q.C., Moheimani, N.R., Mastrangelo, A.J., and Lewis, D.M. (2012). Microalgal biomass for bioethanol fermentation: implications for hypersaline systems with an industrial focus. *Biomass and Bioenergy*, International Conference on Lignocellulosic ethanol 46: 79–88.

Dong-Jie, N., Jing-Yuan, W., Bao-Ying, W., and You-Cai, Z. (2011). Effect of Mo-containing additives on biohydrogen fermentation from cassava's stillage. *International Journal of Hydrogen Energy* 36: 5289–5295.

Doriya, K. and Kumar, D.S. (2018). Solid state fermentation of mixed substrate for l-asparaginase production using tray and in-house designed rotary bioreactor. *Biochemical Engineering Journal* 138: 188–196.

Du, G., Chen, J., Yu, J., and Lun, S. (2001). Continuous production of poly-3-hydroxybutyrate by *Ralstonia eutropha* in a two-stage culture system. *Journal of Biotechnology* 88: 59–65.

Durand, A. (2003). Bioreactor designs for solid state fermentation. *Biochemical Engineering Journal, Solid-State Fermentation* 13: 113–125.

Dürre, P. and Eikmanns, B.J. (2015). C1-carbon sources for chemical and fuel production by microbial gas fermentation. *Current Opinion in Biotechnology*, Chemical biotechnology • Pharmaceutical biotechnology 35: 63–72.

Ecker, D.M., Jones, S.D., and Levine, H.L. (2015). The therapeutic monoclonal antibody market. *MAbs* 7: 9–14.

El-Batal, A.I., ElKenawy, N.M., Yassin, A.S., and Amin, M.A. (2015). Laccase production by *Pleurotus ostreatus* and its application in synthesis of gold nanoparticles. *Biotechnology Reports* 5: 31–39.

Elbeshbishy, E., Dhar, B.R., Nakhla, G., and Lee, H.-S. (2017). A critical review on inhibition of dark biohydrogen fermentation. *Renewable and Sustainable Energy Reviews* 79: 656–668.

Elkhouri, S. and Bakri, Y. (2016). Improvement of amylase production by UV mutagenesis of *Aspergillus flavus* FSS63 under solid state fermentation. *Biological Sciences-PJSIR* 59: 80–84.

Elvira, L., Vera, P., Cañadas, F.J. et al. (2016). Concentration measurement of yeast suspensions using high frequency ultrasound backscattering. *Ultrasonics* 64: 151–161.

Feng, Y., Chen, W., and Chen, F. (2016). A *Monascus pilosus* MS-1 strain with high-yield monacolin K but no citrinin. *Food Science and Biotechnology* 25: 1115–1122.

Finkler, A.T.J., Biz, A., Pitol, L.O. et al. (2017). Intermittent agitation contributes to uniformity across the bed during pectinase production by *Aspergillus niger* grown in solid-state fermentation in a pilot-scale packed-bed bioreactor. *Biochemical Engineering Journal* 121: 1–12.

Fontes Lima, D.M., Moreira, W.K., and Zaiat, M. (2013). Comparison of the use of sucrose and glucose as a substrate for hydrogen production in an upflow anaerobic fixed-bed reactor. *International Journal of Hydrogen Energy* 38: 15074–15083.

Formenti, L.R., Nørregaard, A., Bolic, A. et al. (2014). Challenges in industrial fermentation technology research. *Biotechnology Journal* 9: 727–738.

Freitas, A.C., Baleeiro, F.C.F., Fonseca, R.F. et al. (2015). Bioprocess development to add value to canola cake used as substrate for proteolytic enzyme production. *Food and Bioproducts Processing* 95: 173–182.

García Martín, J.F., Guillemet, L., Feng, C., and Sun, D.-W. (2013). Cell viability and proteins release during ultrasound-assisted yeast lysis of light lees in model wine. *Food Chemistry* 141: 934–939.

Ge, X., Vasco-Correa, J., and Li, Y. (2017). Solid-state fermentation bioreactors and fundamentals. In: *Current Developments in Biotechnology and Bioengineering* (eds. C. Larroche, M.Á. Sanromán, G. Du and A. Pandey), 381–402. Elsevier.

Geiger, E.O. (2014). Statistical methods for fermentation optimization. In: *Fermentation and Biochemical Engineering Handbook* (eds. H.C. Vogel and C.M. Todaro), 415–422. Boston: William Andrew Publishing.

Gernaey, K.V. (2015). A perspective on PSE in fermentation process development and operation. In: *Computer Aided Chemical Engineering* (eds. K.V. Gernaey, J.K. Huusom and R. Gani), 123–130. Elsevier https://doi.org/10.1016/B978-0-444-63578-5.50016-5.

Gernaey, K.V., Lantz, A.E., Tufvesson, P. et al. (2010). Application of mechanistic models to fermentation and biocatalysis for next-generation processes. *Trends in Biotechnology* 28: 346–354.

Gernaey, K.V., Bolic, A., and Svanholm, B. (2012). PAT tools for fermentation processes. *Chimica Oggi / Chemistry Today* 30: 38–42.

Gildemyn, S., Verbeeck, K., Slabbinck, R. et al. (2015). Integrated production, extraction, and concentration of acetic acid from CO_2 through microbial electrosynthesis. *Environmental Science & Technology Letters* 2: 325–328.

Gomes, B.C., Rosa, P.R.F., Etchebehere, C. et al. (2015). Role of homo-and heterofermentative lactic acid bacteria on hydrogen-producing reactors operated with cheese whey wastewater. *International Journal of Hydrogen Energy* 40: 8650–8660.

Gong, Z., Shen, H., Yang, X. et al. (2014). Lipid production from corn stover by the oleaginous yeast *Cryptococcus curvatus*. *Biotechnology for Biofuels* 7: 158.

González-Ramos, D., Gorter de Vries, A.R., Grijseels, S.S. et al. (2016). A new laboratory evolution approach to select for constitutive acetic acid tolerance in *Saccharomyces cerevisiae* and identification of causal mutations. *Biotechnology for Biofuels* 9: 173. https://doi.org/10.1186/s13068-016-0583-1.

Guo, X.M., Trably, E., Latrille, E. et al. (2010). Hydrogen production from agricultural waste by dark fermentation: a review. *International Journal of Hydrogen Energy* 35: 10660–10673.

Guo, H., Geng, B., Liu, X. et al. (2013). Characterization of bacterial consortium and its application in an ectopic fermentation system. *Bioresource Technology* 139: 28–33.

Gupta, S. and Abu-Ghannam, N. (2012). Probiotic fermentation of plant based products: possibilities and opportunities. *Critical Reviews in Food Science and Nutrition* 52: 183–199.

Gustavsson, J., Cederberg, C., and Sonesson, U. (2011). *Global Food Losses and Food Waste: Extent, Causes and Prevention*. Rome: Food and Agriculture Organization of the United Nations.

Guzel-Seydim, Z.B., Kok-Tas, T., Greene, A.K., and Seydim, A.C. (2011). Review: functional properties of Kefir. *Critical Reviews in Food Science and Nutrition* 51: 261–268.

Han, W., Liu, D.N., Shi, Y.W. et al. (2015). Biohydrogen production from food waste hydrolysate using continuous mixed immobilized sludge reactors. *Bioresource Technology* 180: 54–58.

Hao, X., Zhou, M., Yu, H. et al. (2006). Effect of sodium ion concentration on hydrogen production from sucrose by anaerobic hydrogen-producing granular sludge. *Chinese Journal of Chemical Engineering* 14: 511–517.

Henry, O., Kwok, E., and Piret, J.M. (2008). Simpler noninstrumented batch and semi-continuous cultures provide mammalian cell kinetic data comparable to continuous and perfusion cultures. *Biotechnology Progress* 24: 921–931.

Hoche, S., Hussein, M.A., and Becker, T. (2014). Critical process parameter of alcoholic yeast fermentation: speed of sound and density in the temperature range 5–30 °C. *International Journal of Food Science & Technology* 49: 2441–2448.

Hwang, J.-H., Choi, J.-A., Oh, Y.-K. et al. (2011). Hydrogen production from sulfate- and ferrous-enriched wastewater. *International Journal of Hydrogen Energy* 36: 13984–13990.

Idris, A.S.O., Pandey, A., Rao, S.S., and Sukumaran, R.K. (2017). Cellulase production through solid-state tray fermentation, and its use for bioethanol from sorghum stover. *Bioresource Technology* 242: 265–271.

Isikgor, F.H. and Becer, C.R. (2015). Lignocellulosic biomass: a sustainable platform for the production of bio-based chemicals and polymers. *Polymer Chemistry* 6: 4497–4559.

Jain, R. and Naik, S.N. (2018). Adding value to the oil cake as a waste from oil processing industry: production of lipase in solid state fermentation. *Biocatalysis and Agricultural Biotechnology* 15: 181–184.

Jamal, M. (2018). *The CRISPR/Cas System: Emerging Technology and Application*. Caister Academic Press.

Jiang, Y., Liu, J., Jiang, W. et al. (2015). Current status and prospects of industrial bio-production of n-butanol in China. *Biotechnology Advances, Industrial Biotechnology: Tools and Applications* 33: 1493–1501.

Jung, Y.H., Kim, I.J., Kim, H.K., and Kim, K.H. (2013). Dilute acid pretreatment of lignocellulose for whole slurry ethanol fermentation. *Bioresource Technology* 132: 109–114.

Jung, Y.H., Park, H.M., and Kim, K.H. (2015). Whole slurry saccharification and fermentation of maleic acid-pretreated rice straw for ethanol production. *Bioprocess and Biosystems Engineering* 38: 1639–1644.

Junghare, M., Subudhi, S., and Lal, B. (2012). Improvement of hydrogen production under decreased partial pressure by newly isolated alkaline tolerant anaerobe, *Clostridium butyricum* TM-9A: optimization of process parameters. *International Journal of Hydrogen Energy* 37: 3160–3168.

Kalil, M.S., Alshiyab, H.S., and Yusoff, W.M.W. (2008). Effect of nitrogen source and carbon to nitrogen ratio on hydrogen production using *C. acetobutylicum*. *American Journal of Biochemistry and Biotechnology* 4: 393–401.

Kargi, F., Eren, N.S., and Ozmihci, S. (2012). Hydrogen gas production from cheese whey powder (CWP) solution by thermophilic dark fermentation. *International Journal of Hydrogen Energy* 37: 2260–2266.

Khanahmadi, M., Arezi, I., Amiri, M., and Miranzadeh, M. (2018). Bioprocessing of agro-industrial residues for optimization of xylanase production by solid- state fermentation in flask and tray bioreactor. *Biocatalysis and Agricultural Biotechnology* 13: 272–282.

Kilikian, B.V., Afonso, L.C., Souza, T.F.C. et al. (2014). Filamentous fungi and media for cellulase production in solid state cultures. *Brazilian Journal of Microbiology* 45: 279–286.

Kim, J.K., Nhat, L., Chun, Y.N., and Kim, S.W. (2008). Hydrogen production conditions from food waste by dark fermentation with *Clostridium beijerinckii* KCTC 1785. *Biotechnology and Bioprocess Engineering* 13: 499–504.

Kim, E.K., An, S.-Y., Lee, M.-S. et al. (2011). Fermented kimchi reduces body weight and improves metabolic parameters in overweight and obese patients. *Nutrition Research* 31: 436–443.

Kim, I., Seo, Y.H., Kim, G.-Y., and Han, J.-I. (2015). Co-production of bioethanol and biodiesel from corn stover pretreated with nitric acid. *Fuel* 143: 285–289.

Kiyoshi, K., Furukawa, M., Seyama, T. et al. (2015). Butanol production from alkali-pretreated rice straw by co-culture of *Clostridium thermocellum* and *Clostridium saccharoperbutylacetonicum*. *Bioresource Technology* 186: 325–328.

Kosaric, N. and Vardar-Sukan, F. (2015). *Biosurfactants: Production and Utilization – Processes, Technologies, and Economics*. Boca Raton: CRC Press.

Kracke, F., Vassilev, I., and Krömer, J.O. (2015). Microbial electron transport and energy conservation – the foundation for optimizing bioelectrochemical systems. *Frontiers in Microbiology* 6 (575) https://doi.org/10.3389/fmicb.2015.00575.

Kulkarni, P., Loira, I., Morata, A. et al. (2015). Use of non-*Saccharomyces* yeast strains coupled with ultrasound treatment as a novel technique to accelerate ageing on lees of red wines and its repercussion in sensorial parameters. *LWT – Food Science and Technology* 64: 1255–1262.

Kumar, A. and Kanwar, S.S. (2012). Lipase production in solid-state fermentation (SSF): recent developments and biotechnological applications. *Dynamic Biochemistry, Process Biotechnology and Molecular Biology* 6: 13–27.

Kumar, G., Bakonyi, P., Periyasamy, S. et al. (2015). Lignocellulose biohydrogen: practical challenges and recent progress. *Renewable and Sustainable Energy Reviews* 44: 728–737.

Kumar, D., Juneja, A., and Singh, V. (2018). Fermentation technology to improve productivity in dry grind corn process for bioethanol production. *Fuel Processing Technology* 173: 66–74.

Kuo, F.-J., Sheng, C.-T., and Ting, C.-H. (2008). Evaluation of ultrasonic propagation to measure sugar content and viscosity of reconstituted orange juice. *Journal of Food Engineering* 86: 84–90.

Kurosawa, K., Wewetzer, S.J., and Sinskey, A.J. (2014). Triacylglycerol production from corn stover using a xylose-fermenting *Rhodococcus opacus* strain for lignocellulosic biofuels. *Journal of Microbial & Biochemical Technology* 6: 254–259.

Kurosawa, K., Laser, J., and Sinskey, A.J. (2015). Tolerance and adaptive evolution of triacylglycerol-producing *Rhodococcus opacus* to lignocellulose-derived inhibitors. *Biotechnology for Biofuels* 8: 76.

Kyazze, G., Martinez-Perez, N., Dinsdale, R. et al. (2006). Influence of substrate concentration on the stability and yield of continuous biohydrogen production. *Biotechnology and Bioengineering* 93: 971–979.

Lane, J. (2016). Coskata's technology re-emerges as Synata Bio: Biofuels Digest. URL https://www.biofuelsdigest.com/bdigest/2016/01/24/coskatas-technology-re-emerges-as-synata-bio (accessed 2.13.20).

Lang, X. and Zhang, T. (2010). Automatic chain-type solid-state fermentation reaction device and method of automatic solid-state fermentation thereof. Patent number: CN101717716.

Lanza, P., Nicosia, S., Valenti, D. et al. (2012). Materials recovery from WEEE: current situation in Sicily. Presented at the SIDISA 2012-9th Edition of the International Symposium of Sanitary and Environmental Engineering, Politecnico di Milano, DIIAR, Sezione Ambientale-Milano, pp. 1–4.

Lapin, A., Klann, M., and Reuss, M. (2010). Multi-scale spatio-temporal modeling: lifelines of microorganisms in bioreactors and tracking molecules in cells. In: *Biosystems Engineering II: Linking Cellular Networks and Bioprocesses, Advances in Biochemical Engineering / Biotechnology* (eds. C. Wittmann and R. Krull), 23–43. Berlin, Heidelberg: Springer.

Le Borgne, S. (2012). Genetic engineering of industrial strains of *Saccharomyces cerevisiae*. *Methods in Molecular Biology* 824: 451–465.

Lee, C.-L., Wang, J.-J., Kuo, S.-L., and Pan, T.-M. (2006). *Monascus* fermentation of dioscorea for increasing the production of cholesterol-lowering agent – monacolin K and antiinflammation agent – monascin. *Applied Microbiology and Biotechnology* 72: 1254–1262.

Lee, K.-S., Hsu, Y.-F., Lo, Y.-C. et al. (2008). Exploring optimal environmental factors for fermentative hydrogen production from starch using mixed anaerobic microflora. *International Journal of Hydrogen Energy, Asian bioH2 Symposium* 2006 (33): 1565–1572.

Lee, D.-Y., Li, Y.-Y., Oh, Y.-K. et al. (2009). Effect of iron concentration on continuous H_2 production using membrane bioreactor. *International Journal of Hydrogen Energy* 34: 1244–1252.

Lee, D.-J., Show, K.-Y., and Su, A. (2011). Dark fermentation on biohydrogen production: pure culture. *Bioresource Technology, Special Issue: Biofuels-III: Biohydrogen* 102: 8393–8402.

Lemire, J.A., Harrison, J.J., and Turner, R.J. (2013). Antimicrobial activity of metals: mechanisms, molecular targets and applications. *Nature Reviews. Microbiology* 11: 371–384.

Leng, D.E., Katti, S.S., and Atiemo-Obeng, V. (2009). Industrial mixing technology. In: *Albright's Chemical Engineering Handbook* (ed. L. Albright), 615–707. Bacon Raton: CRC Press.

Li, J., Baral, N.R., and Jha, A.K. (2014). Acetone-butanol-ethanol fermentation of corn stover by *Clostridium* species: present status and future perspectives. *World Journal of Microbiology and Biotechnology* 30: 1145–1157.

Licht, F. O. (2018). World ethanol and biofuels report. https://www.agra-net.com/agra/worldethanol-and-biofuels-report/.

Liew, F., Martin, M.E., Tappel, R.C. et al. (2016). Gas fermentation – a flexible platform for commercial scale production of low-carbon-fuels and chemicals from waste and renewable feedstocks. *Frontiers in Microbiology* 7 https://doi.org/10.3389/fmicb.2016.00694.

Lin, C.-Y. and Cheng, C.-H. (2006). Fermentative hydrogen production from xylose using anaerobic mixed microflora. *International Journal of Hydrogen Energy* 31: 832–840.

Lin, C.Y. and Lay, C.H. (2004). Carbon/nitrogen-ratio effect on fermentative hydrogen production by mixed microflora. *International Journal of Hydrogen Energy* 29: 41–45.

Lin, C.-Y., Chang, C.-C., and Hung, C.-H. (2008). Fermentative hydrogen production from starch using natural mixed cultures. *International Journal of Hydrogen Energy* 33: 2445–2453.

Liu, D., Zhu, Y., Beeftink, R. et al. (2004). Chinese vinegar and its solid-state fermentation process. *Food Reviews International* 20: 407–424.

Liu, Y., Li, C., Meng, X., and Yan, Y. (2013). Biodiesel synthesis directly catalyzed by the fermented solid of *Burkholderia cenocepacia* via solid state fermentation. *Fuel Processing Technology* 106: 303–309.

Liu, Z., Ho, S.-H., Sasaki, K. et al. (2016). Engineering of a novel cellulose-adherent cellulolytic *Saccharomyces cerevisiae* for cellulosic biofuel production. *Scientific Reports* 6: 1–10. https://doi.org/10.1038/srep24550.

Liu, H., Hu, H., Jin, Y. et al. (2017). Co-fermentation of a mixture of glucose and xylose to fumaric acid by *Rhizopus arrhizus* RH 7-13-9. *Bioresource Technology* 233: 30–33.

Lo, Y.-C., Chen, W.-M., Hung, C.-H. et al. (2008). Dark H_2 fermentation from sucrose and xylose using H_2-producing indigenous bacteria: feasibility and kinetic studies. *Water Research* 42: 827–842.

Luo, H., Schmid, F., Grbin, P.R., and Jiranek, V. (2012). Viability of common wine spoilage organisms after exposure to high power ultrasonics. *Ultrasonics Sonochemistry* 19: 415–420.

Luttmann, R., Bracewell, D.G., Cornelissen, G. et al. (2012). Soft sensors in bioprocessing: a status report and recommendations. *Biotechnology Journal* 7: 1040–1048.

Mandenius, C.-F. and Brundin, A. (2008). Bioprocess optimization using design-of-experiments methodology. *Biotechnology Progress* 24: 1191–1203.

Marques, N.P., de Cassia Pereira, J., Gomes, E. et al. (2018). Cellulases and xylanases production by endophytic fungi by solid state fermentation using lignocellulosic substrates and enzymatic saccharification of pretreated sugarcane bagasse. *Industrial Crops and Products* 122: 66–75.

Marsh, A.J., Hill, C., Ross, R.P., and Cotter, P.D. (2014). Fermented beverages with health-promoting potential: past and future perspectives. *Trends in Food Science & Technology* 38: 113–124.

Martín-Sampedro, R., Eugenio, M.E., and Villar, J.C. (2012). Effect of steam explosion and enzymatic pre-treatments on pulping and bleaching of *Hesperaloe funifera*. *Bioresource Technology* 111: 460–467.

Mazaheri, D. and Shojaosadati, S.A. (2013). Mathematical models for microbial kinetics in solid-state fermentation: a review. *Iranian Journal of Biotechnology* 11: 156–167.

McNeil, B. and Harvey, L.M. (2008). *Practical Fermentation Technology*. West Sussex, England: Wiley.

Mears, L., Stocks, S.M., Albaek, M.O. et al. (2016). Mechanistic models for process development and optimization of fed-batch fermentation systems. In: *Computer Aided Chemical Engineering* (eds. Z. Kravanja and M. Bogataj), 1311–1316. Elsevier.

Mears, L., Stocks, S.M., Albaek, M.O. et al. (2017). Mechanistic fermentation models for process design, monitoring, and control. *Trends in Biotechnology* 35: 914–924.

Miller, D.H., Lamport, D.T., and Miller, M. (1972). Hydroxyproline heterooligosaccharides in *Chlamydomonas*. *Science* 176: 918–920.

Molino, A., Chianese, S., and Musmarra, D. (2016). Biomass gasification technology: the state of the art overview. *Journal of Energy Chemistry* 25: 10–25.

Mondala, A.H. (2015). Direct fungal fermentation of lignocellulosic biomass into itaconic, fumaric, and malic acids: current and future prospects. *Journal of Industrial Microbiology & Biotechnology* 42: 487–506.

Moreno, A.D., Ibarra, D., Ballesteros, I. et al. (2013). Ethanol from laccase-detoxified lignocellulose by the thermotolerant yeast *Kluyveromyces marxianus* – effects of steam pretreatment conditions, process configurations and substrate loadings. *Biochemical Engineering Journal* 79: 94–103.

Mulinari, J., Venturin, B., Sbardelotto, M. et al. (2017). Ultrasound-assisted hydrolysis of waste cooking oil catalyzed by homemade lipases. *Ultrasonics Sonochemistry* 35: 313–318.

Munoz, L.E.A. and Riley, M.R. (2008). Utilization of cellulosic waste from tequila bagasse and production of polyhydroxyalkanoate (PHA) bioplastics by *Saccharophagus degradans*. *Biotechnology and Bioengineering* 100: 882–888.

Musa, H., Han, P.C., Kasim, F.H. et al. (2017). Turning oil palm empty fruit bunch waste into substrate for optimal lipase secretion on solid state fermentation by *Trichoderma strains*. *Process Biochemistry* 63: 35–41.

Najafpour, G.D. (2015). *Biochemical Engineering and Biotechnology*. Amsterdam: Elsevier.

Neumann, P., Pesante, S., Venegas, M., and Vidal, G. (2016). Developments in pre-treatment methods to improve anaerobic digestion of sewage sludge. *Reviews in Environmental Science and Biotechnology* 15: 173–211.

Nguyen, T.M.P., Lee, Y.K., and Zhou, W. (2012). Effect of high intensity ultrasound on carbohydrate metabolism of bifidobacteria in milk fermentation. *Food Chemistry* 130: 866–874.

Noike, T. (2002). Biological hydrogen production of organic wastes – development of the two-phase hydrogen production process. Presented at the Proceedings of International Symposium on Hydrogen and Methane Fermentation of Organic Waste, pp. 31–39.

Novoa-Díaz, D., Rodríguez-Nogales, J.M., Fernández-Fernández, E. et al. (2014). Ultrasonic monitoring of malolactic fermentation in red wines. *Ultrasonics* 54: 1575–1580.

Ogasawara, H., Mizutani, K., Ohbuchi, T., and Nakamura, T. (2006). Acoustical experiment of yogurt fermentation process. *Ultrasonics* 44: e727–e730.

Oh, Y.-K., Kim, S.H., Kim, M.-S., and Park, S. (2004). Thermophilic biohydrogen production from glucose with trickling biofilter. *Biotechnology and Bioengineering* 88: 690–698.

Ojha, K.S., Mason, T.J., O'Donnell, C.P. et al. (2017). Ultrasound technology for food fermentation applications. *Ultrasonics Sonochemistry* 34: 410–417.

Ong, K.L., Kaur, G., Pensupa, N. et al. (2018). Trends in food waste valorization for the production of chemicals, materials and fuels: case study South and Southeast Asia. *Bioresource Technology, Bioconversion of Food Wastes* 248: 100–112.

Ortiz, G.E., Noseda, D.G., Ponce Mora, M.C. et al. (2016). A comparative study of new *Aspergillus* strains for proteolytic enzymes production by solid state fermentation. *Enzyme Research* https://doi.org/10.1155/2016/3016149.

O-Thong, S., Prasertsan, P., Intrasungkha, N. et al. (2008). Optimization of simultaneous thermophilic fermentative hydrogen production and COD reduction from palm oil mill effluent by *Thermoanaerobacterium*-rich sludge. *International Journal of Hydrogen Energy* 33: 1221–1231.

Pandey, A.K., Edgard, G., and Negi, S. (2016). Optimization of concomitant production of cellulase and xylanase from *Rhizopus oryzae* SN5 through EVOP-factorial design technique and application in Sorghum stover based bioethanol production. *Renewable Energy, Special Issue: New Horizons in Biofuels Production and Technologies* 98: 51–56.

Patel, H. and Gupte, A. (2016). Optimization of different culture conditions for enhanced laccase production and its purification from *Tricholoma giganteum* AGHP. *Bioresources Bioprocessing* 3: 11. https://doi.org/10.1186/s40643-016-0088-6.

Pathania, S., Sharma, N., and Handa, S. (2018). Utilization of horticultural waste (Apple Pomace) for multiple carbohydrase production from *Rhizopus delemar* F2 under solid state fermentation. *Journal, Genetic Engineering & Biotechnology* 16: 181–189.

Patsalou, M., Menikea, K.K., Makri, E. et al. (2017). Development of a citrus peel-based biorefinery strategy for the production of succinic acid. *Journal of Cleaner Production* 166: 706–716.

Pedroso, G.B., Montipó, S., Mario, D.A.N. et al. (2017). Building block itaconic acid from left-over biomass. *Biomass Conversion and Biorefinery* 7: 23–35.

Pereira, R.N. and Vicente, A.A. (2010). Environmental impact of novel thermal and non-thermal technologies in food processing. *Food Research International, Climate Change and Food Science* 43: 1936–1943.

Pirozzi, D., Fiorentino, N., Impagliazzo, A. et al. (2015). Lipid production from *Arundo donax* grown under different agronomical conditions. *Renewable Energy* 77: 456–462.

Pitol, L.O., Finkler, A.T.J., Dias, G.S. et al. (2017). Optimization studies to develop a low-cost medium for production of the lipases of *Rhizopus microsporus* by solid-state fermentation and scale-up of the process to a pilot packed-bed bioreactor. *Process Biochemistry* 62: 37–47.

Planinić, M., Zelić, B., Čubel, I. et al. (2016). Corn forage biological pretreatment by *Trametes versicolor* in a tray bioreactor. *Waste Management & Research* 34: 802–809.

Pollock, J., Ho, S.V., and Farid, S.S. (2013). Fed-batch and perfusion culture processes: economic, environmental, and operational feasibility under uncertainty. *Biotechnology and Bioengineering* 110: 206–219.

Postemsky, P.D., Bidegain, M.A., González-Matute, R. et al. (2017). Pilot-scale bioconversion of rice and sunflower agro-residues into medicinal mushrooms and laccase enzymes through solid-state fermentation with *Ganoderma lucidum*. *Bioresource Technology* 231: 85–93.

Procentese, A., Raganati, F., Olivieri, G. et al. (2017). Pre-treatment and enzymatic hydrolysis of lettuce residues as feedstock for bio-butanol production. *Biomass and Bioenergy* 96: 172–179.

Qureshi, A.S., Zhang, J., and Bao, J. (2015). High ethanol fermentation performance of the dry dilute acid pretreated corn stover by an evolutionarily adapted *Saccharomyces cerevisiae* strain. *Bioresource Technology* 189: 399–404.

Qureshi, A.S., Khushk, I., Ali, C.H. et al. (2016). Coproduction of protease and amylase by thermophilic *Bacillus* sp. BBXS-2 using open solid-state fermentation of lignocellulosic biomass. *Biocatalysis and Agricultural Biotechnology* 8: 146–151.

Raimbault, R. and Germon, J.C. (1976). Procédé d'enrichissement en protéines de produits comestibles solides. French Patent number: 76-06-677.

Rao, Y., Wan, J., Liu, Y. et al. (2018). A novel process for volatile fatty acids production from syngas by integrating with mesophilic alkaline fermentation of waste activated sludge. *Water Research* 139: 372–380.

Rashad, M.M., Abdou, H.M., and Mahmoud, A.E. (2016). Production of some bioactive materials by *Pleurotus ostreatus* from pineapple residues and rice straw via solid state fermentation. *Research Journal of Pharmaceutical, Biological and Chemical Sciences* 7: 2730–2736.

Rastogi, S., Soni, R., Kaur, J., and Soni, S.K. (2016). Unravelling the capability of *Pyrenophora phaeocomes* S-1 for the production of ligno-hemicellulolytic enzyme cocktail and simultaneous bio-delignification of rice straw for enhanced enzymatic saccharification. *Bioresource Technology* 222: 458–469.

Rathore, D., Nizami, A.-S., Singh, A., and Pant, D. (2016). Key issues in estimating energy and greenhouse gas savings of biofuels: challenges and perspectives. *Biofuel Research Journal* 3: 380–393.

Reilly, M., Dinsdale, R., and Guwy, A. (2014). Mesophilic biohydrogen production from calcium hydroxide treated wheat straw. *International Journal of Hydrogen Energy* 39: 16891–16901.

Resa, P., Elvira, L., Montero de Espinosa, F., and Gómez-Ullate, Y. (2005). Ultrasonic velocity in water-ethanol-sucrose mixtures during alcoholic fermentation. *Ultrasonics* 43: 247–252.

Rodríguez-Couto, S. (2018). Solid-state fermentation for laccases production and their applications. In: *Current Developments in Biotechnology and Bioengineering* (eds. A. Pandey, C. Larroche and C.R. Soccol), 211–234. Elsevier.

Roy, M.M., Wang, W., and Bujold, J. (2013). Biodiesel production and comparison of emissions of a DI diesel engine fueled by biodiesel–diesel and canola oil–diesel blends at high idling operations. *Applied Energy* 106: 198–208.

Roy, P., Dutta, A., and Deen, B. (2015). Greenhouse gas emissions and production cost of ethanol produced from biosyngas fermentation process. *Bioresource Technology* 192: 185–191.

Saha, B.C., Kennedy, G.J., Qureshi, N., and Bowman, M.J. (2017). Production of itaconic acid from pentose sugars by *Aspergillus terreus*. *Biotechnology Progress* 33: 1059–1067.

Sahnoun, M., Kriaa, M., Elgharbi, F. et al. (2015). *Aspergillus oryzae* S2 alpha-amylase production under solid state fermentation: optimization of culture conditions. *International Journal of Biological Macromolecules* 75: 73–80.

Salvachúa, D., Mohagheghi, A., Smith, H. et al. (2016). Succinic acid production on xylose-enriched biorefinery streams by *Actinobacillus succinogenes* in batch fermentation. *Biotechnology for Biofuels* 9: 28. https://doi.org/10.1186/s13068-016-0425-1.

Sasaki, K., Tsuge, Y., Sasaki, D. et al. (2015). Mechanical milling and membrane separation for increased ethanol production during simultaneous saccharification and co-fermentation of rice straw by xylose-fermenting *Saccharomyces cerevisiae*. *Bioresource Technology* 185: 263–268.

Schievano, A., Pepé Sciarria, T., Vanbroekhoven, K. et al. (2016). Electro-fermentation – merging electrochemistry with fermentation in industrial applications. *Trends in Biotechnology* 34: 866–878.

Sekhon Randhawa, K.K. and Rahman, P.K.S.M. (2014). Rhamnolipid biosurfactants – past, present, and future scenario of global market. *Frontiers in Microbiology* 5 https://doi.org/10.3389/fmicb.2014.00454.

Sethi, B.K., Jana, A., Nanda, P.K. et al. (2016). Thermostable acidic protease production in *Aspergillus terreus* NCFT 4269.10 using chickling vetch peels. *Journal of Taibah University for Science* 10: 571–583.

Shanmugavel, M., Vasantharaj, S., Yazhmozhi, A. et al. (2018). A study on pectinases from *Aspergillus tamarii*: toward greener approach for cotton bioscouring and phytopigments processing. *Biocatalysis and Agricultural Biotechnology* 15: 295–303.

Shen, L., Zhou, Y., Mahendran, B. et al. (2010). Membrane fouling in a fermentative hydrogen producing membrane bioreactor at different organic loading rates. *Journal of Membrane Science* 360: 226–233.

Sim, D.-W., Song, J., Kim, J.-H. et al. (2018). Enhancing bacterial production of a recombinant cetuximab-Fab by partial humanization and its utility for drug conjugation. *Process Biochemistry* 69: 123–130.

Singh, B. and Satyanarayana, T. (2008). Phytase production by a thermophilic mould *Sporotrichum thermophile* in solid state fermentation and its potential applications. *Bioresource Technology* 99: 2824–2830.

Sipiczki, M. (2011). Diversity, variability and fast adaptive evolution of the wine yeast (*Saccharomyces cerevisiae*) genome – a review. *Annales de Microbiologie* 61: 85–93.

Sivagurunathan, P., Anburajan, P., Kumar, G., and Kim, S.-H. (2016). Effect of hydraulic retention time (HRT) on biohydrogen production from galactose in an up-flow anaerobic sludge blanket reactor. *International Journal of Hydrogen Energy* 41: 21670–21677.

Soccol, C.R. and Vandenberghe, L.P.S. (2003). Overview of applied solid-state fermentation in Brazil. *Biochemical Engineering Journal, Solid-State Fermentation* 13: 205–218.

Soccol, C.R., da Costa, E.S.F., Letti, L.A.J. et al. (2017). Recent developments and innovations in solid state fermentation. *Biotechnology Research and Innovation* 1: 52–71.

Sodhi, H.K., Sharma, K., Gupta, J.K., and Soni, S.K. (2005). Production of a thermostable α-amylase from *Bacillus* sp. PS-7 by solid state fermentation and its synergistic use in the hydrolysis of malt starch for alcohol production. *Process Biochemistry* 40: 525–534.

Somavat, P., Li, Q., de Mejia, E.G. et al. (2016). Coproduct yield comparisons of purple, blue and yellow dent corn for various milling processes. *Industrial Crops and Products* 87: 266–272.

Soni, S. and Dey, G. (2014). Perspectives on global fermented foods. *British Food Journal* 116: 1767–1787.

Souza, L.O., de Brito, A.R., Bonomo, R.C.F. et al. (2018). Comparison of the biochemical properties between the xylanases of *Thermomyces lanuginosus* (Sigma®) and excreted by *Penicillium roqueforti* ATCC 10110 during the solid state fermentation of sugarcane bagasse. *Biocatalysis and Agricultural Biotechnology* 16: 277–284.

Spendley, W., Hext, G.R., and Himsworth, F.R. (1962). Sequential application of simplex designs in optimisation and evolutionary operation. *Technometrics* 4: 441–461.

Spier, M.R., Vandenberghe, L., Medeiros, A.B.P., and Soccol, C.R. (2011). Application of different types of bioreactors in bioprocesses. In: *Bioreactors: Design, Properties and Applications* (eds. P.G. Antolli and Z. Liu), 55–90. Nova science publishers, Inc.

Stanbury, P.F., Whitaker, A., and Hall, S.J. (1995). Media for industrial fermentations. In: *Principles of Fermentation Technology* (eds. P.F. Stanbury, A. Whitaker and S.J. Hall), 93–122. Amsterdam: Pergamon.

Stanbury, P.F., Whitaker, A., and Hall, S.J. (2018). *Principles of Fermentation Technology*. Amsterdam: Elsevier.

Su, H., Liu, G., He, M., and Tan, F. (2015). A biorefining process: sequential, combinational lignocellulose pretreatment procedure for improving biobutanol production from sugarcane bagasse. *Bioresource Technology* 187: 149–160.

Tallapragada, P., Dikshit, R., Jadhav, A., and Sarah, U. (2017). Partial purification and characterization of amylase enzyme under solid state fermentation from *Monascus sanguineus*. *Journal, Genetic Engineering & Biotechnology* 15: 95–101.

Tao, Y., Zhang, Z., and Sun, D.-W. (2014). Experimental and modeling studies of ultrasound-assisted release of phenolics from oak chips into model wine. *Ultrasonics Sonochemistry* 21: 1839–1848.

Teixeira, J.A. and Vicente, A.A. (2017). *Engineering Aspects of Food Biotechnology*. London: Taylor & Francis Ltd. ed.

Theuretzbacher, F., Blomqvist, J., Lizasoain, J. et al. (2015). The effect of a combined biological and thermo-mechanical pretreatment of wheat straw on energy yields in coupled ethanol and methane generation. *Bioresource Technology* 194: 7–13.

Thiry, M. and Cingolani, D. (2002). Optimizing scale-up fermentation processes. *Trends in Biotechnology* 20: 103–105.

Tian, T., Yang, H., Yang, F. et al. (2018). Optimization of fermentation conditions and comparison of flavor compounds for three fermented greengage wines. *Lebensmittel-Wissenschaft & Technologie* 89: 542–550.

Tirpanalan, Ö., Reisinger, M., Smerilli, M. et al. (2015). Wheat bran biorefinery – an insight into the process chain for the production of lactic acid. *Bioresource Technology* 180: 242–249.

Tracy, B.P., Jones, S.W., Fast, A.G. et al. (2012). Clostridia: the importance of their exceptional substrate and metabolite diversity for biofuel and biorefinery applications. *Current Opinion in Biotechnology*, Energy biotechnology • Environmental biotechnology 23: 364–381.

Ueno, Y., Haruta, S., Ishii, M., and Igarashi, Y. (2001). Microbial community in anaerobic hydrogen-producing microflora enriched from sludge compost. *Applied Microbiology and Biotechnology* 57: 555–562.

Utami, T.S., Hariyani, I., Alamsyah, G., and Hermansyah, H. (2017). Production of dry extract extracellular lipase from *Aspergillus niger* by solid state fermentation method to catalyze biodiesel synthesis. *Energy Procedia* 136: 41–46.

Valdez-Vazquez, I., Sparling, R., Risbey, D. et al. (2005). Hydrogen generation via anaerobic fermentation of paper mill wastes. *Bioresource Technology* 96: 1907–1913.

Van Ginkel, S.W. and Logan, B. (2005). Increased biological hydrogen production with reduced organic loading. *Water Research* 39: 3819–3826.

Walker, G.M. (2011). 125th anniversary review: fuel alcohol: current production and future challenges. *Journal of the Institute of Brewing* 117: 3–22.

Walker, G.M. (2014). Fermentation (industrial): media for industrial fermentations. *Encyclopedia of food microbiology*: 769–777. https://doi.org/10.1016/B978-0-12-384730-0.00107-5.

Walker, G.M. and Walker, R.S.K. (2018). Enhancing yeast alcoholic fermentations. *Advances in Applied Microbiology* 105: 87–129.

Walsh, G. (2014). Biopharmaceutical benchmarks 2014. *Nature Biotechnology* 32: 992–1000.

Wang, L. and Chen, H. (2011). Increased fermentability of enzymatically hydrolyzed steam-exploded corn stover for butanol production by removal of fermentation inhibitors. *Process Biochemistry* 46: 604–607.

Wang, M., Han, J., Dunn, J.B. et al. (2012). Well-to-wheels energy use and greenhouse gas emissions of ethanol from corn, sugarcane and cellulosic biomass for US use. *Environmental Research Letters* 7: 045905. https://doi.org/10.1088/1748-9326/7/4/045905.

Wang, K., Tian, Y., Zhou, N. et al. (2018). Studies on fermentation optimization, stability and application of prolyl aminopeptidase from *Bacillus subtilis*. *Process Biochemistry* 74: 10–20.

Wei, D., Liu, X., and Yang, S.-T. (2013). Butyric acid production from sugarcane bagasse hydrolysate by *Clostridium tyrobutyricum* immobilized in a fibrous-bed bioreactor. *Bioresource Technology* 129: 553–560.

Wu, S.-Y., Chu, C.-Y., and Shen, Y.-C. (2012). Effect of calcium ions on biohydrogen production performance in a fluidized bed bioreactor with activated carbon-immobilized cells. *International Journal of Hydrogen Energy* 37: 15496–15502.

Wu, D., Li, Q., Wang, D., and Dong, Y. (2013). Enzymatic hydrolysis and succinic acid fermentation from steam-exploded corn stalk at high solid concentration by recombinant *Escherichia coli*. *Applied Biochemistry and Biotechnology* 170: 1942–1949.

Xi, Y., Dai, W., Xu, R. et al. (2013). Ultrasonic pretreatment and acid hydrolysis of sugarcane bagasse for succinic acid production using *Actinobacillus succinogenes*. *Bioprocess and Biosystems Engineering* 36: 1779–1785.

Xie, H., Shen, H., Gong, Z. et al. (2012). Enzymatic hydrolysates of corn stover pretreated by a N-methylpyrrolidone–ionic liquid solution for microbial lipid production. *Green Chemistry* 14: 1202–1210.

Xing, D., Ren, N., Wang, A. et al. (2008). Continuous hydrogen production of auto-aggregative *Ethanoligenens harbinense* YUAN-3 under non-sterile condition. *International Journal of Hydrogen Energy, Asian bioH2 Symposium* 2006 (33): 1489–1495.

Xu, Q., Li, S., Fu, Y. et al. (2010). Two-stage utilization of corn straw by *Rhizopus oryzae* for fumaric acid production. *Bioresource Technology* 101: 6262–6264.

Xu, J., Wang, X., Hu, L. et al. (2015). A novel ionic liquid-tolerant *Fusarium oxysporum* BN secreting ionic liquid-stable cellulase: consolidated bioprocessing of pretreated lignocellulose containing residual ionic liquid. *Bioresource Technology* 181: 18–25.

Yang, Y., Foulquié-Moreno, M.R., Clement, L. et al. (2013). QTL analysis of high thermotolerance with superior and downgraded parental yeast strains reveals new minor QTLs and converges on novel causative alleles involved in RNA processing. *PLoS Genetics* 9: e1003693. https://doi.org/10.1371/journal.pgen.1003693.

Yang, X., Geng, B., Zhu, C. et al. (2018). Fermentation performance optimization in an ectopic fermentation system. *Bioresource Technology* 260: 329–337.

Yazid, N.A., Barrena, R., and Sánchez, A. (2016). Assessment of protease activity in hydrolysed extracts from SSF of hair waste by and indigenous consortium of microorganisms. *Waste Management* 49: 420–426.

Yu, H., Zhu, Z., Hu, W., and Zhang, H. (2002). Hydrogen production from rice winery wastewater in an upflow anaerobic reactor by using mixed anaerobic cultures. *International Journal of Hydrogen Energy* 27: 1359–1365.

Yu, X., Zheng, Y., Dorgan, K.M., and Chen, S. (2011). Oil production by oleaginous yeasts using the hydrolysate from pretreatment of wheat straw with dilute sulfuric acid. *Bioresource Technology* 102: 6134–6140.

Zhang, Y. and Shen, J. (2006). Effect of temperature and iron concentration on the growth and hydrogen production of mixed bacteria. *International Journal of Hydrogen Energy* 31: 441–446.

Zhang, M.-L., Fan, Y.-T., Xing, Y. et al. (2007). Enhanced biohydrogen production from cornstalk wastes with acidification pretreatment by mixed anaerobic cultures. *Biomass and Bioenergy* 31: 250–254.

Zhang, Y., Sun, W., Wang, H., and Geng, A. (2013). Polyhydroxybutyrate production from oil palm empty fruit bunch using *Bacillus megaterium* R11. *Bioresource Technology* 147: 307–314.

Zhang, Y., Chen, X., Luo, J. et al. (2014). An efficient process for lactic acid production from wheat straw by a newly isolated *Bacillus coagulans* strain IPE22. *Bioresource Technology* 158: 396–399.

Zhang, Q.-A., Shen, Y., Fan, X.-H., and Martín, J.F.G. (2016). Preliminary study of the effect of ultrasound on physicochemical properties of red wine. *CyTA Journal of Food* 14: 55–64.

Zhang, B.-B., Xing, H.-B., Jiang, B.-J. et al. (2018). Using millet as substrate for efficient production of monacolin K by solid-state fermentation of *Monascus ruber*. *Journal of Bioscience and Bioengineering* 125: 333–338.

Zhao, Q.-B. and Yu, H.-Q. (2008). Fermentative H_2 production in an upflow anaerobic sludge blanket reactor at various pH values. *Bioresource Technology* 99: 1353–1358.

Zhao, L., Cao, G.-L., Wang, A.-J. et al. (2013). Simultaneous saccharification and fermentation of fungal pretreated cornstalk for hydrogen production using *Thermoanaerobacterium thermosaccharolyticum* W16. *Bioresource Technology* 145: 103–107.

Zhao, G., Ding, L.-L., Pan, Z.-H. et al. (2019). Proteinase and glycoside hydrolase production is enhanced in solid-state fermentation by manipulating the carbon and nitrogen fluxes in *Aspergillus oryzae*. *Food Chemistry* 271: 606–613.

Zheng, P., Fang, L., Xu, Y. et al. (2010). Succinic acid production from corn stover by simultaneous saccharification and fermentation using *Actinobacillus succinogenes*. *Bioresource Technology* 101: 7889–7894.

Zhou, P., Elbeshbishy, E., and Nakhla, G. (2013). Optimization of biological hydrogen production for anaerobic co-digestion of food waste and wastewater biosolids. *Bioresource Technology* 130: 710–718.

3

Culture Condition Changes for Enhancing Fermentation Processes

Marina Al Daccache[1], Nicolas Louka[1], Richard G. Maroun[1], and Dominique Salameh[2]

[1] *Faculté des Sciences, Centre d'Analyses et de Recherche, UR TVA, Laboratoire CTA, Université Saint-Joseph, Beyrouth, Lebanon*
[2] *Faculté des Sciences, Centre d'Analyses et de Recherche, UR - EGP, Laboratoire E2D, Université Saint-Joseph, Beyrouth, Lebanon*

3.1 Introduction

In practice, at any scale of production, every fermentation reaction is initiated with microorganism's cultivation. Monitoring microbiological cultures and understanding different cultivation methods should be the strong suit of every microbiologist (Monod 1949). It is known that the constitution and performance of microorganisms are directly linked to their growth environment. Eminent microbiologists have noted the importance of the media and culture conditions. Therefore, possessing deep knowledge of the fermentation process and metabolic activities will lead to better control of the growth conditions and media formulation.

The term "fermentation" has received different meanings and definitions over the past decades. In a microbiological dimension, fermentation is widely associated with the anaerobic or oxygen-starved environment. In parallel, respiration is known to occur under an oxygen-rich atmosphere. In a much wider consideration, the term "bioreaction" can be used for any reaction in which the raw material (substrate) is transformed by the action of a microorganism (or a cell culture used as a microbial culture). Its biochemical meaning describes fermentation as a bioreaction producing energy by the metabolism of organic compounds. Therefore, the result of bioreaction is the accomplishment of all or part of the metabolism of the microbial cell. In a complete definition, metabolism is the set of reactions that allow organisms to obtain, from the elements of the medium, the molecules and the energy necessary to synthesize the functional or structural macromolecules. It ensures the reactions of growth, production, and maintenance. Metabolism consists of two main categories: catabolism and anabolism.

Fermentation Processes: Emerging and Conventional Technologies, First Edition.
Edited by Mohamed Koubaa, Francisco J. Barba, and Shahin Roohinejad.

Catabolism is an oxidative process that results in the degradation of substrates; catabolic reactions provide the subunits and the energy required for the synthesis reactions of macromolecules. They produce reduced pyridine nucleotides which must be reoxidized to ensure the continuity of the process. Anabolism groups together the constructive metabolic processes where the energy liberated by catabolism is consumed to synthesize complex molecules.

The chemical definition of the fermentation reaction can be considered as the most complete definition. It is defined as an incomplete oxidation–reduction reaction in which the donor and the electron acceptor are both organic compounds produced, in general, from the degradation of the same initial substrate (Teusink et al. 1998). From another side, respiration is a complete redox reaction in which the electron donor is generally an organic compound and the final electron acceptor is an inorganic compound, usually oxygen. Taking into account the chemical definition exposed above, several complex parameters and factors may interfere and must be considered when designing the optimized culture medium. Such prerequisites could be mentioned as electrochemical conditions, thermodynamic laws, chemical mass balance, stoichiometry conditions, medium composition, and many other physical parameters such as agitation that regulates the overall kinetics.

It should be noted that the presence of oxygen is not enough to ensure that the microorganism will undergo the respiratory metabolism, and similarly, a cell can function in fermentative metabolism even in the presence of oxygen. In fact, microorganisms prefer respiratory route, the most energetic pathway, but the metabolic fate of the substrate depends largely on yeast growth conditions. Explicitly, some microorganisms can switch from respiration to fermentation in the presence of large amounts of sugars even when oxygen is abundant (Piskur et al. 2006; Pronk et al. 1996). Therefore, the naming of the fermentation is based on the end product to mitigate any ambiguity of the process conditions. The following are the major groups of commercially important fermentation processes depending on the end products generated by the microorganism: fermentations that produce biomass, microbial enzymes, microbial metabolites, recombinant products, and finally those that convert a specific compound into a structurally related one, but more valuable.

Taking all that complexity into consideration, the medium formulation seems to be an essential point influencing the reactions behind the fermentation. It should fulfill the elemental requirements for cell biomass and metabolite production and must supply adequate energy for biosynthesis and cell maintenance. Culture condition changes and the medium composition effects on the fermentation process will be discussed in this chapter.

3.2 Culture Media Used for Fermentation

3.2.1 The Culture Media Purpose

Precise formulation is needed to optimize the most appropriate medium for a specific fermentation process. Some transversal requirements must exist in all media. The survival of microorganisms requires water, a source of energy, carbon, nitrogen, and some mineral elements. The formula robustness of a media is scale dependent. Even if growth prerequisites are satisfied on a small scale, it may be poorly reproducible at a larger volume reactor. It must be remembered that the design of the medium may be dependent on fermentation purposes. For a large-scale fermentation, the choice of nutrient must be made to meet many criteria such as:

- Reaching the maximum yield of product per gram of substrate consumed;
- Reaching the maximum rate of product formation;
- Reproduction of a constant end product quality;
- Satisfying practical conditions of manipulation during media preparation and sterilization; and
- Causing as minimum as possible complications during the production process (i.e. aeration, agitation, purification, etc.).

The medium selection is of great interest since it will also affect the fermenter design. It was reported that the optimal quantities of nitrogen presented some dissimilarities with the type of fermenter used. Because of direct interaction between the medium components and the process conditions, another factor to be considered is the scalability of the fermentation. A laboratory medium may not be convenient in a large fermenter when the gas-transfer pattern is lower. Also, a high viscosity could limit the agitation effectiveness on a larger scale. Besides, the medium may also affect the pH variation, foam accumulation, the redox potential, and even the morphological form of the microorganism. Consequently, a list of criteria may be taken into consideration before choosing the appropriate medium for the desired bioprocess, for example:

1) The elemental composition of the microorganism used;
2) The energy source required by the cells;
3) The purpose of the bioprocess: biomass production, product formation, or both;
4) The impacts due to the medium formulation (pH, viscosity, etc.);
5) The process design: batch, fed-batch, continuous flows, the geometry of the reactor, etc.;
6) The requirement of different processes such as filtration, centrifugation, purification, etc. and the medium limitation on these activities; and
7) The sterilization methods needed and their availability.

3.2.2 Media Types

Cells can be cultured in solid, semi-solid, or liquid media. The most common growth media for microorganisms are nutrient broths (liquid nutrient medium) or lysogeny broth (LB) medium. Liquid media are often mixed with agar and poured via a sterile media dispenser into Petri dishes to solidify. These agar plates provide a solid medium on which microbes may be cultured. They remain solid, as very few bacteria can decompose agar (the exception being some species in the genera: *Cytophaga*, *Flavobacterium*, *Bacillus*, *Pseudomonas*, and *Alcaligenes*).

Three media categories can be broadly found based on their composition: synthetic, semi-synthetic, and complex.

3.2.2.1 Synthetic Media

Synthetic media are composed of a defined concentration of each chemical component. These media are generally simple, containing a carbon/energy source, a nitrogen source, and inorganic salts. The medium complexity will increase depending on the microorganism growth requirements.

Mainly synthetic media are used for research and in laboratory. Synthetic media tend to be expensive especially if they contain specific components such as vitamins and growth factors needed in their pure form. They are completely soluble and offer the least downstream processing challenges in an industry context or purification/recovery challenges.

3.2.2.2 Semi-synthetic Media

Semi-synthetic media contain a large percentage of chemically defined elements, including one or more poorly specified component of variable with controlled composition, for example, yeast extract. Such media are also useful in research and laboratory situations for an organism requiring an expensive pure substrate. Such components have specific nutritional needs and a rich source of many unknown nutrients.

3.2.2.3 Complex Media

Complex media are composed of substances from plant or animal origin. So, these materials may change from batch to batch where the composition can be influenced by time of year, location of origin, and small changes in production protocols. A simple example of a complex medium is the famous LB "lysogeny broth" popular for cultivating microbes. Even though this medium is commonly used and very suitable for the researcher, it has many drawbacks when employed for physiological studies in batch culture. LB medium includes a digest of casein and yeast extract with small amounts of amino acids, mostly peptides of varying lengths, and just small quantities of carbohydrates. Thus, the medium supports only limited exponential growth in the early phase after low inoculation.

3.2.2.4 Defined Mineral Media

A medium having an excess concentration of all nutrients and one of them limiting the cell growth, because of stoichiometry inadequacy, is known as a defined mineral medium. In this kind of media, the growth-limiting factor is usually the carbon/energy source. In such a medium, the growth of the cultures can occur at maximum specific growth rate (μ_{max}) until the consumption of the limiting factor at levels that affect the growth rate kinetically. If cells must be used later, they will be selected during the exponential growth phase before the start of kinetic restrictions. Conventional batch cultivation does not allow cell growth at different growth rates. The common way to obtain different μ_{max} is to change the limiting factor. Consequently, the cell's growth physiology will be affected either due to the changes in the growth rates or due to the changes in the medium composition. For example, the dependence of cell size on the specific growth rate in *Salmonella* was reported using 22 different media (Schaechter et al. 1959). From another side, different specific growth rates can be obtained in the same medium by lowering the concentration of the limiting substrate.

3.2.3 Culture Media: A Quantitative Approach

Chemistry laws, such as mass conservation, energy conservation, and especially stoichiometry, are also applicable for the bioreactions. In 1975, Pirt established a quantitative method for microbiology. He provided an explanation consisting of basic concepts and mathematical descriptions of cultivation methods and processes of cellular growth (Pirt 1975). The next stage will be modeling the metabolism in accordance with the chemistry principle which states that nothing is lost nor is created; everything is transformed (Figure 3.1).

In another way of understanding the bioreaction, each substrate catabolized through the bioreaction will bring energy to maintain the microorganisms, first to produce new molecules through anabolism, and finally to participate in producing bigger molecules synthetizing new microorganisms to ensure cell population growth (Eq. (3.1)).

$$\Delta S = \Delta S_{maintenance} + \Delta S_{metabolite\ production} + \Delta S_{multiplication} \tag{3.1}$$

It is then important, when accounting yields of biomass production, to consider the bioreaction specificities (Eq. (3.2)).

$$\frac{1}{Y_{X/S}} = \frac{\Delta S_{maintenance}}{\Delta X} + \frac{\Delta S_{metabolite\ production}}{\Delta X} + \frac{\Delta S_{multiplication}}{\Delta X} \tag{3.2}$$

where $Y_{X/S}$ represents the biomass yield ($g_{biomass}/g_{substrate}$).

From the separate measurements of the substrates (input) and products (output), it is possible to write the global equation of the reaction. In this case (writing according to the input/output mass balance), the hypothesis is that this equation is valid throughout the reaction. However, there are also more complex cases where the balance equation only partially reflects the reality as for example when one is dealing with a reaction scheme in several stages.

It is common in biological reactions that the intracellular regulatory mechanisms (see above) cause changes in the rates of one of the reaction steps, or even the absolute ending of a reaction chain. The substrate flow then moves to another metabolic branch.

To be able to describe the cellular growth as well as its metabolic activities, we must assimilate them to chemical reactions. It is then necessary to establish a relation between the raw formula of the cell in dry matter, the incoming substrates, those accumulated, and the products formed. If one is interested only in growth by neglecting the formation of

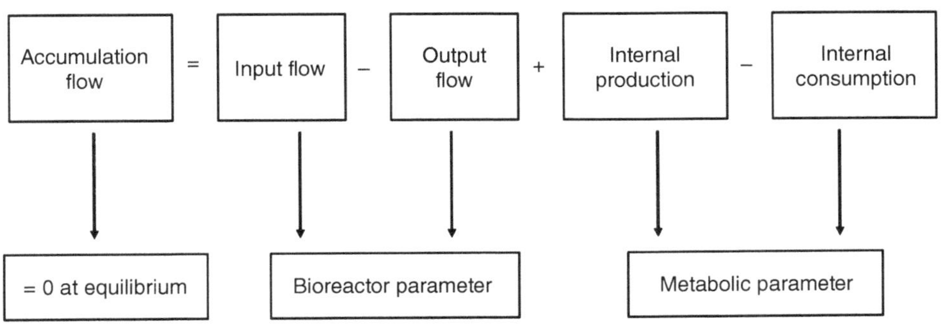

Figure 3.1 Schematic representation of metabolism modeling. *Source:* Salameh et al. (2013).

products, one can translate the phenomenon of the microbial growth by the chemical equation (Eq. (3.3)):

$$AC_aH_bO_c + BO_2 + CC_\alpha H_\beta O_\gamma N_\delta + DNH_4OH \rightarrow EC_\alpha H_\beta O_\gamma N_\delta + FCO_2 + GH_2O + QKJ \quad (3.3)$$

where A, B, C, D, E, F, and G are the stoichiometric coefficients; a, b, and c represent the composition of the carbon substrate; α, β, γ, and δ represent the composition of the biomass (which depends on the nature of the substrate, the type of microorganism, and the physiological state of the cells); and $C_\alpha H_\beta O_\gamma N_\delta$ corresponds to the gross formula of the cellular dry matter.

This equation must be expressed quantitatively, which is crucial to maintain component wastage as minimal as possible. So, it is possible to calculate the amount of nutrients needed to produce a certain quantity of biomass. Also, it should be possible to evaluate the initial substrate quantities to obtain defined product yields, taking into consideration that some components may be only needed for the formation of the products and are not required for biomass production. To find a solution for this, balance equation of growth yield factors for elements, $Y_{X/S}$, can be derived from the elemental composition of the biomass. Thus, a stoichiometric link between the quantity of nutrients provided in a medium and the biomass produced from it may be established. The growth yield factors have been demonstrated and used for many microorganisms (Egli 2009; Pirt 1975). Not only the nutrients refer to the growth yield factors, but it similarly encloses the electron donors and acceptors serving as energy sources and electron acceptors (Egli 2009). Table 3.1 may serve as a guide showing the minimal quantities of the elements present in a culture medium.

Another point concerning the use of an appropriate media is the nutrient stoichiometrically limiting growth in the cultivation system. The nutrient will define the maximum quantity of the biomass in the system. In fact, the carbon substrate has a dual function: biosynthesis and energy generation. Thus, the carbon needs for biomass synthesis under aerobic conditions may be assessed from the cellular yield coefficient (Y), which is well defined according to Eq. (3.4):

$$Y = \frac{Quantity\ of\ cell\, dry\, matter\ produced}{Quantity\ of dry\ carbon\ substrate\ utilized} \quad (3.4)$$

This nutrient will determine the necessity of the other sources of nutrient that should be provided in the medium. An appropriate medium may contain an excess of all nonlimiting nutrients and typical excess factors. The design of a medium should also take into consideration the several elements needed for an optimal culture.

3.2.4 Culture Media: A Compositional Approach

It is extremely important to understand the role of the different compositional parts of a culture media to be able to reach its optimum formulation under the scope of microbial metabolism, energy flows, and fermentation purpose. The ratio between the energy chemical sources and the water content is a key parameter in substrate concentration optimization.

3.2.4.1 Water

Water is a major component found in all fermentation media, which is also needed for auxiliary services such as cooling, heating, rinsing, and cleaning. The pH value, as well as

Table 3.1 Average elemental composition of bacteria, yeasts, and fungi (by % dry weight), and content of major cellular constituents of Gram-negative cells (by % dry weight).

Elemental constituents in dry biomass	Bacteria	Yeats	Fungi
Carbon	50–53	45–50	40–63
Hydrogen	7	7	—
Nitrogen	12–15	7.5–11	7–10
Phosphorus	2–3	0.8–2.6	0.4–4.5
Sulfur	0.2–1	0.01–0.24	0.1–0.5
Potassium	1.0–4.5	1.0–4.0	0.2–2.5
Sodium	0.5–1	0.01–0.1	0.02–0.5
Calcium	0.01–1.1	0.1–0.3	0.1–1.4
Magnesium	0.1–0.5	0.1–0.5	0.1–0.5
Chloride	0.5	—	—
Iron	0.02–2	0.01–0.5	0.1–0.2
Cellular constituents	Gram-negative cells[a]	Cellular constituents	Gram-negative cells[a]
Protein	55	Phospholipids	9
RNA	21	Glycogen	3
DNA	3	Polyhydroxybutyrate (PHB)	—
Peptidoglycan	3	Polyphosphate	—

[a] Cells growing with an excess of nutrients at μ_{max} in batch culture.
Source: Aiba et al. (1973); Egli (2009); Herbert (1976); Lilly (1965); Luria (1960).

the dissolved salts, should be considered in water quality assessment. Therefore, the importance of water quality will depend on the bioprocess. Nowadays, the water problem is solved due to the large range of water purification systems such as treatments using deionization techniques, reverse osmosis, and other methods including adding salts or pH adjustments. Specifically, the mineral elements found in water are important for the brewing process affecting the positioning of breweries and the sorts of beer produced. Hard water rich in $CaSO_4$ is suitable for English Burton bitter beers, while the production of dark beers requires water with high carbonate content.

3.2.4.2 Energy Sources

The oxidation of medium elements in regular fermentation or light sources in photofermentation generates energy for the cell's growth. Nevertheless, since the major industrial microorganisms are chemoorganotrophs, the usual source of energy will be the carbon source (e.g. carbohydrates, lipids, and proteins). Besides, some microorganisms can use hydrocarbons or methanol (Bauchop and Elsden 1960).

3.2.4.3 Carbon Sources

The metabolization rate of the carbon source may influence the production of biomass and the primary or secondary metabolites. High concentrations of rapidly metabolized sugars induce fast growth and low productivity of secondary metabolites. The choice of the carbon source will depend on the main product of a fermentation process, especially if this product will be formed directly from its dissimilation.

3.2.4.4 Examples of Commonly Used Carbon Sources

3.2.4.4.1 Carbohydrates

The starch obtained from maize grain is the most broadly available carbohydrate. The direct usage of maize and other cereals may also be adopted. A variety of glucose preparations may be obtained by starch hydroxylation by dilute acids and enzymes, for example, hydrolyzed cassava starch was used as a carbon source for glutamic acid production in Japan (Minoda 1986). Furthermore, starch particles and glucose are considered as a cheap medium for ethanol production (Bawa et al. 2010). Malt, sucrose, lactose, and xylose may be considered as other carbon sources.

3.2.4.4.2 Glucose

Most cell lines can consume glucose. This substrate is widely available, reliable, easily stored, and handled. However, there are also several drawbacks of using glucose:

1) The occurrence of "Crabtree effect" if glucose is oversupplied in the initial stages of growth
2) The loss of some substrates due to the Maillard reaction (if glucose is sterilized with a nitrogen source)

3.2.4.4.3 Sucrose

It is often used in research laboratories, but not all microorganisms can metabolize sucrose. Many forms and grades of sucrose are found commercially: pure granulated forms to complex molasses solutions. However, sucrose has several drawbacks: it can caramelize when oversterilized. Molasses can be quickly contaminated, and the sterilization can significantly reduce sucrose availability.

3.2.4.4.4 Lactose

A limited range of microorganisms such as *Escherichia coli* can only use lactose, and it is slowly metabolized. Sucrose is useful for some commercial processes when the complex substrate whey is used. It is a by-product of the dairy industry containing around 50 g of lactose per liter as whey and 4% w/w nitrogen. Some fermentation processes initially use glucose for faster growth and then add lactose after glucose is exhausted.

3.2.4.4.5 Oils and Fats

Vegetable oils such as olive, maize, cottonseed, linseed, and soybean may also be considered as carbon substrates. Several factors favor the use of oils instead of carbohydrates. Typical oil generates twice more energy than glucose on a weight basis. Oil-based fed-batch fermentations allow the usage of the maximum working capacity of a vessel. They are also characterized by their antifoam properties. Due to the diminution of the surface tension, when microemulsion of rapeseed oil was used for oxytetracycline, higher biomass was produced, and oxytetracycline production was increased (Papapanagiotou et al. 2005). Glycerol is used for the fermentation where the substrate's purity is important. The successful usage of soybean oil for secondary metabolite production by *Streptomyces hygroscopicus* was reported (Junker et al. 1998).

3.2.4.4.6 Hydrocarbons and Their Derivatives

n-alkanes were used for the production of organic acids, amino acids, vitamins and cofactors, nucleic acids, antibiotics, enzymes, and proteins (Fukui and Tanaka 1980). For biomass production, methane, methanol, and *n*-alkanes were usually adopted (Drozd 1987; Hamer 1979; Jobling 1990; Levi et al. 1979). The disadvantages of using hydrocarbons as substrates are related to the cost, purity, and the process dimension (Drozd 1987).

3.2.4.4.7 Other Carbon Sources

Biodiesel generated by animal fats and vegetable oils gives almost 10% by weight glycerol, which could be consumed as a carbon and energy source in microbial fermentations aiming to produce valuable products such as 1,3-propanediol, dihydroxyacetone, and ethanol (da Silva et al. 2009).

Many agricultural waste products, by-products, and residues may also be used as fermentation substrates. Palm oil mill effluent (POME), a waste from palm oil production, was used for acetone–butanol fermentation (Somrutai et al. 1996). Besides, Rajagopalan et al. have reported the use of wheat bran and sunflower oil cake for α-amylase production (Rajagopalan and Krishnan 2009).

3.2.4.5 Nitrogen Sources

Most of the microorganisms can assimilate inorganic or organic sources of nitrogen. Inorganic nitrogen could be ammonia gas, ammonium salts, or nitrates (Hutner 1972). Several points must be taken into consideration before selecting the nitrogen *Source:*

1) Defining the processing purpose: biomass production, intracellular or extracellular production, or both;
2) Deciding whether the nitrogen source would be the limiting factor; and
3) Defining the effect on the nitrogen source on the pH of the medium.

3.2.4.5.1 Ammonia

Ammonia is usually used in industrial processes and rarely used in research laboratories due to the liquid volatility and the handling problems associated with it. Ammonia is readily dissolved in the medium and immediately available to the cells, in addition to its cheapness.

3.2.4.5.2 Nitrogen-Based Salts

Ammonium sulfate ($(NH_4)_2SO_4$) and ammonium chloride (NH_4Cl) are usual nitrogen sources. Organisms use ammonium, and the usage of this salt is a cheap and convenient method to respond to cell requirements. Furthermore, the analysis of the nitrogen content will be easy and achieved by a simple assay and an ammonia probe. It is smart to consider the other requirements of the organism to choose an adequate salt.

3.2.4.5.3 Complex Nitrogen Sources

Several nitrogen sources can be considered as complexes such as yeast extract, soya powder (8% w/w nitrogen), and corn steep liquor (4% w/w). In addition to the nitrogen elements, these sources may contain essential nutrients such as carbon, vitamins, and minerals. A significant batch-to-batch variability may be remarked.

3.2.4.6 Minerals

Certain mineral elements are required for microorganism's growth and metabolism (Hughes and Poole 1991). Essential components are needed, such as magnesium, phosphorus, potassium, sulfur, calcium, and chlorine. Other elements such as cobalt, copper, iron, manganese, molybdenum, and zinc are also needed, but they can be found as impurities in the major ingredients (Table 3.2.).

3.2.4.7 Chelators

The preparation of some media will induce the formation of an insoluble precipitate due to insoluble metal phosphates. This problem can be solved by adding small quantities of chelating agents such as ethylene diamine tetraacetic acid, citric acid, polyphosphates, etc. into the medium. It is important to note that a chelating agent should not cause growth inhibition of the microorganism.

3.2.4.8 Growth Factors

The growth factors required are vitamins, but specific amino acids, fatty acids, or sterols may also be needed. Some natural carbon and nitrogen sources normally found in media formulations comprise some of the desired growth factors (Atkinson and Mavituna 1991). A smart blending of materials can eliminate vitamin deficiency.

Table 3.2 Description of the typical concentration ranges of mineral components.

Component	Range ($mg\,ml^{-1}$)
KH_2PO_4	0.01–4.0
$MgSO_4{\cdot}7H_2O$	0.01–20.0
KCl	0.5–12.0
$CaCO_3$	0.08–17.0
$FeSO_4{\cdot}4H_2O$	0.01–0.1
$MnSO_4{\cdot}H_2O$	0.0001–2.0
$ZnSO_4{\cdot}8H_2O$	0.0001–1.0
$CuSO_4{\cdot}5H_2O$	0.003–0.01
$Na_2MoO_4{\cdot}2H_2O$	0.005–2.5
$CaSO_4$	0.1–2.0
NH_4Cl	1–3.0

3.2.4.9 Buffers

To obtain optimal productivity, pH should be controlled. A compound may be added to serve either as a buffer or as a nutrient source. The incorporation of calcium carbonate adjusts the media pH to around 7.0. A pH decrease can cause carbonate decomposition. Furthermore, phosphates may play a major role in buffering. However, a high phosphate concentration may influence the production of several secondary metabolites. A balance between carbon and nitrogen sources may be another way for pH control. It may also be regulated externally by adding ammonia or sodium hydroxide and sulfuric acid.

3.2.4.10 Precursors and Metabolic Regulators to Media

Not all the components of a fermentation medium should regulate the production of the product. Some have a major role in supporting the growth of the microorganisms. Those additives involve mainly precursors, inhibitors, and inducers.

3.2.4.11 Precursors and Inhibitors

Some substances, known as "precursors," are added in a fermentation medium. They are then incorporated in some reactions to produce the desired end product. By adding inhibitors to fermentations, more of a specific product may be formed, or a metabolic intermediate could be accumulated. In a defined medium, it is easy to avoid known inhibitors, but in complex media, some components such molasses may contain some inhibitors.

3.2.5 Impact of Culture Conditions on Fermentation Processes

A controlled and successful fermentation process requires the creation of a balance between the best growing medium, under optimal conditions at the required rate with the necessary oxygen supply. These conditions seem utopian but require an understanding of the purpose of the fermentation in question and the nature of the microorganism used.

Therefore, to obtain optimal fermentation as a specific growth or production rate, it is essential to adjust and control different external parameters.

3.2.5.1 The Temperature

Temperature is a key factor influencing the activities of microorganisms. For example, each yeast cell can grow at very specific temperatures (Ribereau-Gayon et al. 2006). Fleet suggests that different temperatures are suitable for different strains, ranging from 10 to 30°C (Fleet 1992). More precisely, the growth rate varies from one strain to another depending on the temperature. Bordeaux oenologists have devoted a great deal of research to this subject, which has been summarized by Ribereau-Gayon et al. (Ribereau-Gayon et al. 1975). The temperature profoundly affects the respiratory and fermentative intensity of the yeasts: the intensity of the fermentation doubles for each temperature increase of 10°C. It reaches its maximum at 35°C and begins to decrease when the temperature reaches 40°C.

3.2.5.2 The pH

High medium acidity can induce slow or blocked fermentations, but high pH can also cause more severe damages to fermentation. Another undesirable consequence of a low pH that can be noted is the formation of volatile acidity by the yeast.

3.2.5.3 The Cell Concentration

The concentration of yeast cells and their aggregation could have an inhibitory effect on the metabolism of yeast. For example, a high concentration leads to competition over the available nutrients (Ciftci et al. 1983).

3.2.5.4 The Carbon Dioxide

Carbon dioxide is a major product of fermentation. Each mole of ethanol produced is coupled to one mole of CO_2. This CO_2 is at the same time an activator and a yeast inhibitor. Under normal conditions, CO_2 is released and does not inhibit fermentation. However, an increase in this gas will be inhibitory, which is common during pressure fermentations. A slight internal pressure in the tank is enough to slow down the fermentation.

3.2.5.5 The Ethanol

Ethanol produced during fermentation is toxic to yeasts and may inhibit their function (Carlsen 1991). Besides, ethanol could denature certain enzymes (e.g. hexokinases, alcohol dehydrogenase, etc.) or disrupt their active sites. There is a link between the inhibitory effect of ethanol and structural changes in the plasma membrane. This could lead to an increase in ion permeability with the destabilization of the balance of transmembrane charges. Thus, the ethanol produced by fermentation slows the assimilation of nitrogen (Ribereau-Gayon et al. 2006).

3.3 Metabolic Approaches

If Pirt provided quantitative and mathematical concepts of cultivation techniques focused on the extracellular factors (e.g. nutrient availability, pH, temperature, etc.) (Pirt 1975), Monod highlighted the importance of the cells' response to the environment by regulating their physiological activities (Monod 1949).

The composition of sugars in the medium and the availability of oxygen represent the two main environmental conditions that have a significant impact on the metabolic physiology of yeasts.

Three commonly observed effects are associated with the types of energy generation processes involved in sugar metabolism and/or oxygen availability: Pasteur, Crabtree, and Custer effects (Deken 1966; Pronk et al. 1996).

3.3.1 Pasteur Effect

Pasteur was the first to compare yeast growth under aerobic and anaerobic conditions. Under low glucose concentration conditions, yeasts use sugars either by respiration or by fermentation. The aeration induces an increase in the formed biomass (total and per unit of degraded sugar) and, in parallel, a decrease in the production of alcohol and the consumption of sugar. Pasteur then holds that respiration inhibits fermentation. The fact that two enzymes compete to catalyze the respiration or fermentation of pyruvate may explain this inhibition. Pyruvate decarboxylase is involved in the fermentation pathway. It is characterized by a low affinity for pyruvate, unlike the enzyme pyruvate dehydrogenase.

Besides, oxidative phosphorylation consumes a lot of adenosine diphosphate (ADP) and inorganic phosphate, which migrate to the mitochondria. This results in a deficiency of ADP and inorganic phosphate in the cytoplasm. This deficit can limit phosphorylation and thus slow down the glycolytic flux. The inhibition of glycolysis enzymes by adenosine triphosphate (ATP) largely explains the Pasteur effect.

3.3.2 Crabtree Effect

In media containing high concentrations of glucose, for example, in grape must, *Saccharomyces cerevisiae* metabolizes sugars only by fermentation. Even in the presence of oxygen, breathing is impossible. This effect was discovered by Crabtree on tumor cells and is known by several names: catabolic repression by glucose, contrary effect Pasteur, and Crabtree effect (Crabtree 1929). The yeasts show the following signs during this effect: degeneration of mitochondria, decrease of the proportion of cellular sterols and fatty acids, and repression of the synthesis of mitochondrial enzymes of the Krebs cycle and constituents of the respiratory chain.

The Crabtree effect has been divided into short- and long-term effects, mainly based on continuous culture studies where, under steady-state conditions, the growth rate can be manipulated experimentally (Petrik et al. 1983). The short-term effect is defined as the ability to trigger alcoholic fermentation in case of sudden glucose excess, while the long-term effect is characterized by respiratory fermentative metabolism observed specifically with discontinuous cultures.

3.3.3 Custer Effect

The Custer effect is the inhibition of alcoholic fermentation in the absence of oxygen. This effect characterizes yeasts belonging to the genera *Brettanomyces* and *Dekkera* that ferment glucose to ethanol and acetic acid under aerobic conditions (van Dijken et al. 1986; Wijsman et al. 1984). However, when switching to anaerobic conditions, fermentation is strongly inhibited. This phenotype can be fully saved by reintroducing oxygen into the culture medium or by the addition of H^+ acceptors such as acetoin (reduced to 2,3-butanediol) or other aliphatic carbonyl compounds (Scheffers 1966; Wijsman et al. 1984). Reducing stress can cause this effect. Yeasts exhibiting this effect seem to be in some way incapable of establishing the redox balance by the production of glycerol or other strongly reduced compounds.

Cell metabolism is made from several parallel and branched reactions catalyzed by different enzymes. Enzyme production represents one of the important metabolism aspects. Furthermore, it is important to know that microorganism never synthesizes a single product. As a result, slow kinetics, modest yields, and complex purification operations may occur. By highlighting this, we can see that biological reaction must be controlled under specific conditions to fulfill its purpose. The biological reaction can be modeled then as presented in Figure 3.2.

This means that, when occurring, a bioreaction consumes substrates, to produce products. These will be new microorganisms, referring fermentation theory models and/or other smaller macromolecules or regular molecules. The regulation of the bioreaction

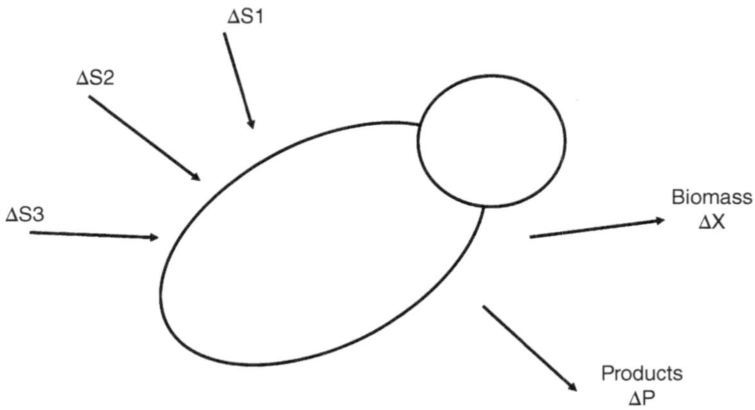

Figure 3.2 Schematic representation of bioreaction. *Source:* Salameh et al. (2013).

could be understood, independently of the physical parameters under the scope of the metabolism.

A simple equation linking the enzymatic reaction rate to the rate-limiting substrate concentration is described by the Michaelis–Menten equation given below (Eq. (3.5)):

$$v = \frac{K.ES}{K_M + S} \tag{3.5}$$

where v = the reaction velocity; K = the rate constant; E = the total amount of enzyme; K_M = the Michaelis–Menten (or saturation) constant; and S = the substrate concentration.

The Michaelis–Menten equation is based on several principals as follows:

- The substrate concentration is constant during the entire reaction where the measured rate is still equal to the initial rate and where the substrate concentration is by far greater than the enzyme concentration;
- The product conversion back into the substrate is negligible since no product accumulation is observed during the initial rate measurement; and
- The substrate fixation on the specific site of the enzyme is faster in comparison with the dissociation of the complex enzyme-product.

Besides, Monod proposed a relationship (Eq. (3.6)) between the specific growth rate and the limiting substrate concentration:

$$\mu = \frac{\mu_{max} S}{(K_s + S)} \tag{3.6}$$

where μ_{max} is the maximum specific growth rate (h^{-1}); S is the concentration of the limiting substrate concentration, and K_S is the saturation constant.

The specific growth rate represents an important parameter influencing the physiological state and the cellular composition. As discussed earlier, a great dependency is observed between the growth rate and the substrate concentration. However, a relation also exists between the growth rate and carbon sources. A link between the growth rate and the cellular composition, the gene expression, regulators concentration, and the metabolites are discussed.

3.3.4 Oxygen Requirements

Oxygen, a non-added component to the medium, represents an important element in many processes. Furthermore, oxygen availability may control the growth rate and metabolite production. The medium design may affect the oxygen availability in several ways, as follows:

1) *Fast metabolism:* A medium rich in fast-metabolized sugars will lead to high oxygen demand. So, reducing the initial concentration of the substrates and adding additional quantities during the fermentation process could overcome an oxygen limitation due to fast metabolism. Another way to overcome this is by changing the medium composition and adding higher carbohydrates (e.g. lactose, starch, etc.) and proteins that cannot be metabolized fast and do not demand a large specific oxygen uptake rate (Johnson 1946).
2) *Rheology:* Many components present in the medium can influence the culture viscosity and consequently affect its behavior with respect to aeration and agitation. Polymers such as starch and polysaccharides may influence the rheological aspect of the fermentation broth (Tuffile and Pinho 1970).
3) *Antifoams:* The usage of some antifoams may reduce the oxygen transfer rate. Foaming can lead to cell removal from the medium. It can induce cell autolysis and release of proteins, which increases foam formation.

3.3.5 Oxygen Function in Fermentation

Oxygen acts as a final electron acceptor in the redox reactions. Also, it can interfere like an inducer or repressor of some enzyme reactions or, in some cases, as a substrate. If oxygen is considered as a substrate during the respiration and stoichiometry is respected, 192 g of oxygen is necessary for the complete oxidation of 180 g of glucose. Oxygen is not soluble in water in high quantities, so it is impossible to provide the culture all the oxygen needed from the beginning of the fermentation process. Therefore, cultures are supplied with oxygen during the fermentation process through aeration or broth agitation to respond to the oxygen demand of the microorganism. Many researchers have tried to establish an equation based on the stoichiometry of oxygen conversion and conversion of carbon and nitrogen sources of carbon into biomass to determine the oxygen demand of fermentation (Nielsen and Villadsen 2013). Based on these equations, we can understand that the microbial demand for oxygen will depend on the carbon source in the medium. For a reduced source of carbon, higher oxygen demand is observed. Nevertheless, we can conclude from the equations that high carbon content of hydrocarbon substrates is equivalent to high yield factors (grams of biomass per gram of substrate consumed). The usage of such substrates will be based on the balance between the benefit of high biomass yield and the disadvantage of high oxygen need and heat generation. Some equations consider only biomass production (Darlington 1964) for the estimation of oxygen demands, while others include product formation (Cooney 1979). However, the provision of oxygen for fermentation cannot be simply based on an overall estimation of demand. Cell metabolism is influenced by the concentration of dissolved oxygen in the medium. The effect of dissolved oxygen concentration on the specific oxygen uptake rate (Q_{O2}, millimoles of oxygen consumed per gram dry weight of cells per hour) has been demonstrated to be of the Michaelis–Menten type (Figure 3.3).

The specific oxygen uptake rate increases proportionally with the dissolved oxygen concentration until attaining the critical dissolved oxygen concentration. Above this value, no increase in the oxygen uptake rate is observed. Thus, preserving the dissolved oxygen concentration greater than the critical level assures a maximum biomass production. A disturbance in the cells' metabolism is observed when the dissolved oxygen concentration decreases below the critical level. In parallel, an oxygen concentration greater by far than the critical level may stimulate only the product formation with no significant influence on the biomass production. Knowing that, it is the fermentation technologist's function to control the oxygen concentration in a way to produce more biomass or to starve the microorganism to produce more products.

The effects of the dissolved oxygen concentration on fermentation are well demonstrated by amino acid biosynthesis by *Brevibacterium flavum*. The critical dissolved oxygen concentration for *B. flavum* is 0.01 mg l^{-1}. The production of members of the glutamate (e.g. glutamine, proline, and arginine) and aspartate (e.g. lysine, threonine, and isoleucine) families of amino acids was affected detrimentally by levels of oxygen satisfaction below 1.0, whereas optimum production of phenylalanine, valine, and leucine occurred at oxygen

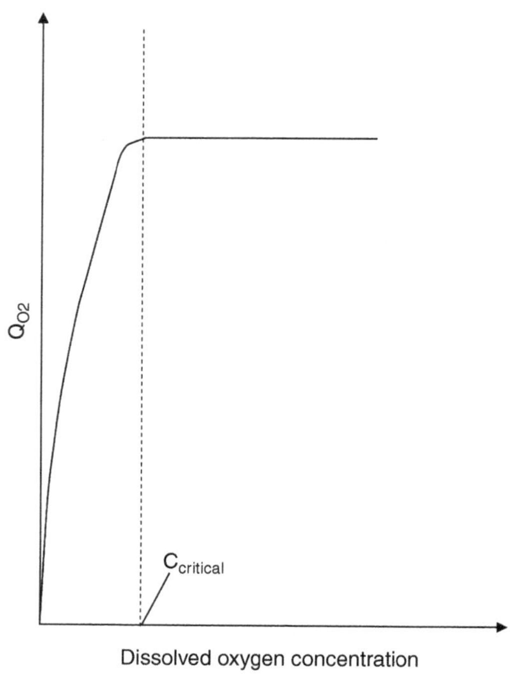

Figure 3.3 Dissolved oxygen concentration increases in a solution during a period of aeration. *Source:* Salameh et al. (2013).

satisfaction levels of 0.55, 0.60, and 0.85, respectively. The glutamate and aspartate families are produced from tricarboxylic acid (TCA) cycle intermediates. Phenylalanine, valine, and leucine are produced from glycolysis intermediates. The excess of oxygen should increase the liberation of TCA cycle intermediates, and oxygen limitation should reduce the glucose oxidation via the TCA cycle, thereby more intermediates should be available for the biosynthesis of phenylalanine, valine, and leucine. The metabolic disruption results in higher production of pyruvate derived amino acids.

Oxygen can be provided to microbial cultures depending on the culture scale. Usually, oxygen is supplied in the form of air, a cheap source of gas. In a laboratory scale, cultures are aerated by shake flask in a controlled environmental chamber. Microbial pilot- and industrial-scale fermentations are performed in stirred, aerated vessels. Some fermenters are designed in a way that suitable oxygen transfer is provided without agitation.

Oxygen transfer from air to the cell during fermentation was described by several steps as follows (Bartholomew et al. 1950):

1) Oxygen is diffused from the bulk gas phase inside the air bubble to the gas–liquid interface;
2) Oxygen is transferred through the gas–liquid interface;
3) Dissolved oxygen is diffused through a relatively still "boundary layer" close to the air bubble to the bulk fermentation medium;
4) Dissolved oxygen is transferred from the external edge of the stagnant boundary layer, through the fermentation medium to the suspended organism – be it single bacterial or animal cells, aggregates of cells, dispersed mycelium, mycelial aggregates, or pellets;
5) The dissolved oxygen is diffused through another relatively stagnant boundary layer to the cell or mycelium surface, or through aggregates or pellets to the cell or mycelium surface;
6) The dissolved oxygen is accepted by the fermentation organism across its cell membrane; and
7) The dissolved oxygen is transferred within the cell to the enzyme site.

The rate of oxygen transfer from the air bubble to the liquid phase may be described by Eq. (3.7):

$$\frac{\mathrm{d}C_L}{\mathrm{d}t} = K_{La}\left(C^* - C_L\right) \tag{3.7}$$

where C_L is the concentration of dissolved oxygen in the fermentation broth (mmol l^{-1}), t is time (h), dC_L/dt is the change in oxygen concentration over a time, K_L represents the summation of the oxygen transfer resistances from gas to liquid, and the difference between the saturated dissolved oxygen concentration and the actual concentration in the fermentation broth (C^*-C_L) is the "driving force" across the resistances.

Measuring both K_L and "*a*," the gas–liquid interface area, in fermentation is extremely difficult. So, the two terms are normally combined in the term K_{La}, the volumetric mass-transfer coefficient. It is used to determine the aeration capacity of a fermenter: the larger the K_{La}, the greater the aeration capacity of the system. It depends on the fermenter design and operating conditions. Furthermore, it will be influenced by the aeration rate, agitation rate, and impeller design. They may reduce the resistances to transfer and affect "*a*" by changing the number, size, and residence time of air bubbles. It is appropriate to use K_{La} as an index of fermenter performance. In fact, the oxygen transfer rate does not depend on the dissolved oxygen concentration, but the fact that the oxygen transfer rate is a key criterion in fermentation cannot be denied. The dissolved oxygen concentration mirrors the balance between the supply of dissolved oxygen by the fermenter and the oxygen demand of the organism. The K_{La} of the fermenter should be the optimum oxygen concentration for the product production preserved in solution during the fermentation process.

3.4 Conclusions

The changes in culture condition can enhance the fermentation process. Detailed investigations about the microorganism used and the purpose of the process are crucial to designing an appropriate medium that meets the basic requirements. Several medium types are commonly used, but the medium selection should be based on the fermenter design, the process conditions, and the scalability of the fermentation. A suitable medium may hold an excess of all nonlimiting nutrients and typical excess factors. The design of a medium should also take into consideration the several elements needed for an optimal culture. The composition of sugar and the availability of oxygen are among the major factors to be considered since they have a significant impact on yeast metabolism. Overall, the key factors to consider when designing the medium are its effects on the strain selection, the fermentation scale-up, and the target behind the considered fermentation.

References

Aiba, S., Humphrey, A.E., and Millis, N. (1973). *Biochemical Engineering*. New York: Press, Academic.

Atkinson, B. and Mavituna, F. (1991). Process biotechnology. In: *Biochemical Engineering and Biotechnology Handbook*, 43–81. New York: Stockton Press.

Bartholomew, W.H., Karow, E.O., Sfat, M.R., and Wilhelm, R.H. (1950). Oxygen transfer and agitation in submerged fermentations. Mass transfer of oxygen in submerged fermentation of Streptomyces griseus. *Ind. Eng. Chem.* 42: 1801–1809.

Bauchop, T. and Elsden, S.R. (1960). The growth of micro-organisms in relation to their energy supply. *Microbiology* 23: 457–469.

Bawa, N., Bear, D., Hill, G. et al. (2010). Fermentation of glucose and starch particles using an inexpensive medium. *J. Chem. Technol. Biotechnol.* 85: 441–446.

Carlsen, N. (1991). Effects of alcohols on the respiration and fermentation of aerated suspensions of baker's yeast. *J. Gen. Microbiol.* 137: 2879–2883.

Ciftci, T., Constantinides, A., and Wang, S.S. (1983). Optimization of conditions and cell feeding procedures for alcohol fermentation. *Biotechnol. Bioeng.* 25: 2007–2023.

Cooney, C. (1979). Conversion yields in penicillin production: theory vs. practice. *Process Biochem.* 14: 31–33.

Crabtree, H.G. (1929). Observations on the carbohydrate metabolism of tumours. *Biochem. J.* 23: 536–545.

Darlington, W.A. (1964). Aerobic hydrocarbon fermentation – a practical evaluation. *Biotechnol. Bioeng.* 6: 241–242.

Deken, R.H.D. (1966). The Crabtree effect: a regulatory system in yeast. *J. Gen. Microbiol.* 44: 149–156.

van Dijken, J.P., van den Bosch, E., Hermans, J.J. et al. (1986). Alcoholic fermentation by "non-fermentative" yeasts. *Yeast* 2: 123–127.

Drozd, J.W. (1987). Hydrocarbons as feedstocks for biotechnology. In: *Carbon Substrates in Biotechnology* (eds. J.D. Stowell, A.J. Beardsmore, C.W. Keevil and J.R. Woodward), 119–138. Oxford: IRL Press.

Egli, T. (2009). Microbial nutrition. In: *Desk Encyclopedia of Microbiology* (ed. M. Schaechter), 308–324. Oxford, UK; San Diego, CA: Academic Press/Elsevier.

Fleet, G. (1992). Spoilage yeasts. *Crit. Rev. Biotechnol* 12: 1–44.

Fukui, S. and Tanaka, A. (1980). *Production of Useful Compounds from Alkane Media in Japan, in: Advances in Biochemical Engineering*, vol. 17, 1–35. Berlin, Heidelberg: Springer Berlin Heidelberg.

Hamer, G. (1979). Biomass from natural gas. In: *Economic Microbiology, Microbial Biomass*, vol. 4 (ed. A.H. Rose), 315–360. London: Academic Press.

Herbert, D. (1976). Stoichiometric aspects of microbial growth. In: *Continuous culture 6, applications and new fields* (eds. A.C.R. Dean, D.C. Ellwood, C.G.T. Evans and J. Melling), 1–30. London: Society of Chemical Industry.

Hughes, M.N. and Poole, R.K. (1991). Metal speciation and microbial growth--the hard (and soft) facts. *J. Gen. Microbiol.* 137: 725–734.

Hutner, S.H. (1972). Inorganic nutrition. *Annu. Rev. Microbiol.* 26: 313–346.

Jobling, A. (1990). Bioprotein manufacture: a critical assessment. *Br. Polym. J.* 23: 362.

Johnson, M.J. (1946). Metabolism of penicillin – producing molds. *Ann. N. Y. Acad. Sci.* 48: 57–66.

Junker, B., Mann, Z., Gailliot, P. et al. (1998). Use of soybean oil and ammonium sulfate additions to optimize secondary metabolite production. *Biotechnol. Bioeng.* 60: 580–588.

Levi, J.D., Shennan, L., and Ebbon, G.P. (1979). Biomass from liquid n-alkanes. In: *Economic Microbiology', Microbial Biomass*, vol. 4 (ed. A.H. Rose), 361–419. London: Academic Press.

Lilly, V.G. (1965). The chemical environment for growth. 1. Media, macro- and micro-nutrients. In: *The fungi—An advance treatise 1. The fungal cell* (eds. G.C. Ainsworth and A.S. Sussman), 465–478. New York: Academic Press.

Luria, S.E. (1960). The bacterial protoplasm: composition and organisation. In: *The bacteria: a treatise on structure and function*, vol. 1 (eds. I.C. Gunsalus and R.Y. Stanier), 1–34. New York: Academic Press, Inc.

Minoda, Y. (1986). Raw materials for amino acid fermentation. *Prog. Ind. Microbiol.* 24: 51–66.

Monod, J. (1949). The growth of bacterial cultures. *Annu. Rev. Microbiol.* 3: 371–394.

Nielsen, J. and Villadsen, J. (2013). *Bioreaction Engineering Principles*. Bioreact. Eng. Princ, Springer International Publishing.

Papapanagiotou, P.A., Quinn, H., Molitor, J.-P. et al. (2005). The use of phase inversion temperature (PIT) microemulsion technology to enhance oil utilisation during *Streptomyces rimosus* Fed-batch fermentations to produce oxytetracycline. *Biotechnol. Lett.* 27: 1579–1585.

Petrik, M., Kappeli, O., and Fiechter, A. (1983). An expanded concept for the glucose effect in the yeast Saccharomyces uvwum: Involvement of short- and long-term regulation. *J. Gen. Microbiol.* 129: 43–49.

Pirt, S.J. (1975). *Principles of Microbe and Cell Cultivation*. New York: Wiley.

Piskur, J., Rozpedowska, E., Polakova, S. et al. (2006). How did *Saccharomyces* evolve to become a good brewer? *Trends Genet.* 22: 183–186.

Pronk, J.T., Steensma, H.Y., and Van Dijken, J.P. (1996). Pyruvate metabolism in *Saccharomyces cerevisiae*. *Yeast* 12: 1607–1633.

Rajagopalan, G. and Krishnan, C. (2009). Optimization of agro-residual medium for α-amylase production from a hyper-producing *Bacillus subtilis* KCC103 in submerged fermentation. *J. Chem. Technol. Biotechnol.* 84: 618–625.

Ribereau-Gayon, P., Lafon-Lafourcade, S., and Bertrand, A. (1975). La débourbage des moûts de vendage blanche. *Conn. Vigne Vin.* 9: 117–139.

Ribereau-Gayon, P., Glories, Y., Maujeam, A., and Dubourdieu, D. (2006). *Handbook of Enology Volume 2 the Chemistry of Wine Stabilization and Treatments*, 2e. Wiley.

Salameh, D., Strehaiano, P., and Maalouf, M. (2013). *Manuel de microbiologie industrielle*. Saint-Joseph University.

Schaechter, M.O., Maale, O., and Kjeldgaard, N.O. (1959). Dependency of medium and temperature on cell size and chemical composition during balanced growth of *Salmonella Typhimurium*. *J. Gen. Microbiol.* 19: 592–606.

Scheffers, W.A. (1966). Stimulation of fermentation in yeasts by acetoin and oxygen. *Nature* 210: 533–534.

da Silva, G.P., Mack, M., and Contiero, J. (2009). Glycerol: a promising and abundant carbon source for industrial microbiology. *Biotechnol. Adv.* 27: 30–39.

Somrutai, W., Takagi, M., and Yoshida, T. (1996). Acetone-butanol fermentation by *Clostridium aurantibutyricum* ATCC 17777 from a model medium for palm oil mill effluent. *J. Ferment. Bioeng.* 81: 543–547.

Teusink, B., Diderich, J.A., Westerhoff, H.V. et al. (1998). Intracellular glucose concentration in derepressed yeast cells consuming glucose is high enough to reduce the glucose transport rate by 50%. *J. Bacteriol.* 180: 556–562.

Tuffile, C.M. and Pinho, F. (1970). Determination of oxygen-transfer coefficients in viscous streptomycete fermentations. *Biotechnol. Bioeng.* 12: 849–871.

Wijsman, M.R., van Dijken, J.P., van Kleeff, B.H., and Scheffers, W.A. (1984). Inhibition of fermentation and growth in batch cultures of the yeast *Brettanomyces intermedius* upon a shift from aerobic to anaerobic conditions (Custers effect). *Antonie Van Leeuwenhoek* 50: 183–192.

4

Emerging Technologies and Their Mechanism of Action on Fermentation

Krystian Marszałek[1,2], Łukasz Woźniak[3], Artur Wiktor[4], Justyna Szczepańska[3], Sylwia Skąpska[3], Dorota Witrowa-Rajchert[4], Jorge A. Saraiva[5], Jose M. Lorenzo[6], and Francisco J. Barba[7]

[1] *Department of Fruit and Vegetable Product Technology, Prof. Wacław Dąbrowski Institute of Agricultural and Food Biotechnology, Warsaw, Poland*
[2] *Department of Food Technology and Human Nutrition, Institute of Food Technology and Nutrition, College of Natural Science, University of Rzeszow, Rzeszow, Poland*
[3] *Department of Fruit and Vegetable Product Technology, Institute of Agricultural and Food Biotechnology, Warsaw, Poland*
[4] *Faculty of Food Sciences, Department of Food Engineering and Process Management, Warsaw University of Life Sciences (WULS-SGGW), Warsaw, Poland*
[5] *QOPNA, Chemistry Department, University of Aveiro, Aveiro, Portugal and LAQV-REQUIMTE, Department of Chemistry, University of Aveiro, Aveiro, Portugal*
[6] *Centro Tecnológico de la Carne, ParqueTecnológico de Galicia, Ourense, Spain*
[7] *Faculty of Pharmacy, Preventive Medicine and Public Health, Food Science, Toxicology and Forensic Medicine Department, Universitat de València, València, Spain*

4.1 Introduction

Nowadays, consumer requirements for foods are constantly changing, with growing preference for fresh, natural, or minimally processed foods. Therefore, food processing should be designed to maintain or even enhance the nutritional and sensory quality of raw materials. Innovative approaches such as high hydrostatic pressure (HHP), ultrasound (US), pulsed electric fields (PEF), and microwave radiation (MW) have been shown as attractive technologies for food technologists and engineers not only due to their food preservation capabilities but also because of their potential applications for achieving new functional effects.

4.2 HHP Processing

HHP is an innovative mild-thermal technology commonly used for the inactivation of pathogenic and spoilage microorganisms as well as tissue enzymes while retaining the nutritional value and freshness of food (Barba et al. 2015a; Kultur et al. 2018). Recently, HHP has been used in a wide range of food processes and mechanisms, including pasteurization, freezing, enzyme inactivation, protein unfolding, shucking of shellfish, cell wall

Fermentation Processes: Emerging and Conventional Technologies, First Edition.
Edited by Mohamed Koubaa, Francisco J. Barba, and Shahin Roohinejad.

opening, cold gelatinization of starch, or even for enhancing microbial growth at mild process conditions (e.g. enhancing fermentation processes) (Barba et al. 2012, 2015a). All of these applications require specific process parameters (Barba et al. 2018; Mota et al. 2018).

Nowadays, HHP is a worldwide known technique used in the food industry for preserving and extending the shelf life of fruit and vegetables (F&V), meat, seafood, and ready-to-eat products. F&V products enclose over 40% of the HPP food market, especially purees, juices, and beverages (Barba et al. 2015a; Huang et al. 2017; Misra et al. 2017). The biggest challenge in the use of this technique is extending the shelf life of food products by controlling native tissue enzymes, which generally exhibit much higher resistance to the combined action of pressure and time under pressure than microorganism (Barba et al. 2012, 2014, 2015a; Marszałek et al. 2016a, 2017a). A newest trend is also the application of HHP to carry out probiotic bacteria fermentation to produce fermented foods (Mota et al. 2018).

More than 300 HHP equipment (Figure 4.1) are available worldwide in the food industry, mainly in North America and Mexico (67%), but also in Europe (18%), Asia (8%), Latin America (3%), Oceania (3%), and Africa (1%) (Huang et al. 2017). Avure Technologies (Västerås, Sweden), Hiperbaric (Burgos, Spain), and UHDE High Pressure Technologies (Hagen, Germany) are the main companies for industrial-scale equipment production, with horizontal and vertical vessel systems available for batch HHP equipment. The new trend is the production of continuous or semi-continuous HHP equipment. Avure Technologies offers semi-continuous systems for the treatment of liquid beverages such as juices. A newest solution offered by Hiperbaric allows beverages to be processed in bulk, with the largest available processing capacity to date, allowing the processing of 10 000 $L.h^{-1}$ (Barba et al. 2016).

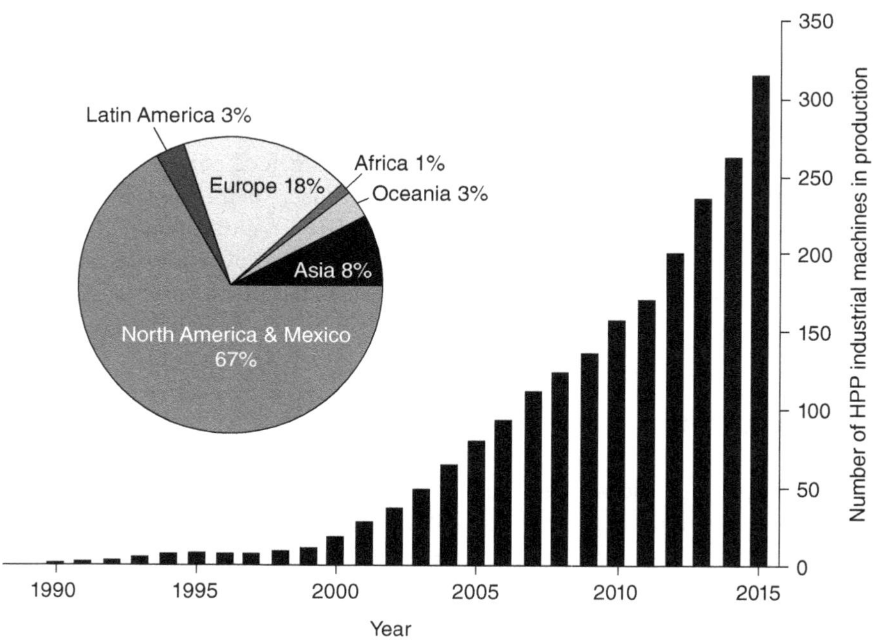

Figure 4.1 Worldwide trend of HHP food production. *Source:* Huang et al. (2017). © 2017, Elsevier.

In batch systems, products have to be prepackaged and HPP-treated for a predefined time, temperature, and pressure to obtain the desired results. Batch process can be used for all kinds of food: liquids, semi-solid, and solid. F&V juices, drinks, smoothies, or purees are generally prepacked in plastic bottles of different shapes, designed to fill the pressure chamber with the maximum number of products. Meat and ready-to-eat products are prepackaged at vacuum conditions for elimination of air bubbles.

Commercial applications of HHP in food industry use pressures in the range from 100 MPa up to 600 MPa at room temperature, whereas laboratory equipment can achieve up to 1400 MPa, connected with cooling or heating system. These pressures are extremely high in relation to other pressures observed in nature but not the highest existing in the world.

Industrial applications with the highest volume of 687 l and maximum pressure up to 310 MPa or 525 l with maximum pressure up to 600 MPa generally work at ambient temperature or with mild heating up to 50°C. Semi-industrial applications with volumes below 150 l reach temperatures of up to 95°C and pressures of up to 700 MPa (Barba et al. 2016, 2018; Horita et al. 2018). Some laboratory-scale equipment can reach pressures and temperatures up to 900 MPa and 110°C, respectively, but the volume of the HPP chamber does not exceed 5 l. The highest pressure applied for food was 1400 MPa at temperature up to 110°C in a chamber with a volume of 35 ml (Reineke et al. 2008).

High pressure can be generated by different ways: an external pump, a moving piston, or by heating the medium in a closed chamber. There are three principles that underline the application and effects of HHP: Pascal's law, adiabatic heat of compression, and Le Chatelier–Braun principle (Barba et al. 2018).

The first is the Pascal's law, also well known as the isostatic rule, which states that the transmittance of pressure is uniform and instantaneous in the whole system under pressure (Figure 4.2). The HPP equipment immediately generates the stable pressure in the whole product volume regardless of its size, shape, and consistency (Barba et al. 2018). Increase in the pressure occurs at a much faster rate than that in the case of heat conductive processes.

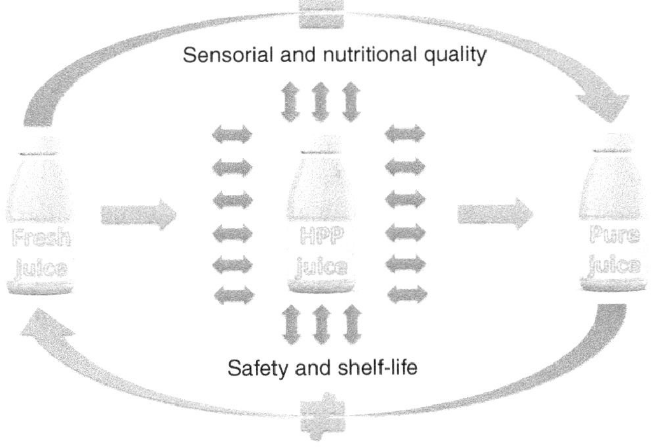

Figure 4.2 Application of high-pressure processing.

Batch processes consist of three stages: increase of the pressure, pressurization using a low compressibility medium (e.g. water), and depressurization. During the first stage, the liquid medium is pumped into the pressure chamber using a high-pressure pump or pumps. The second stage is associated with the second rule according to which temperature is increasing as a consequence of the adiabatic heat of compression. This heating is specific to each food constituent and kind of transmitting fluids. For example, the temperature of water increases in a range from 3 to 5°C per each 100 MPa, whereas temperature of silicone oil can increase over 20°C per 100 MPa. Oils and fats in food have the largest heat of compression values ranging from 3 to 9°C per each 100 MPa due to a different molecular structure of lipids compared with water. This heat of compression should be taken into account, together with the initial temperature and target pressure, in designing food processing. The heat transfer among pressure-transmitting fluid, through the product, and to the high-pressure chamber strongly influences the effect of microbial and enzyme inactivation as well as final product quality (Barba et al. 2018).

Most foods contain a considerable amount of water, thus fairly resembling the pressure-transmitting water and minimizing the adiabatic heating thermal effect. Consequently, fatty foods may undergo a higher temperature rise in comparison to F&V products (Barba et al. 2016, 2018). The initial temperature of the food product, the kind of pressure-transmitting fluid, and temperature as well as the pressure vessel initial temperature must be designed and equilibrated before food processing in order to avoid high uncontrolled temperature fluctuations. The adiabatic heating is completely reversible after pressure release, but the temperature does not return to the initial value, due to the heat transfer to the high-pressure chamber and temperature equilibration during holding time (particularly for longer pressure holding times). Although the temperature increase is relatively small, it can significantly affect microorganisms, especially at the mild temperatures used in the experiments. After holding time, the depressurization is applied; the pressure falls down immediately, and the temperature decreases according to the adiabatic heating rules (generally below initial temperature, particularly for longer pressure holding times). The level of pressure, treatment time, and depressurization time depend on the equipment capabilities, specific food types, and expected final results. Once the desired pressure is achieved, no additional energy is required, and pressure is maintained during holding time. The temperature of the treatment chamber should be controlled externally or internally to achieve temperature control (Barba et al. 2018).

The third rule is the Le Chatelier–Braun principle that concerns the influence of high pressure on biomolecules. This principle states that under equilibrium conditions, any chemical reaction, phase transformation, molecular transformation, etc. is affected by a decrease in volume due to pressure increase. Generally, high pressure can influence large molecules (polymers), including proteins, enzymes, and polysaccharides, due to changes in noncovalent bonds, such as hydrogen, ionic, and hydrophobic, which are more sensitive to pressure changes than covalent bonds. Pressure values up to 1200 MPa have no effect upon covalent bonds but can lead to permanent changes in hydrogen and ionic bonds as well as in hydrophobic interactions responsible for the secondary and tertiary structure of proteins (Balny et al. 1992).

Smaller molecules such as amino acids, vitamins, pigments, or flavor substances are generally not sensitive to pressure (Barba et al. 2015a; Butz et al. 2003; Cao et al. 2011).

Vitamins A, C, B_1, B_2, and E in F&V do not undergo any significant changes as a result of HHP but can change due to mild processing temperature and during storage time (Sancho et al. 1999). HHP coupled with process temperature over 20°C can have a negative effect on the content of anthocyanin pigments and vitamin C, whereas the process run at temperature below 20°C has no or only slight impact on their level (Corrales et al. 2008; Garcia-Palazon et al. 2004; Marszałek et al. 2015a, 2017b). The color of fruit products treated by HHP can be even more intensive than before pressurization and thus more attractive. This phenomenon is generally explained by the extraction of anthocyanin pigments and flavonols from the tissues to the intercellular juice (Otero and Préstamo 2009; Stinco et al. 2019).

The principles presented above justify why low-molecular-weight compounds, responsible for nutritional and sensory characteristic, are not affected by pressure whereas higher molecules are pressure sensitive (Barba et al. 2012, 2015a). Chemical or physical reactions resulting in lower volume of the reaction products are accelerated by HHP, whereas reactions causing an increase in the total volume of the products are retarded (Barba et al. 2018, 2016).

The products treated by pressures of 10–1000 MPa suffer changes in their volume that reach from 4 to 15%, depending on the pressure applied, as well as the structure and composition of food (Barba et al. 2016, 2018). It is worthy to note that these changes are not permanent because the impact of pressure on product's structure is generally reversible. In addition, if the food structure does not contain air (that is highly compressible), the isostatic action of pressure from all directions with the same intensity generally does not induce shearing forces that could result in tissue damage (Dervisi et al. 2001).

The biggest advantage of HHP is the possibility of reducing tissue enzymes and the inactivation of microorganisms by damaging the structure of their cell proteins while maintaining the nutritional quality and fresh-like flavor of the products. However, in some cases, microbial growth and enzymatic activity can be enhanced by stress conditions under HHP, thus being a useful tool in fermentation processes (Barba et al. 2015a; Marszałek et al. 2017a, 2019; Mota et al. 2018).

Another shortcoming of this technique is that although most microorganisms, including spoilage and pathogenic ones, could be inactivated effectively at commercially available pressures, most of the plant tissue enzymes, especially from oxidoreductases group, remain active (Barba et al. 2014; Szczepańska et al. 2020). Considering that pressurization up to 600 MPa resulted in water compression even up to 15% of volume, the packaging material should be flexible in a way to both withstand this compression and retain its original geometry upon decompression. In addition, packaging material migration to the food products must be avoided. Plastic pouches, containers, semi-rigid trays with at least one flexible interface, and bottles are usually used for HHP processing, while metal cans and glassware are obviously not (Barba et al. 2018).

There are two major limitations to the further development of HHP processes. First is the level of pressure possible to achieve at industrial scale. The current research on the tissue enzyme inactivation shows that pressurization below 600 MPa is not enough for enzyme inactivation. Decimal reduction time (*D-value*), temperature and pressure required for 90% reduction of *D-value* (z_t*-value* and z_p*-value*), activation energy (*Ea*), and activation volume (*Va*), calculated for oxidoreductive enzymes, indicate that the inactivation of tissue enzymes is not comprehensive enough to ensure the reliability of HHP at available pressures as an alternative to thermal processing. Secondly, developing a continuous pressure food processor still remains an engineering challenge (Barba et al. 2016). Semi-continuous processing can be achieved using multiple sequential chambers connected in series. In this case, while some chambers are under pressurization, others are unloaded or loaded (Barba et al. 2016, 2018).

The current cost of using HHP treatment for food preservation depends on many factors, including equipment costs and its amortization, production capacity, and running costs (labor and energy) (Barba et al. 2016). The production rate (process flow: kg or l of product per year) is determined by the cycle time and the volumetric packaging efficiencies in the pressure chamber. The better the packing density, the lower the cost of production per kg or l. The cost of preserving foods using HHP is calculated around 0.3 $€.kg^{-1}$ and is much higher in relation to conventional pasteurization (0.05 $€.kg^{-1}$), but often significantly lower in comparison to well-known traditional methods such as drying (5 $€.kg^{-1}$), smoking (2 $€.kg^{-1}$), or brining (1 $€.kg^{-1}$) (Heinz and Buckow 2010). Nowadays, many consumers accept higher costs of HHP products due to their better quality.

4.3 Ultrasound

The term *ultrasound* describes the acoustic waves of frequencies exceeding the hearing range of human ear. The highest frequency that can be heard by people varies by age and sex, although the borderline value is established at 20 kHz (Figure 4.3). Due to their features, the US found numerous applications in medicine, chemistry, and industry; in many cases, they are used as routine operations rather than an innovation or a novel alternative (Awad et al. 2012; Roselló-Soto et al. 2015a; Zhu et al. 2017a, 2017b).

For instance, reflection of US is used for nondestructive testing of materials. The homogenous structure of dense materials is a good conductor of waves. Any disruption in the internal structure reflects part of the waves, and therefore being possible to find such changes by the implementation of the US. Figure 4.4 presents the principles of using wave reflection to test materials. The sound velocities in biological tissues are similar to those of water (i.e. 1500 $m.s^{-1}$). Surprisingly, air is a poor supporter of sound wave transmission due to the low concentration of molecules and high compressibility.

Cavitation is the formation of vapor cavities as a result of forces occurring in fluids. Cavitation occurs during the sonication or decrease of pressure below vapor pressure of the liquid. Two types of cavitation can be distinguished. In stable (non-inertial) cavitation, bubbles oscillate in

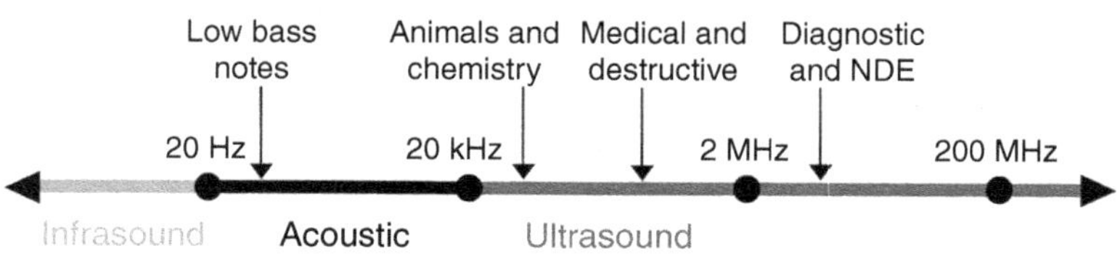

Figure 4.3 Frequency ranges of ultrasound and corresponding applications (Wikipedia. Wikimedia Foundation, Inc.).

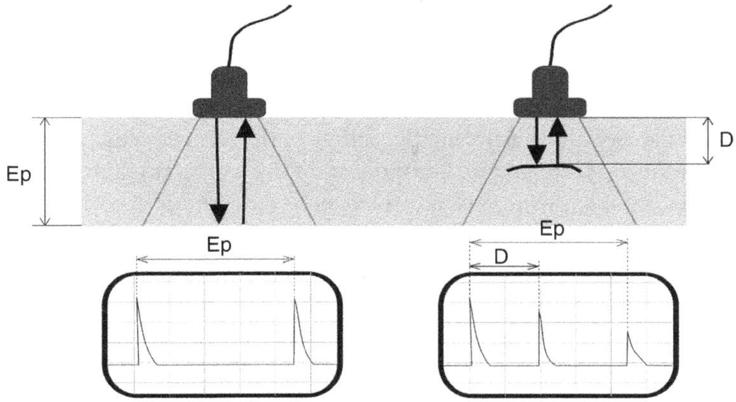

Figure 4.4 The utilization of ultrasound reflection in testing of materials. The presence of an additional indication as a result of presence of the defect in the internal structure (Wikipedia. Wikimedia Foundation, Inc.).

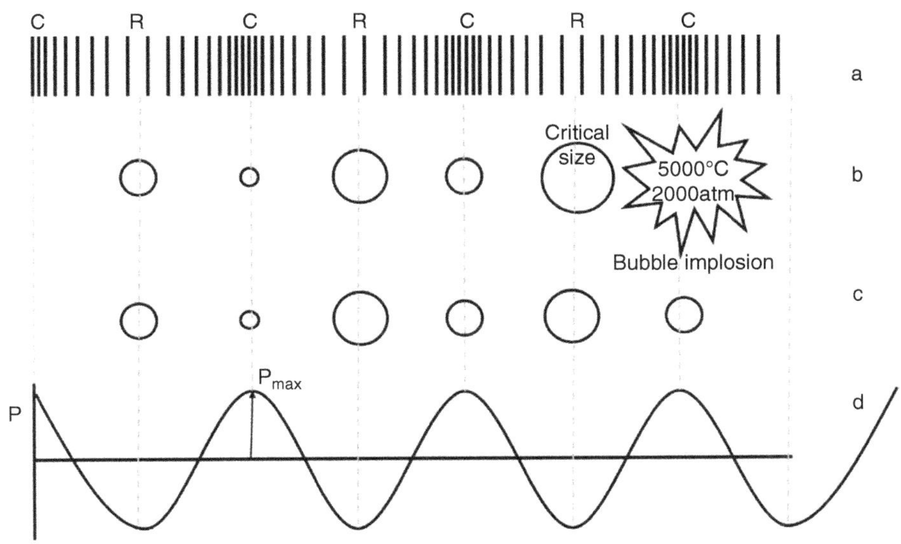

Figure 4.5 The propagation of ultrasonic wave (a) compression and refraction, (b) transient ultrasonic cavitation, (c) stable ultrasonic cavitation, and (d) variations of the pressure. *Source:* Vyas and Ting (2018). Licensed under CC-BY-4.0.

size, but no rupture occurs. This type is beneficial in processes where improvement of mass transfer is required, e.g. cleaning. The use of higher-pressure oscillations leads to transient (inertial) cavitation. The bubbles implode producing shockwaves and local release of energy along with high temperature (up to 5000°C) and pressure (up to 200 MPa). Both the propagation of an ultrasonic wave and cavitation are presented in Figure 4.5.

The possibility of occurrence of a phenomenon known as *sonofusion* (bubble fusion) has been proposed by some scientists, although it is seen as a matter of controversy. It claims that during the collapse of a cavitation bubble, the temperature is high enough to start a nuclear fusion on a microscale. The effect was observed during the cavitation of deuterated acetone; however, a reliability of the experiment was later undermined (Shapira and Saltmarsh 2002; Taleyarkhan et al. 2002).

The use of the US is associated with transferring the energy to the sample, which results in an increase of the temperature of the sample. The increase of the temperature is equal in the whole volume of the sample, which is advantageous concerning the practical implementations of US.

The practical implementation of ultrasonic technologies needs equipment capable of generating sound waves of selected frequency and power. The main constituents of most of the devices are an electrical generator, transducer, and emitter. The transducer is a central element in ultrasonic devices. It is responsible for converting the energy to a sound wave with the desired parameters. The most often used type of transducers is based on the piezoelectricity, an effect observed for the first time in 1880 by Jacques and Pierre Curie. The piezoelectricity occurs in certain solid materials, as a result of application of the mechanical stress to a crystal structure. Other types of transducers are magnetostrictive (using magnetic energy) and liquid-driven, which rely on pure mechanical energy. The emitters (known also as ultrasonic reactors or cells) have two main forms: baths (where sample is placed) and horns (which are placed inside the sample) (Koubaa et al. 2015; Roselló-Soto et al. 2015a, 2015b).

The wide range of parameters used in ultrasonic technologies led to a practical division of the processes into two groups, depending on frequency and power of the US used. The low-intensity US uses high frequencies (100 kHz – 1 MHz) and low power ($<1\,W{\cdot}cm^{-2}$); they are typically used for the nondestructible analysis of the internal structure of the samples and patients. The high-intensity US typically use lower frequencies (16–100 kHz) and much larger power ($10–1000\,W.cm^{-2}$) and may affect the structure or the properties of the samples.

US found numerous applications in the industry, medicine, navigation, and everyday life. Selected applications are presented in Figure 4.6. Due to volume restrictions, only selected uses of US related to the food industry will be discussed in this chapter.

Sonication is a general term describing the utilization of US as a separate industrial process or as a method of enhancement of other processes. For example, US was used for thawing of meat samples (Miles et al. 1999); the utilization of this technique minimized the problem of surface heating. Sonication also found its application in the enhancement of processes such as freezing and crystallization. The cavitation processes generate bubbles that act as nuclei for crystal growth initiation. This method has been tested for potato freezing (Sun and Li 2003); the authors have reported decreased processing times as well as better structure of the product after thawing.

The phenomena occurring during the action of high-intensity US are very efficient for cleaning purposes. US was found to be two times more efficient in biofilm removal than swabbing of food processing equipment (Oulahal-Lagsir et al. 2000). Both the increased convection and cavitation are also helpful during cleaning of membranes (Muthukumaran et al. 2007).

Ultrasound-assisted extraction (UAE) consists of an enhancement of the solvent extraction processes allowing an increase in the extraction rate. UAE have been applied for the

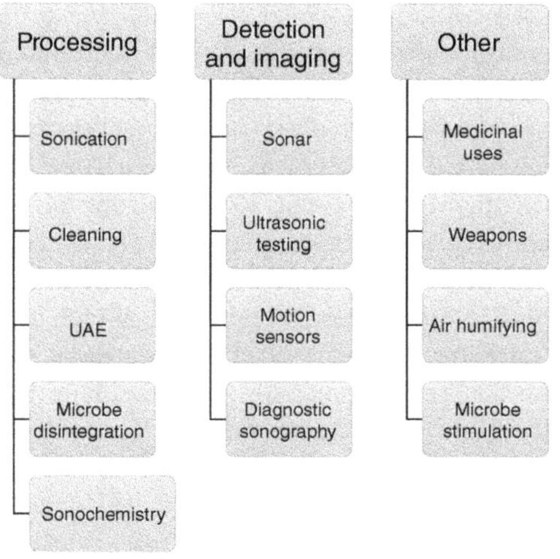

Figure 4.6 The practical applications of ultrasound.

isolation of numerous plant components, using a wide range of conditions and solvents, in both batch and continuous processes (Vilkhu et al. 2008). Coupling UAE with another novel method of extraction known as *supercritical fluid extraction* was previously described (Riera et al. 2004). The process was implemented for the isolation of almond oil and resulted in increasing the yield by 30%. The applications of UAE are not limited to biological samples. Richter et al. described the implementation of this technique for the purification of soils contaminated by aliphatic and polycyclic aromatic hydrocarbons (Richter et al. 2006). US was also used for enhancing pressurized solvent extraction resulting in a decreased processing time from 20 minutes to 10 minutes.

US also proved to be an efficient method of inactivation of tissue enzymes present in food samples, therefore positively influencing the quality of the stored products. Examples of the enzymes affected by US include horseradish peroxidase and pectin methylesterase from orange (Vercet et al. 2001), watercress peroxidase (Cruz et al. 2006), polyphenol oxidases, proteases, and lipases from various sources (Raso and Barbosa-Cánovas 2003). The inactivation of enzymes is a result of the denaturation of proteins or chemical changes due to oxidative stress. The structure of enzymes impacts its vulnerability to US; some enzymes such as amylase are resistant to inactivation (Mason et al. 1996).

Low-intensity US is used as a fast and nondestructive method of product testing. Coupland used US for testing the texture of cheese and vegetables as well as ripeness of fruits (Coupland 2004). Benedito et al. applied this method for the evaluation of various food products, including meat, fish, milk, and oils (Benedito et al. 2002). US testing was also used for the evaluation of ripening of tofu cheese (Ting et al. 2009).

US can have a lethal effect on the living species; this feature is a basis of microorganism disintegration applications. The inactivation of microbes present in samples is critical in food technology, but it is also important in other tasks such as wastewater management. US frequencies of above 18 kHz can inactivate microorganisms, although it is usually not enough to ensure safety of the product. Therefore, US should be used together with other preservation factors, such as heat, pressure, or chemical agents (Koubaa et al. 2016). Raso et al. achieved good reduction of microbial populations using US, pressure, and heat along with low water activity (Raso and Barbosa-Cánovas 2003). Knorr et al. showed that combination of US with direct steam injection was capable to inactivate *Bacillus stearothermophilus* and *Escherichia coli*, although *Lactobacillus acidophilus* was more resistant to such treatment (Knorr et al. 2004). The antimicrobial mode of action of US is linked to the damage to cell wall structures, which can explain differences in resistance against US of various bacterial strains. The damage is caused by cavitation bubbles: their implosions create very high temperatures and pressures, which are responsible for the disruption of the cell walls (Earnshaw et al. 1995). US can also be used for enhancing the fermentation processes. Low-intensity sonication can stimulate the growth of yeast and lactic acid bacteria and, therefore, increase the rate of bioprocesses (Al Daccache et al. 2020b; Huang et al. 2017; Sulaiman et al. 2011).

4.4 Pulsed Electric Fields

PEF, next to HHP and US treatment, is one of the most promising innovative alternative food processing technologies. All over the world, at least 69 research units in 25 countries study different applications of PEF treatment (Miklavcic 2012; Misra et al. 2017). The main advantage of PEF treatment is its nonthermal character. It means that it is effective below the temperature range traditionally used in the case of thermal processes.

PEF treatment depends on applying external electric fields to material enclosed between two electrodes and the subsequent application of series of short pulses characterized by very high intensity (0.1–65 $kV\cdot cm^{-1}$) (Buckow et al. 2013; Gabrić et al. 2018; Geveke et al. 2015; Puértolas and Barba 2016; Puértolas et al. 2016). The application of an external electric field to a biological cell can lead to a phenomenon called electroporation (Figure 4.7). In a simplified explanation, electroporation can be described as the electrically induced perforation of a cell membrane, as a result of the application of the external electric field.

There are some theories that attempt to explain in detail the mechanism of electroporation. One of the most popular explanations of electrically induced perforation is that of a transient aqueous pore model. Cell membrane semi-permeability is linked with the presence of specific structures, which are necessary to perform transport of chemical compounds between the outer and the inner parts of the cell. Moreover, because of the multilayer structure and specific composition of the cell membrane, which is mainly constituted of lipids and proteins, the transmembrane pores can be formed. These structures are called hydrophobic pores, which have very small radius and are metastable (Mason 1999; Tsong 1991). They are formed spontaneously due to the local temperature fluctuations and changes of lipids properties. Due to the asymmetrical distribution of ions, and therefore the uneven polarization of the two sides of the cell membrane, the high voltage application will result in the flow of electric current. In turn, the specific structure of the cell membrane causes that this element has both electric capacity and electrical resistance so that the current flowing through it will cause heating of the cell membrane, in accordance with Joule's law. The increment of temperature, being a local phenomenon in relation to the

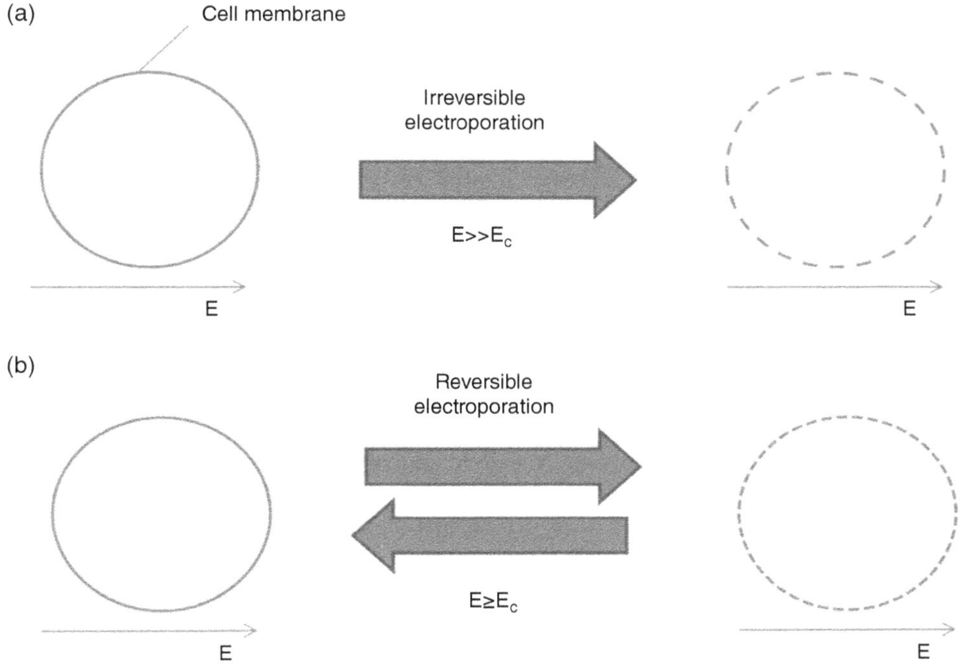

Figure 4.7 Schematic presentation of (a) irreversible and (b) reversible electroporation phenomenon. E = applied electric field intensity (kV·cm^{-1}); Ec = critical electric field intensity (kV·cm^{-1}).

entire cell membrane continuity, will lead to conformation changes of lipids, which are part of the cell membrane. These regrouping and rearrangement processes will cause formation of metastable hydrophobic pores. If the size of the applied external electric field is sufficient, new structures called hydrophilic pores, with a radius and stability much greater than the hydrophobic pores, may be formed in the cell membrane. Hydrophilic pores can also be created by the growth of already existing hydrophobic pores. It is assumed that the change of pore character occurs when the diameter of the pore reaches critical dimensions. This critical diameter is generally equal to the thickness of the cell membrane (Mason 1999). The diameter of the pores formed will affect the possibilities of transmembrane transport of molecules, which can largely determine the intensification of processes based on mass exchange, for example, osmotic dehydration (Wiktor et al. 2014). As a result of the formation of pores with a more stable character and due to the accumulation of charges on both sides of the membrane and the applied electric field, an electroosmotic flow may also occur in the biological cell (Barba et al. 2015b, 2015c; Puértolas and Barba 2016). It is also worth noting that experiments carried out using molecular dynamics modeling show that the higher the transmembrane potential (TMP) resulting from an applied external electric field, the easier the hydrophilic pores are formed. However, it should be noted that the formation of more stable pores in the cell membrane occurs when the TMP reaches a critical value (TMPc). For most biological materials, the TMPc oscillates between 0.7 and 2.2 V (Angersbach et al. 1999).

Another theory of electroporation mechanism, which is very similar to the theory of hydrophobic and hydrophilic pores, is based on an electromechanical model. According to this theory, placing a biological cell in an external electric field leads to the reorganization of the distribution of charges on both sides of the cell membrane, which leads to the occurrence of viscoelastic stresses through electrostatic forces of the same sign. If the external intensity of the electric field is sufficiently large (critical), these stresses will lead to an increase in the permeability of the cell membrane. This model explains very easily the changes taking place in the cell membrane under the influence of the applied electric field, which does not change the fact that the hydrophilic and hydrophobic pore model is a more precise theory and explains a larger number of dependencies (Buckow et al. 2013).

Despite many controversies, it is assumed that the application of PEF, causing a significant increase in the membrane potential above the critical value (TMPc$\gg$TMP), leads to excessive growth of the cell pores, permanent disruption of the cell membrane, and thereby cell death (Figure 4.7). Therefore, this phenomenon is called irreversible electroporation (IRE). In turn, if the applied PEF energy leads to an increase of the TMP to a level close to TMPc, slightly greater or slightly smaller, it is observed that the cell is able to restore the cell membrane continuity and reseal the pores formed after some time from its perforation. This phenomenon is called reversible electroporation (RE) (Vorobiev and Lebovka 2008). The "closure" of the resulting pores will also occur in the case of IRE, but this phenomenon will last very long. Identification of the occurrence of the phenomenon of electroporation is assessed mostly by indirect methods. The most frequently used method depends on measurement of electrical properties of material subjected to the action of PEF (Kotnik et al. 2015). Another, less popular, method is the use of cell staining, thanks to which an intracellular juice leak can be observed (Faridnia et al. 2015).

Electroporation efficiency depends on many factors, which are linked to both processing conditions and properties of the treated system (Wiktor and Witrowa-Rajchert 2016). Within the processing parameters, the following factors play very important roles: electric field intensity, specific energy input, pulse geometry, pulse width, number of pulses, frequency temperature, and the design of the treatment chamber. In turn, among the properties of the treated material, the role of some can be emphasized, including cell size and geometry, electrical conductivity, pH, structure and porosity, water content and water activity and, in the case of microorganisms, growth phase, strain, and type of microorganisms.

Nevertheless, based on the Schwan equation (Eq. (4.1)):

$$TMP = -0,75 f\, E d_c cos\theta \tag{4.1}$$

where TMP corresponds to the transmembrane potential (V), d_c corresponds to the cell diameter (m), E corresponds to the external electric field intensity (V.m^{-1}), θ corresponds to the angle between the vector of electric field and any point on the periphery of the cell, and f corresponds to a constant dependent on the form and electrical properties of the treated biological cell. It can be concluded that the most important factors are electric field intensity and the diameter of the treated cell.

As aforementioned, in order to achieve the perforation of the cell membrane, induced TMP needs to reach at least critical value, and that can be done by the application of an external electric field with a critical value of electric field intensity (E_c).The value of the electric field intensity will also influence the technological effect that can be achieved. In general, higher intensities (5–65 kV·cm^{-1}) of the electric field are used to inactivate microorganisms (Barba et al. 2015b), and lower intensities (0.6–1.5 kV·cm^{-1}) of the electric field are used to enhance unit operations, based on mass and/or heat transfer (Amami et al. 2007; Lebovka et al. 2007). Considering the impact of electric field intensity on microorganisms, it can be stated that there is a linear relationship between the electric field strength and the microbial survival, if the critical value of electric field intensity is exceeded (Heinz et al. 1999). Below the E_c, the impact of PEF on microorganisms is ambiguous, and it can cause either RE or proliferation stimulation, as described further.

The size and shape of a biological cell also affects the electroporation efficiency as shown by the Schwan equation. The smaller the cell diameter, the higher the electric field strength should be applied to increase the membrane potential above the critical value. For this reason, in order to make IRE of microorganisms, an electric field with a higher field strength should be applied than in the case of processing of plant-derived materials with cell diameter larger than the diameter of microorganisms (Barba et al. 2015b). The shape of the cell is also a factor influencing the efficiency of electroporation not only because it directly affects the value of the membrane potential. Considering that the perforation of the cell membrane located in the external electric field is most likely to occur at its poles, it can be said that the shape of the cell will indirectly affect the orientation of the cell relative to the external electric field and consequently the efficiency of electroporation. It is worth noting that the membrane potential of large diameter cells is much more dependent on their shape (Barba et al. 2015b).

Studies carried out so far also indicate that the sensitivity to PEF depends on the group of the microorganism, its strain, form type (vegetative or spores), or growth phase. Generally, it can be assumed that vegetative cells are sensitive to the action of PEF, while spores are resistant. For instance, yeast is a group of microorganisms that is much easier to inactivate than bacteria (Barba et al. 2015b). Research conducted by Wouters et al. proved that bacteria are much easier to inactivate with PEF if they are at the beginning of the logarithmic phase (Wouters et al. 2001). It is also important to note that the sensitivity of microorganisms to PEF and high temperature does not correlate with each other (i.e. temperature-sensitive microorganisms may be very resistant to PEF) (Toepfl et al. 2005).

The ability to carry out PEF treatments in a way that leads to RE or IRE gives great opportunity to achieve very different technological goals and makes this technology flexible and versatile. In the case when the main purpose of treatment is linked to the cell death (e.g. microbial inactivation) or the rupture of the cellular structure (e.g. enhancement of heat- and mass transfer–based processes), PEF parameters should be set in order to cause IRE (Evrendilek 2017; Geveke et al. 2009; Yu et al. 2017). In turn, RE will be applied when the objective depends on introduction of "matter" inside the cell without causing it severe injury or death or in order to provoke stress reaction (Mercer and Armenta 2011; Soliva-Fortuny et al. 2009). For instance, PEF can improve the production of secondary metabolites, what has been proven based on the research performed on *Taxus chinensis* cells (Ye et al. 2004).

Despite electroporation, electric fields can stimulate cell proliferation and affect the metabolic pathways (Al Daccache et al. 2020a, 2020c; Fiedler et al. 1995; Hunt et al. 2009).

The mechanism of such effect of PEF treatments is rather ambiguous. However, there are some theories, which suggest that this phenomenon is linked with the change of enzyme activity and metabolic pathway alterations (Mota et al. 2018). Such an explanation is consistent with papers describing the effect of PEF on the metabolism of potato or apple tissue (Dellarosa et al. 2016; Galindo et al. 2009), and with articles concerning changes of activity of enzymes subjected to PEF (Ohshima et al. 2007; Tian et al. 2016).

Fermentation depends on the utilization of enzymatic systems of microbiota to perform chemical transformation of different compounds. Therefore, the aforementioned consequences of PEF treatments clearly suggest that this is a promising technology to modify the progress of fermentation processes and to shape its course and progress.

4.5 Microwaves

Electromagnetic waves are those of electrical and magnetic energy, which move together through space and which are widely present in the environment. Electromagnetic radiation includes radio waves, microwaves, infrared radiation, visible light, ultraviolet radiation, X-rays, and gamma rays. Microwave wavelength interval ranges from 1 mm to 1 m with frequencies between 300 MHz and 300 GHz (Datta and Ni 2002; Kalla and Devaraju 2017).

The first applications of microwaves were for radar detection and military communication. In the second half of the twentieth century, Dr. Percy Spencer observed increasing temperature of a water-rich substance under MW. This observation led to the application of microwave energy to food industry as an emerging fast heating technique. Dr. Spencer is considered as the inventor of the application of microwave in the food industry (Fito et al. 2005; Kalla and Devaraju 2017). All types of electromagnetic waves, including microwaves, can result in absorption, reflection, or dispersion, whereas a material rich in dipolar molecules such as water can only absorb microwaves. Higher frequency of electromagnetic fields generates higher oscillation of water particles, thereby transferring the microwave energy to heat. Heating time generally depends on the water content of the product, microwave frequency, and power of the resonance chamber. Higher frequency waves result in shorter heating time due to faster oscillation of water particles in the products. Low-frequency MW shows an inverse relationship. Microwaves at frequency of 2.45 GHz cause faster heating of products than at frequency 915 MHz. This is because waves are shorter, but their penetration is limited to 2.5 cm, whereas 915 MHz waves lead to slower heating, which is more uniform and with higher depth of penetration (12.9 cm) (Vadivambal and Jayas 2010). The absorption of microwaves takes place due to the effect of ionic conductivity, which depends on the acceleration of the ions and oscillation of the dipoles in the alternating electric field. In variable electromagnetic fields, dipoles of water oscillate, whereas ions move parallel to the direction of the field. Higher frequency of this process causes a higher number of collisions between the molecules, transmitting kinetic energy to them. Frequency of 2.45 GHz means almost 2.5 billion oscillations per seconds. The simultaneous occurrence of both mechanisms contributes to friction and leads to increase of the temperature.

Magnetron is the most widely used source of microwaves for industrial and domestic applications. The power of a single magnetron generally oscillates between 300 and

3000 W. Magnetron is built-in into a generator, where microwaves are generated, because magnetron converts the supplied power into microwave energy and causes emission of high-frequency radiant energy (Vollmer 2003). The magnetron is a vacuum tube with a central electron-emitting cathode of highly negative potential, which is surrounded by a structured anode. The main operation principle of the magnetron is based on the motion of electrons under influence of the combined action of an electric and a magnetic field. The anode structure forms cavities, and, as a result, the intended microwave resonant frequency is created. This frequency depends on the cavities used. Due to the magnetic field, the emitted electrons are accelerated at first radially and later they move spirally. If the electric and magnetic fields are strong enough, the electrons cannot reach the anode. The cavities of the anode affect electrons, which are speeding up or slowing down. Single electrons become electron bunches, and they move around the cathode at the microwave frequencies. At the end, oscillations of the resonant cavities are formed, and they get into a waveguide (Stolyarov 2016), which are metal tubes. They can be of any shape, but the most popular have a rectangular cross section. The size of waveguides defines a minimum frequency by appropriate boundary conditions, below which waves do not propagate.

Waveguides carry microwaves from the magnetron directly to the product through the heating chamber. Domestic microwave ovens have reflecting cavity walls that form several modes of microwaves. This phenomenon leads to the maximization of heating efficiency (Orsat et al. 2017). The microwaves are reflected inside the oven chamber; therefore, domestic microwave ovens have to be equipped with dense grid for safety reasons. Industrial equipment systems have applicators, which may be installed in continuous or batch systems (Figure 4.8). Microwave applicators may be designed to allow controlled interaction under a diversity of ambient conditions, such as various humidity and conditions ranging from vacuum to high pressure (Rajpurohit and Chhibber 2016).

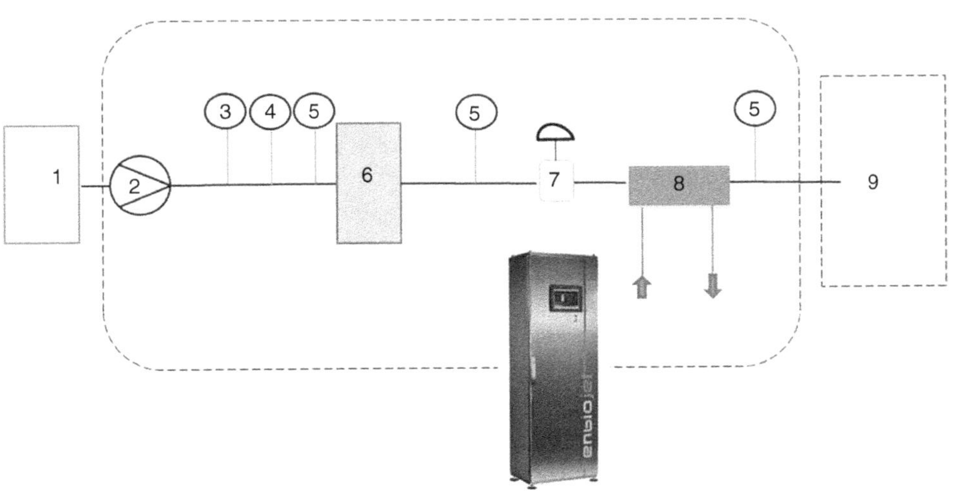

Figure 4.8 Schematic diagram of *Enbio Jet®* continuous microwave heating system. 1: product tank, 2: pump, 3: flow measurement, 4: pressure measurement, 5: temperature measurement, 6: microwave chamber, 7: back pressure modules, 8: cooling exchanger, and 9: laminar chamber. *Source:* Marszałek et al. (2015b). © 2015, Springer Nature.

Applicators can be divided into two types, depending on the type of field configurations: single-mode and multimode applicators. Single-mode applicators are used in high-efficiency microwaves. More complicated construction such as single-mode applicators may consist of resonant cavities (Asmussen et al. 1987; Orsat et al. 2017; Rajpurohit and Chhibber 2016).

Multimode applicators are suitable for bulk processing because they can have large dimensions, and it is possible to use rotating dishes, mode stirrers, or other solutions, which lead to improvements in the heating homogeneity. Due to the large size of multimode applicators, a large number of standing-wave modes can exist near or at the operating frequency inside the cavity. Furthermore, multimode applicators can work in batch or continuous systems (Asmussen et al. 1987; Mandal and Yogesh-Mohan 2007).

MW can be used in many traditional and emerging technological processes such as cooking, baking, pasteurization, sterilization, thawing, tempering, and drying. Cooking is one of the major applications of microwave, and it has become one of the most adapted methods of food heating in the world. The main reason for the popularity of microwave ovens is the significant reduction in product preparation time as compared to conventional cooking or baking (Marconi et al. 2000).

Microwave heating results in nonuniform cooking, and products may contain cold or hot spots, which may cause a lack of microbiological food safety. When the shape of food is flat, and the sizes of food are small, rather even heating through overall volume is possible. Nowadays, there have been better and better construction solutions, minimizing the problems of uneven cooking, and enabling proper processing, especially in the food industry. To eliminate inhomogeneous heating of the product in the food industry, the most common solution is the microwave oven tunnel. This solution allows the heated product to move inside. The microwave radiators placed on the tunnel walls emit microwaves in a controlled manner, and the heated food moves inside the tunnel in bulk (Kalla and Devaraju 2017).

It is proved that microwave heating does not induce the Maillard reaction due to the short time and low temperature of microwaved food, resulting in slow rates of Maillard compound formation (Yeo and Shibamoto 1991). On the other hand, the results of Meißner and Erbersdobler's study showed that microwave heating of milk did not cause significant differences in the formation of Maillard products between microwave and traditional cooking (Meißner and Erbersdobler 1996).

The main factors, which affect microwave cooking, include the size and shape of the products, their composition, density and physical state, as well as the time and temperature of the heating process. Microwave cooking method causes minimal loss of moisture when compared to traditional heating systems (Das and Rajkumar 2013; Puligundla et al. 2013).

Microwave heating can be also used for cake or bread baking. However, it was found that baking a cake using microwave affected its textural properties negatively due to the nonuniform heating (Megahey et al. 2005). Moreover, microwaves do not have the ability to induce crust browning, which takes place in a conventional baking. A possible solution was to bake (e.g. the loaf of bread) in a conventional oven in a metal pan and, at the end of baking, put it in microwave oven (Kalla and Devaraju 2017; Sánchez-Pardo et al. 2008).

Various types of pans, such as glass, Teflon, or metal, are used to avoid uneven product heating. The metal pans proved to be good for heat conduction and shows high heat transfer coefficient but caused uneven baking, resulting in poor microwave penetrability

(Kalla and Devaraju 2017). In recent years, there has been an increased interest in the use of microwaves combined with other heating systems to increase the quality of baked products and reduce the processing time. Using microwave-assisted heating turned out to be more effective in two cycles. The first cycle is an internal cooking using traditional method, whereas the second is cooking using microwaves. Another combined technology can be halogen lamp–microwave-assisted heating. This way of baking brings additional advantages such as crisping and browning effects (Sevimli et al. 2005). Microwaves can be also used as a part of donut frying process, resulting in lower fat uptake (Sevimli et al. 2005).

An example of a microwave system for pasteurization and sterilization is MicVac, which is a single-mode microwave equipment working at a frequency of 915 MHz. This method is suitable and used for ready-to-eat meals. The MicVac method consists of four steps. The first step is filling the tray with ingredients, which is by machine or by hand. The second step is based on the application of a covering film equipped with a valve. Afterwards, the sealed product is cooked and pasteurized in a microwave tunnel. During this stage, the valve opens to let out steam and oxygen, and then valve closes and forms a vacuum. The last step consists of a cooling process, and after that the fresh-made meals are ready for storage and distribution for consumption. MicVac products have high quality, and the shelf life period was 30–45 days when products were stored at 8°C and 90 days when products were stored at 4°C (MicVac 2018).

Another solution of microwave pasteurization is the use of a continuous flow process. This technique is designed for some liquid products, such as milk, fruit juices, purée, or ciders. The raw fluid is pumped through a series of coils placed inside the microwave chamber. Next, the fluid passes through the holding section, which allows a controllable time of microwave heating. At the end, fluid is chilled and stored. The advantages of continuous flow microwave heating are the rapid rise in temperature and the possibility of fast cooling liquid in the flow. Furthermore, the equipment does not require steam installation (Ahmed and Ramaswamy 2007;Marszałek et al. 2016b). Microwave absorption causes heat production, and in this way, it inactivates microorganisms and enzymes; therefore, microwave heating at proper conditions can result in a pasteurization effect (Marszałek et al. 2015b). Microwave pasteurization operates at a temperature between 60 and 82°C, whereas microwave sterilization is carried out at a higher temperature range, between 110 and 130°C (Ahmed and Ramaswamy 2007).

Thawing is the transformation of a product from frozen state to an unfrozen state until the product reaches 0°C and has no residual ice (Archer et al. 2008). It is important that a lower temperature, from −5 to −2°C, be applied for tempered products, which are destined for mechanical chopping and are easily diced or separated. In this temperature range, the product remains firm. Hence, these two processes– thawing and tempering– are not the same process because tempering constitutes only the initial phase of a total thawing process. Microwave tempering can be an alternative to microwave thawing because often complete thawing is not necessary (Fellows 2009; Kumar 2015; Taher and Farid 2001;Yarmand and Rad 2011).

The conventional thawing in water is often a time-consuming process. Water has a much lower thermal conductivity compared to ice, so the heat transfer rate is reduced as the outer layer of water in thawing product increases in thickness. It is a disadvantage because thawing time is required to be as short as possible to minimize excessive drip loss, dehydration, and microbial growth. A possible solution is using microwave thawing. However, the use of microwave thawing leads to nonuniform heating because, in some parts of product, thawing is faster than in the others due to uneven power distributions in the liquid area. This phenomenon is mostly seen in large frozen foods, such as meat or fish blocks. This uncontrolled heating can be avoided by lengthening the defrosting time and reducing power or by using pulsed microwaves, which result in more even distribution of heat in the product (Rakesh et al. 2009; Sevimli et al. 2005; Taher and Farid 2001). Another solution, which reduces uneven thawing, is a continuous movement of the product inside the chamber so that each part of the product is briefly heated in the microwave field. In order to diminish the moisture loss during microwave thawing, a water spray is applied and a high humidity is ensured (Akhtar et al. 2013; Fellows 2009; Rakesh et al. 2009).

The mechanism of drying with the use of microwaves is quite different from conventional drying. In traditional drying, moisture is primitively flashed off from the surface of material, and the remaining water is transferred slowly to the surface. A material can be heated by convection, conduction, and/or radiation. In microwave drying, the temperature rises faster than in traditional drying because heat is generated directly in the interior of material and causes a faster temperature rise. Mass transfer in microwave drying is due to pressure gradient, which is established when vapor is being quickly generated within the material. Additionally, microwaves can penetrate the surface of dried solid foods, thereby reaching moisture inside the material (Gowen et al. 2006; Kassem et al. 2011).

Microwave drying may cause scorching due to excessive temperature along the edges or corners of foods, especially during the final steps of drying. In order to keep product quality and increase drying rate, it is necessary to combine microwave drying with other techniques. The key role in microwave-assisted drying is played by the dielectric food properties. Malafronte et al. proved it by using a mathematical model to simulate the combined action of convective and microwave-assisted drying (Malafronte et al. 2012). Drying methods using microwave are divided into microwave-assisted air-drying, microwave-assisted vacuum drying, and microwave-assisted freeze-drying (Chandrasekaran et al. 2013).

Microwave-assisted air-drying is a method where microwave heating can be combined with hot air at three different stages of drying. At the first stage, microwave heating is used at the beginning of the dehydration process, resulting in rapid heating of the interior of product. At the next stage, vapor is forced outside because drying rate increases, and in this way a more stable temperature profile is formed. This phenomenon creates "puffing," which means a porous structure. At the last stage of drying, vapor is forced outside, and the drying rate starts to fall (Shaheen et al. 2012; Zhang et al. 2006). Generally, microwave-assisted air-drying is more useful at the last stage of drying of products. This type of drying increases the rehydration capacity and does not cause shrinking of the products (Chandrasekaran et al. 2013).

Microwave-assisted vacuum drying is another type of drying process. Lewis equation, Fick's law, and Page's model were applied to describe drying processes. From these equations, it can be concluded that the changes in moisture ratio follow the exponential or empirical function. High-energy water molecules diffuse to the product surface and evaporate due to the decreased pressure. This low pressure causes water vapor to concentrate at the surface, and the boiling point of water is reduced. Thus, using high vacuum leads to improvement in the evaporation of water and its volatilization from the drying product. Furthermore, this type of drying prevents oxidation because vacuum drying takes place without air. Moisture diffusivity in microwave-assisted vacuum drying is increased in comparison to microwave-assisted hot-airdrying(Therdthai and Zhou 2009; Zhang et al. 2006).

During freeze-drying, the frozen water does not pass through the liquid phase because it is directly transferred to vapor phase. Freeze-drying using microwaves can be used in different ways. The first way is freeze-drying accompanied with microwaves, whereas the second way is microwave drying applied after freeze-drying. In the first type, microwave field is used to supply the heat of sublimation. In the second type of the process, it is divided into two steps: freeze-drying and microwave or vacuum microwave-drying (Abdelwahed et al. 2006; Duan et al. 2010; Zhang et al. 2006).

Microwave-assisted extraction (MAE) is a process of using microwave energy to heat the extractant (mainly liquid organic solvents) in contact with a sample in order to facilitate separation of compounds from the matrix into the solvent. This innovative approach of extraction causes heating of the whole sample simultaneously without heating the vessel. The boiling point of the sample-solvent mixture is reached very rapidly, and the extraction time is shorter. Moreover, MAE uses 10 times lower solvent volumes than conventional extraction techniques (SparrEskilsson and Björklund 2000).

MAE is divided into two systems: with closed vessels or with open vessels. The first system performs extraction under controlled pressure and temperature and is known as pressurized MAE, whereas the second system performs extraction under atmospheric pressure and is known as focused MAE or focused microwave-assisted Soxhlet extraction. In the latter one, only a part of the extraction vessel, which contains the sample, is irradiated with microwaves (Camel 2000; Luque-García and Luque De Castro 2004; Priego-Capote and Luque De Castro 2005). A higher temperature is achieved during MAE with closed vessels than in open vessel systems. Increased pressure inside the closed vessel allows rising of the boiling point of the solvents. Another advantage of closed vessel system is that there is no loss of volatile substances, but if the temperature inside the vessel increases rapidly, a separation of the most volatile solutes into the headspace may happen, thus promoting their possible loss. However, closed vessel systems have some disadvantages, such as impossibility of addition of solvents or reagents during multistep extraction process (Camel 2000; Mandal and Yogesh-Mohan 2007).

MAE with open vessels is more suitable for thermolabile species because lower temperatures are used compared to closed vessel systems. Moreover, the addition of solvents or reagents is possible at any time during extraction, and excess of solvent is easy to remove (Camel 2000; Luque-García and Luque De Castro 2004; Priego-Capote and Luque De Castro 2005).

As shown, microwaves can be successfully applied for many food processes, especially for heating purposes. Many processes can be assisted by microwaves, for example, drying and extraction. In order to design a household and industry microwave oven, knowledge of dielectric, physical, and chemical properties of processed foods is necessary. An important advantage of microwave processes is the fact that it decreases operation time compared to conventional operations, what is important from an economic point of view as well as regarding the quality of obtained processed food.

4.6 Conclusions

Consumer demand for high-quality foods that are fresh tasting and nutritious has created considerable interest in the development of new food processing techniques. Innovative alternative technologies have been advanced to replace conventional heat treatments. In addition, due to consumer demand for food with fresh-like taste, crisp texture, and natural color, the development of new food processing technologies had grown dramatically over the past decade. In this regard, developments in nonthermal technologies have been advanced by both industry and academia in an attempt to meet the challenge of producing safe processed food of a high quality. In general, these innovative food processing technologies can affect biotechnological processes such as fermentation. Several innovative approaches can be used to modulate food fermentation processes, obtaining products with novel and improved characteristics. In addition, the combination of these new technologies could lead to develop new and optimized fermentation method under nonconventional conditions, which can decrease the time and cost. Overall, fermentation under unconventional conditions is still under scientific consideration, but the newest publications indicate the high potential of these techniques within the next years, due to the potential biotechnological application. Therefore, it is expected that further work will be conducted in this field focusing on addressing these issues, as well as evaluating the impact of these technologies on other fermentation processes with industrial relevance.

Acknowledgments

Part of this chapter was supported by Project Number 2015/17/D/NZ9/02079 of the National Science Centre, Poland. FCT/MEC for the financial support to the LAQV-REQUIMTE and QOPNA research Units (UIDB/50006/2020 and FCT UID/QUI/00062/2019, respectively), through national funds and where applicable co-financed by the FEDER, within the PT2020 Partnership Agreement.

References

Abdelwahed, W., Degobert, G., Stainmesse, S., and Fessi, H. (2006). Freeze-drying of nanoparticles: formulation, process and storage considerations. *Adv. Drug Deliv. Rev.* 58: 1688–1713.

Ahmed, J. and Ramaswamy, S.H. (2007). Microwave Pasteurization and Sterilization of Foods. In: *Handbook of Food Preservation* (ed. M.S. Rahman), 691–713. New York: CRC Press, Taylor and Francis Group.

Akhtar, S., Khan, M.I., and Faiz, F. (2013). Effect of thawing on frozen meat quality: a comprehensive review. *Pak. J. Food Sci.* 23: 198–211.

Al Daccache, M., Koubaa, M., Maroun, R.G. et al. (2020a). Pulsed electric field-assisted fermentation of *Hanseniaspora* sp. yeast isolated from Lebanese apples. *Food Res. Int.* 129: 108840. https://doi.org/10.1016/j.foodres.2019.108840.

Al Daccache, M., Koubaa, M., Salameh, D. et al. (2020b). Ultrasound-assisted fermentation for cider production from Lebanese apples. *Ultrason. Sonochem.* 63: 104952. https://doi.org/10.1016/j.ultsonch.2019.104952.

Al Daccache, M., Koubaa, M., Salameh, D. et al. (2020c). Control of the sugar/ethanol conversion rate during moderate pulsed electric field-assisted fermentation of a *Hanseniaspora* sp. strain to produce low-alcohol cider. *Innov. Food Sci. Emerg. Technol.* 59: 102258. https://doi.org/10.1016/j.ifset.2019.102258.

Amami, E., Fersi, A., Vorobiev, E., and Kechaou, N. (2007). Osmotic dehydration of carrot tissue enhanced by pulsed electric field, salt and centrifugal force. *J. Food Eng.* 83: 605–613.

Angersbach, A., Heinz, V., and Knorr, D. (1999). Electrophysiological model of intact and processed plant tissues: cell disintegration criteria. *Biotechnol. Prog.* 15: 753–762.

Archer, M., Edmonds, M., and George, M. (2008). Seafood Thawing. *Research and Development Department* SR598: 41.

Asmussen, J., Lin, H.H., Manring, B., and Fritz, R. (1987). Single-mode or controlled multimode microwave cavity applicators for precision materials processing. *Rev. Sci. Instrum.* 58: 1477–1486.

Awad, T.S., Moharram, H.A., Shaltout, O.E. et al. (2012). Applications of ultrasound in analysis, processing and quality control of food: a review. *Food Res. Int.* 48: 410–427.

Balny, C., Hayashi, R., Heremans, K., and Masson, P. (1992). *High Pressure and Biotechnology*, 1e. France: John LibbeyEurotex.

Barba, F.J., Esteve, M.J., and Frígola, A. (2012). High pressure treatment effect on physicochemical and nutritional properties of fluid foods during storage: a review. *Compr. Rev. Food Sci. Food Saf.* 11: 307–322.

Barba, F.J., Criado, M.N., Belda-Galbis, C.M. et al. (2014). *Stevia rebaudiana* Bertoni as a natural antioxidant/antimicrobial for high pressure processed fruit extract: processing parameter optimization. *Food Chem.* 148: 261–267.

Barba, F.J., Terefe, N.S., Buckow, R. et al. (2015a). New opportunities and perspectives of high pressure treatment to improve health and safety attributes of foods. A review. *Food Res. Int.* 77 (Part 4): 725–742.

Barba, F.J., Parniakov, O., Pereira, S.A. et al. (2015b). Current applications and new opportunities for the use of pulsed electric fields in food science and industry. *Food Res. Int.* 77 (Part 4): 773–798.

Barba, F.J., Galanakis, C.M., Esteve, M.J. et al. (2015c). Potential use of pulsed electric technologies and ultrasounds to improve the recovery of high-added value compounds from blackberries. *J. Food Eng.* 167 (Part 4): 38–44.

Barba, F.J., Orlien, V., Mota, M.J. et al. (2016). Implementation of emerging technologies. In: *Innovation Strategies in the Food Industry* (ed. C.M. Galanakis), 117–148. Academic Press.

Barba, F.J., Ahrné, L., Xanthakis, E. et al. (2018). Innovative technologies for food preservation. In: *Innovative Technologies for Food Preservation: Inactivation of Spoilage and Pathogenic Microorganisms* (eds. F.J. Barba, A.S. Sant'Ana, V. Orlien and M. Koubaa), 25–51. Academic Press, Elsevier.

Benedito, J., Carcel, J.A., Gonzalez, R., and Mulet, A. (2002). Application of low intensity ultrasonics to cheese manufacturing processes. *Ultrasonics* 40: 19–23.

Buckow, R., Ng, S., and Toepfl, S. (2013). Pulsed electric field processing of orange juice: a review on microbial, enzymatic, nutritional, and sensory quality and stability. *Compr. Rev. Food Sci. Food Saf.* 12: 455–467.

Butz, P., Fernández García, A., Lindauer, R. et al. (2003). Influence of ultra high pressure processing on fruit and vegetable products. *J. Food Eng.* 56: 233–236.

Camel, V. (2000). Microwave-assisted solvent extraction of environmental samples. *TrAC Trends Anal. Chem.* 19: 229–248.

Cao, X., Zhang, Y., Zhang, F. et al. (2011). Effects of high hydrostatic pressure on enzymes, phenolic compounds, anthocyanins, polymeric color and color of strawberry pulps. *J. Sci. Food Agric.* 91: 877–885.

Chandrasekaran, S., Ramanathan, S., and Basak, T. (2013). Microwave food processing– a review. *Food Res. Int.* 52: 243–261.

Corrales, M., Toepfl, S., Butz, P. et al. (2008). Extraction of anthocyanins from grape by-products assisted by ultrasonics, high hydrostatic pressure or pulsed electric fields: a comparison. *Innov. Food Sci. Emerg. Technol.* 9: 85–91.

Coupland, J.N. (2004). Low intensity ultrasound. *Food Res. Int.* 37: 537–543.

Cruz, R.M.S., Vieira, M.C., and Silva, C.L.M. (2006). Effect of heat and thermosonication treatments on peroxidase inactivation kinetics in watercress (*Nasturtium officinale*). *J. Food Eng.* 72: 8–15.

Das, A.K. and Rajkumar, V. (2013). Effect of different fat level on microwave cooking properties of goat meat patties. *J. Food Sci. Technol.* 50: 1206–1211.

Datta, A.K. and Ni, H. (2002). Infrared and hot-air-assisted microwave heating of foods for control of surface moisture. *J. Food Eng.* 51: 355–364.

Dellarosa, N., Tappi, S., Ragni, L. et al. (2016). Metabolic response of fresh-cut apples induced by pulsed electric fields. *Innov. Food Sci. Emerg. Technol.* 38 (Part B): 356–364.

Dervisi, P., Lamb, J., and Zabetakis, I. (2001). High pressure processing in jam manufacture: effects on textural and colour properties. *Food Chem.* 73: 85–91.

Duan, X., Zhang, M., Mujumdar, S., and Wang, R. (2010). Trends in microwave-assisted freeze drying of foods. *Dry. Technol.* 28: 444–453.

Earnshaw, R.G., Appleyard, J., and Hurst, R.M. (1995). Understanding physical inactivation processes: combined preservation opportunities using heat, ultrasound and pressure. *Int. J. Food Microbiol.* 28: 197–219.

Evrendilek, G.A. (2017). Impacts of pulsed electric field and heat treatment on quality and sensory properties and microbial inactivation of pomegranate juice. *Food Sci. Technol. Int.* 23: 668–680.

Faridnia, F., Burritt, D.J., Bremer, P.J., and Oey, I. (2015). Innovative approach to determine the effect of pulsed electric fields on the microstructure of whole potato tubers: use of cell viability, microscopic images and ionic leakage measurements. *Food Res. Int.* 77: 556–564.

Fellows, P.J. (2009). *Food Processing Technology-Principle and Practices*, 3e. Oxford, UK: Woodhead Publishing Limited.

Fiedler, U., Gröbner, U., and Berg, H. (1995). Electrostimulation of yeast proliferation. *Bioelectrochem. Bioenerg.* 38: 423–425.

Fito, P., Chiralt, A., and Martin, M.E. (eds.) (2005). *Novel Food Processing Technologies*. New York: CRC Press.

Gabrić, D., Barba, F., Roohinejad, S. et al. (2018). Pulsed electric fields as an alternative to thermal processing for preservation of nutritive and physicochemical properties of beverages: a review. *J. Food Process Eng.* 41: e12638. https://doi.org/10.1111/jfpe.12638.

Galindo, F.G., Dejmek, P., Lundgren, K. et al. (2009). Metabolomic evaluation of pulsed electric field-induced stress on potato tissue. *Planta* 230: 469–479.

Garcia-Palazon, A., Suthanthangjai, W., Kajda, P., and Zabetakis, I. (2004). The effects of high hydrostatic pressure on β-glucosidase, peroxidase and polyphenoloxidase in red raspberry (*Rubusidaeus*) and strawberry (*Fragaria×ananassa*). *Food Chem.* 88: 7–10.

Geveke, D.J., Gurtler, J., and Zhang, H.Q. (2009). Inactivation of Lactobacillus plantarum in apple cider, using radio frequency electric fields. *J. Food Prot.* 72: 656–661.

Geveke, D.J., Aubuchon, I., Zhang, H.Q. et al. (2015). Validation of a pulsed electric field process to pasteurize strawberry purée. *J. Food Eng.* 166: 384–389.

Gowen, A., Abu-Ghannam, N., Frias, J., and Oliveira, J. (2006). Optimisation of dehydration and rehydration properties of cooked chickpeas (*Cicer arietinum* L.) undergoing microwave-hot air combination drying. *Trends Food Sci. Technol.* 17: 177–183.

Heinz, V. and Buckow, R. (2010). Food preservation by high pressure. *J. Verbrauch.Lebensm.* 5: 73–81.

Heinz, V., Phillips, S.T., Zenker, M., and Knorr, D. (1999). Inactivation of *Bacillus subtilis* by high intensity pulsed electric fields under close to isothermal conditions. *Food Biotechnol.* 13: 155–168.

Horita, C.N., Baptista, R.C., Caturla, M.Y.R. et al. (2018). Combining reformulation, active packaging and non-thermal post-packaging decontamination technologies to increase the microbiological quality and safety of cooked ready-to-eat meat products. *Trends Food Sci. Technol.* 72: 45–61.

Huang, H.-W., Wu, S.-J., Lu, J.-K. et al. (2017). Current status and future trends of high-pressure processing in food industry. *Food Control* 72: 1–8. https://doi.org/10.1016/j.foodcont.2016.07.019.

Hunt, R.W., Zavalin, A., Bhatnagar, A. et al. (2009). Electromagnetic biostimulation of living cultures for biotechnology, biofuel and bioenergy applications. *Int. J. Mol. Sci.* 10: 4515–4558.

Kalla, M.A. and Devaraju, R. (2017). Microwave energy and its application in food industry: areveiw. *Asian J. Dairy Food Res.* 36: 37–44.

Kassem, A.S., Shokr, A.Z., El-Mahdy, A.R. et al. (2011). Comparison of drying characteristics of Thompson seedless grapes using combined microwave oven and hot air drying. *J. Saudi Soc. Agric. Sci.* 10: 33–40.

Knorr, D., Zenker, M., Heinz, V., and Lee, D.-U. (2004). Applications and potential of ultrasonics in food processing. *Trends Food Sci. Technol.* 15: 261–266.

Kotnik, T., Frey, W., Sack, M. et al. (2015). Electroporation-based applications in biotechnology. *Trends Biotechnol.* 33: 480–488.

Koubaa, M., Roselló-Soto, E., ŠicŽlabur, J. et al. (2015). Current and new insights in the sustainable and green recovery of nutritionally valuable compounds from *Stevia rebaudiana* Bertoni. *J. Agric. Food Chem.* 63: 6835–6846.

Koubaa, M., Mhemdi, H., Barba, F.J. et al. (2016). Oilseed treatment by ultrasounds and microwaves to improve oil yield and quality: an overview. *Food Res. Int.* 85: 59–66.

Kultur, G., Misra, N.N., Barba, F.J. et al. (2018). Effect of high hydrostatic pressure on background microflora and furan formation in fruit purée based baby foods. *J. Food Sci. Technol.* 55: 985–991.

Kumar, Y. (2015). Application of microwave in food drying. *Int. J. Eng. Stud. Tech. Approach* 1: 9–24.

Lebovka, N.I., Shynkaryk, N.V., and Vorobiev, E. (2007). Pulsed electric field enhanced drying of potato tissue. *J. Food Eng.* 78: 606–613.

Luque-García, J.L. and Luque De Castro, M.D. (2004). Focused microwave-assisted Soxhlet extraction: devices and applications. *Talanta* 64: 571–577.

Malafronte, L., Lamberti, G., Barba, A.A. et al. (2012). Combined convective and microwave assisted drying: experiments and modeling. *J. Food Eng.* 112: 304–312.

Mandal, V. and Yogesh-Mohan, S.H. (2007). Microwave assisted extraction: an innovative and promising extraction tool for medicinal plant research. *Pharmacogn. Rev.* 1: 7–18.

Marconi, E., Ruggeri, S., Cappelloni, M. et al. (2000). Physicochemical, nutritional, and microstructural characteristics of chickpeas (*Cicer arietinum* L.) and common beans (*Phaseolus vulgaris* L.) following microwave cooking. *J. Agric. Food Chem.* 48: 5986–5994.

Marszałek, K., Mitek, M., and Skąpska, S. (2015a). The effect of thermal pasteurization and high pressure processing at cold and mild temperatures on the chemical composition, microbial and enzyme activity in strawberry purée. *Innov. Food Sci. Emerg. Technol.* 27: 48–56.

Marszałek, K., Mitek, M., and Skąpska, S. (2015b). Effect of continuous flow microwave and conventional heating on the bioactive compounds, colour, enzymes activity, microbial and sensory quality of strawberry purée. *Food Bioprocess Technol.* 8: 1864–1876.

Marszałek, K., Woźniak, Ł., and Skąpska, S. (2016a). The application of high pressure–mild temperature processing for prolonging the shelf-life of strawberry purée. *High Pressure Res.* 36: 220–234.

Marszałek, K., Woźniak, Ł., Skąpska, S., and Mitek, M. (2016b). A comparative study of the quality of strawberry purée preserved by continuous microwave heating and conventional thermal pasteurization during long-term cold storage. *Food Bioprocess Technol.* 9: 1100–1112.

Marszałek, K., Woźniak, Ł., Skąpska, S., and Mitek, M. (2017a). High pressure processing and thermal pasteurization of strawberry purée: quality parameters and shelf life evaluation during cold storage. *J. Food Sci. Technol.* 54: 832–841.

Marszałek, K., Woźniak, Ł., Kruszewski, B., and Skąpska, S. (2017b). The effect of high pressure techniques on the stability of anthocyanins in fruit and vegetables. *Int. J. Mol. Sci.* 18 https://doi.org/10.3390/ijms18020277.

Marszałek, K., Doesburg, P., Starzonek, S. et al. (2019). Comparative effect of supercritical carbon dioxide and high pressure processing on structural changes and activity loss of oxidoreductive enzymes. *Journal of CO2 Utilization* 29: 46–56. https://doi.org/10.1016/j.jcou.2018.11.007.

Mason, W.T. (1999). *Fluorescent and Luminescent Probes for Biological Activity: APractical Guide to Technology for Quantitative Real-Time Analysis*. Academic Press.

Mason, T.J., Paniwnyk, L., and Lorimer, J.P. (1996). The uses of ultrasound in food technology. *Ultrason. Sonochem.* 3: S253–S260.

Megahey, E.K., McMinn, W.A.M., and Magee, T.R.A. (2005). Experimental study of microwave baking of Madeira cake batter. *Food Bioprod. Process.* 83: 277–287.

Meißner, K. and Erbersdobler, H.F. (1996). Maillard reaction in microwave cooking: comparison of early Maillard products in conventionally and microwave-heated milk. *J. Sci. Food Agric.* 70: 307–310.

Mercer, P. and Armenta, R.E. (2011). Developments in oil extraction from microalgae. *Eur. J. Lipid Sci. Technol.* 113: 539–547.

MicVac (2018). https://www.micvac.com/retail-solutions/method.

Miklavcic, D. (2012). Network for development of electroporation-based technologies and treatments: COST TD1104. *J. Membr. Biol.* 245: 591–598.

Miles, C.A., Morley, M.J., and Rendell, M. (1999). High power ultrasonic thawing of frozen foods. *J. Food Eng.* 39: 151–159.

Misra, N.N., Koubaa, M., Roohinejad, S. et al. (2017). Landmarks in the historical development of twenty first century food processing technologies. *Food Res. Int.* 97: 318–339.

Mota, M.J., Lopes, R.P., Koubaa, M. et al. (2018). Fermentation at non-conventional conditions in food- and bio-sciences by the application of advanced processing technologies. *Crit. Rev. Biotechnol.* 38: 122–140.

Muthukumaran, S., Kentish, S.E., Stevens, G.W. et al. (2007). The application of ultrasound to dairy ultrafiltration: the influence of operating conditions. *J. Food Eng.* 81: 364–373.

Ohshima, T., Tamura, T., and Sato, M. (2007). Influence of pulsed electric field on various enzyme activities. *J. Electrost.* 65: 156–161.

Orsat, V., Raghavan, G.S.V., and Krishnaswamy, K. (2017). *The Microwave Processing of Foods.* Montreal, QC, Canada: McGill University.

Otero, L. and Préstamo, G. (2009). Effects of pressure processing on strawberry studied by nuclear magnetic resonance. *Innov. Food Sci. Emerg. Technol.* 10: 434–440.

Oulahal-Lagsir, N., Martial-Gros, A., Bonneau, M., and Blum, L.J. (2000). Ultrasonic methodology coupled to ATP bioluminescence for the non-invasive detection of fouling in food processing equipment – validation and application to a dairy factory. *J. Appl. Microbiol.* 89: 433–441.

Priego-Capote, F. and Luque De Castro, M.D. (2005). Focused microwave-assisted Soxhlet extraction: a convincing alternative for total fat isolation from bakery products. *Talanta* 65: 81–86.

Puértolas, E. and Barba, F.J. (2016). Electrotechnologies applied to valorization of by-products from food industry: Main findings, energy and economic cost of their industrialization. *Food Bioprod. Process.* 100: 172–184.

Puértolas, E., Koubaa, M., and Barba, F.J. (2016). An overview of the impact of electrotechnologies for the recovery of oil and high-value compounds from vegetable oil industry: energy and economic cost implications. *Food Res. Int.* 80: 19–26.

Puligundla, P., Abdullah, S.A., Choi, W. et al. (2013). Potentials of Microwave Heating technology for select food processing applications – a brief overview and update. *J. Food Process. Technol.* 4: 1–9.

Rajpurohit, D.S. and Chhibber, R. (2016). Design optimization of two input multimode applicator for efficient microwave heating. *Int. J. Adv. Microw. Technol.* 1: 68–73.

Rakesh, V., Datta, A.K., Amin, M.H.G., and Hall, L.D. (2009). Heating uniformity and rates in a domestic microwave combination oven. *J. Food Process Eng.* 32: 398–424.

Raso, J. and Barbosa-Cánovas, G.V. (2003). Nonthermal preservation of foods using combined processing techniques. *Crit. Rev. Food Sci. Nutr.* 43: 265–285.

Reineke, K., Mathys, A., Heinz, V., and Knorr, D. (2008). Temperature control for high pressure processes up to 1400 MPa. *J. Phys. Conf. Ser.* https://doi.org/10.1088/1742-6596/121/14/142012.

Richter, P., Jiménez, M., Salazar, R., and Maricán, A. (2006). Ultrasound-assisted pressurized solvent extraction for aliphatic and polycyclic aromatic hydrocarbons from soils. *J. Chromatogr. A* 1132: 15–20.

Riera, E., Golás, Y., Blanco, A. et al. (2004). Mass transfer enhancement in supercritical fluids extraction by means of power ultrasound. *Ultrason. Sonochem.* 11: 241–244.

Roselló-Soto, E., Galanakis, C.M., Brnčić, M. et al. (2015a). Clean recovery of antioxidant compounds from plant foods, by-products and algae assisted by ultrasounds processing. Modeling approaches to optimize processing conditions. *Trends Food Sci. Technol.* 42: 134–149.

Roselló-Soto, E., Parniakov, O., Deng, Q. et al. (2015b). Application of non-conventional extraction methods: toward a sustainable and green production of valuable compounds from mushrooms. *Food Eng. Rev.* 8: 214–234.

Sánchez-Pardo, M.E., Ortiz-Moreno, A., Mora-Escobedo, R. et al. (2008). Comparison of crumb microstructure from pound cakes baked in a microwave or conventional oven. *LWT – Food Sci. Technol.* 41: 620–627.

Sancho, F., Lambert, Y., Demazeau, G. et al. (1999). Effect of ultra-high hydrostatic pressure on hydrosoluble vitamins. *J. Food Eng.* 39: 247–253.

Szczepańska, J., Barba, F.J., Skąpska, S., and Marszałek, K. (2020). High pressure processing of carrot juice: Effect of static and multi-pulsed pressure on the polyphenolic profile, oxidoreductases activity and colour. *Food Chem.* 307: 125549. https://doi.org/10.1016/j.foodchem.2019.125549.

Sevimli, K.M., Sumnu, G., and Sahin, S. (2005). Optimization of halogen lamp-microwave combination baking of cakes: a response surface methodology study. *Eur. Food Res. Technol.* 221: 61–68.

Shaheen, M.S., El-Massry, K.F., El-Ghorab, A.H., and Anjum, F.M. (2012). Microwave applications in thermal food processing. In: *The Development and Application of Microwave Heating* (ed. W. Cao), 3–16. United Kingdom: IntechOpen.

Shapira, D. and Saltmarsh, M. (2002). Nuclear fusion in collapsing bubbles-is it there? An attempt to repeat the observation of nuclear emissions from sonoluminescence. *Phys. Rev. Lett.* 89: 104302. https://doi.org/10.1103/PhysRevLett.89.104302.

Soliva-Fortuny, R., Balasa, A., Knorr, D., and Martín-Belloso, O. (2009). Effects of pulsed electric fields on bioactive compounds in foods: a review. *Trends Food Sci. Technol.* 20: 544–556.

SparrEskilsson, C. and Björklund, E. (2000). Analytical-scale microwave-assisted extraction. *J. Chromatogr. A.* 902: 227–250.

Stinco, C.M., Szczepańska, J., Marszałek, K. et al. (2019). Effect of high-pressure processing on carotenoids profile, color, microbial and enzymatic stability of cloudy carrot juice. *Food Chem.* 299: 125112. https://doi.org/10.1016/j.foodchem.2019.125112.

Stolyarov, O.I. (2016). Matching the microwave-oven chamber to the magnetron. *J. Commun. Technol. Electron.* 61: 190–196.

Sulaiman, A.Z., Ajit, A., Yunus, R.M., and Chisti, Y. (2011). Ultrasound-assisted fermentation enhances bioethanol productivity. *Biochem. Eng. J.* 54: 141–150.

Sun, D.-W. and Li, B. (2003). Microstructural change of potato tissues frozen by ultrasound-assisted immersion freezing. *J. Food Eng.* 57: 337–345.

Taher, B.J. and Farid, M.M. (2001). Cyclic microwave thawing of frozen meat: experimental and theoretical investigation. *Chem. Eng. Process.* 40: 379–389.

Taleyarkhan, R.P., West, C.D., Cho, J.S. et al. (2002). Evidence for nuclear emissions during acoustic cavitation. *Science* 295: 1868–1873.

Therdthai, N. and Zhou, W. (2009). Characterization of microwave vacuum drying and hot air drying of mint leaves (*Menthacordifolia* Opiz ex Fresen). *J. Food Eng.* 91: 482–489.

Tian, M., Fang, T., Du, M., and Zhang, F. (2016). Effects of pulsed electric field (PEF) treatment on enhancing activity and conformation of α-amylase. *Protein J.* 35: 154–162.

Ting, C.-H., Kuo, F.-J., Lien, C.-C., and Sheng, C.-T. (2009). Use of ultrasound for characterising the gelation process in heat induced $CaSO_4{\cdot}2H_2O$ tofu curd. *J. Food Eng.* 93: 101–107.

Toepfl, S., Heinz, V., and Knorr, D. (2005). Overview of pulsed electric field processing of foods. In: *Emerging Technologies for Food Processing* (ed. D.-W. Sun), 69–97. London: Academic Press.

Tsong, T.Y. (1991). Electroporation of cell membranes. *Biophys. J.* 60: 297–306.

Vadivambal, R. and Jayas, D.S. (2010). Non-uniform temperature distribution during microwave heating of food materials-a review. *Food Bioprocess Technol.* 3: 161–171.

Vercet, A., Burgos, J., Crelier, S., and Lopez-Buesa, P. (2001). Inactivation of proteases and lipases by ultrasound. *Innov. Food Sci. Emerg. Technol.* 2: 139–150.

Vilkhu, K., Mawson, R., Simons, L., and Bates, D. (2008). Applications and opportunities for ultrasound assisted extraction in the food industry — A review. Food Innovation: Emerging Science, Technologies and Applications (FIESTA) Conference 9, 161–169.

Vollmer, M. (2003). Physics of the microwave oven. *Phys. Educ.* 39: 74–81.

Vorobiev, E. and Lebovka, N. (eds.) (2008). *Electrotechnologies for Extraction from Plant Foods and Biomaterials*. New York: Springer.

Vyas, S. and Ting, Y.-P. (2018). A review of the application of ultrasound in bioleaching and insights from sonication in (bio)chemical processes. *Resources* 7: 3. https://doi.org/10.3390/resources7010003.

Wiktor, A. and Witrowa-Rajchert, D. (2016). Pulsed electric fields as pretreatment for subsequent food process operations. In: *Handbook of Electroporation* (ed. D. Miklavčič), 1–16. Cham: Springer International Publishing https://doi.org/10.1007/978-3-319-26779-1_178-1.

Wiktor, A., Śledź, M., Nowacka, M. et al. (2014). Pulsed electric field pretreatment for osmotic dehydration of apple tissue: experimental and mathematical modeling studies. *Dry. Technol.* 32: 408–417.

Wouters, P.C., Alvarez, I., and Raso, J. (2001). Critical factors determining inactivation kinetics by pulsed electric field food processing. *Trends Food Sci. Technol.* 12: 112–121.

Yarmand, M. and Rad, A.H. (eds.) (2011). *Microwave Heating*. United Kingdom, UK: InTech.

Ye, H., Huang, L.-L., Chen, S.-D., and Zhong, J.-J. (2004). Pulsed electric field stimulates plant secondary metabolism in suspension cultures of *Taxus chinensis*. *Biotechnol. Bioeng.* 88: 788–795.

Yeo, H.C.H. and Shibamoto, T. (1991). Chemical comparison of flavours in microwaved and conventionally heated foods. *Trends Food Sci. Technol.* 2: 329–332.

Yu, Y., Jin, T.Z., and Xiao, G. (2017). Effects of pulsed electric fields pretreatment and drying method on drying characteristics and nutritive quality of blueberries. *J. Food Process. Preserv.* 41: e13303. https://doi.org/10.1111/jfpp.13303.

Zhang, M., Tang, J., Mujumdar, A.S., and Wang, S. (2006). Trends in microwave-related drying of fruits and vegetables. *Trends Food Sci. Technol.* 17: 524–534.

Zhu, Z., Guan, Q., Koubaa, M. et al. (2017a). HPLC-DAD-ESI-MS2 analytical profile of extracts obtained from purple sweet potato after green ultrasound-assisted extraction. *Food Chem.* 215: 391–400.

Zhu, Z., Wu, Q., Di, X. et al. (2017b). Multistage recovery process of seaweed pigments: investigation of ultrasound assisted extraction and ultra-filtration performances. *Food Bioprod. Process.* 104: 40–47.

5

Biomass Fractionation Using Emerging Technologies

Mahdi Irani[1], Alireza Rafati[2], Seyedeh-Sara Hashemi[3], Francisco J. Barba[4], Mohamed Koubaa[5], and Shahin Roohinejad[3]

[1]*Department of Food Science and Technology, Ferdowsi University of Mashhad (FUM), Mashhad, Iran*
[2]*Division of Pharmacology & Pharmaceutical Chemistry, Sarvestan Branch, Islamic Azad University, Sarvestan, Iran*
[3]*Burn & Wound Healing Research Center, Shiraz University of Medical Science, Shiraz, Iran*
[4]*Faculty of Pharmacy, Preventive Medicine and Public Health, Food Science, Toxicology and Forensic Medicine Department, Universitat de València, València, Spain*
[5]*ESCOM, UTC, EA 4297 TIMR, Compiègne, France*

5.1 Introduction

One of the most plentiful materials in the environment is lignocellulosic biomass such as agricultural residues, agro-industrial residues, wood and paper waste, and food residues. These materials are mainly generated in the form of post- or pre-harvest agricultural waste as well as food industry waste, of which most of them are disposed of to the ecosystem or burned. Food products such as fruits and vegetables contain different components including sugars, fibers, proteins, pigments, antioxidants, as well as organic and mineral compounds (Chemat et al. 2017). For instance, soluble and insoluble carbohydrates derived from agricultural or food residues represent potential biomass for producing biofuels such as biogas and ethanol (Putnik et al. 2017).

Different conventional extraction methods can be applied to extract carbohydrates from agricultural and food waste. Generally, these methods offer high production efficiency but require long processing time and high energy consumption (Chemat et al. 2017). Therefore, the use of nonconventional technologies has been more and more applied because of its gentler impact on the ecosystem. Moreover, these technologies significantly reduce the processing time and improve the quality and productivity of carbohydrate extraction, compared to the nonconventional ones. The applications of nonconventional methods for biomass fractionation, including the ultrasound (García et al. 2011; Germec et al. 2016; Subhedar et al. 2018), microwave (Diaz et al. 2015; Mai et al. 2014; Yuan and Macquarrie 2015), pulsed electric fields (PEFs) (Almohammed et al. 2016; Pothula et al. 2014; Redondo et al. 2018; Vorobiev and Lebovka 2016), enzyme-assisted (de Camargo et al. 2016; Huynh et al. 2014; Khan et al. 2010), and supercritical fluids (SCFs)

Fermentation Processes: Emerging and Conventional Technologies, First Edition.
Edited by Mohamed Koubaa, Francisco J. Barba, and Shahin Roohinejad.

(Cardenas-Toro et al. 2014; Prado et al. 2014, 2016), have been studied. Many scientific reports, book chapters, and review papers have reviewed the nonconventional extraction techniques of food and agricultural ingredients (Barba et al. 2017; Koubaa et al. 2016, 2018); however, information about the use of these technologies for the recovery of carbohydrates toward their use as fermentation substrates remains insufficiently reviewed. In this chapter, a comprehensive review of emerging technologies for biomass fractionation, to use in fermentation technology, including the application of ultrasonic, microwave, PEF, enzyme-assisted and SCF technologies, has been presented.

5.2 Ultrasound Application for Biomass Fractionation

Ultrasounds are sound waves having a frequency in the range of 18–100 kHz. Similar to other waves, it can come up with compression and expansion when it passes through a medium. This system produces vapor bubbles that collapse in the medium according to the cavitation phenomenon. The conversion of the kinetic energy of motion to the heating of bubble contents creates a large amount of energy. The application of ultrasound has been developed based on generating temperature, pressure, and heating–cooling rate of bubbles, reaching about 5000 K, 1000 atm, and $<10^{10}\,K{\cdot}s^{-1}$, respectively (Suslick and Doktycz 1990), which facilitates the extraction of organic and inorganic compounds from plant matrices (Herrera and De Castro 2005).

Generally, ultrasonic waves increase the mass transfer and solvent access to the plant cell materials. Two types of physical phenomena were suggested to explain the extraction mechanism using ultrasounds, namely (i) the diffusion across the cell wall and (ii) the rinsing of the cell contents of the breaking walls (Mason et al. 1996). Typically, there are two types of ultrasonic devices: ultrasonic cleaning bath and ultrasonic probe (horn system). The first one is usually used for solid dispersion into the solvent, which reduces the particle size of solids and increases their solubility, and the second, more powerful, consists of applying ultrasound directly to the reaction container and is usually applied for a small number of samples. To have efficient and effective extraction, the parameters of moisture content, milling degree, particle size, and solvent type should be taken into account. Moreover, pressure, temperature, time, and frequency of sonication are elements that control the ultrasonic performance and should be optimized. The advantages of ultrasonic treatment are mainly the decrease in the extraction time as well as the energy and solvent consumption. Besides, the application of ultrasounds can lead to a better mixing, accelerating the transfer of energy, reducing the thermal gradients, allowing selective fractionation, increasing productivity, and reducing the processing steps (Chemat et al. 2008).

In detail, the application of ultrasounds to lignocellulosic biomass, having a rigid structure, results in destroying their waxy layers and helps to remove them from the structure (Yunus et al. 2010). Also, the ultrasound-assisted application reduces the particle size of lignocellulosic biomass, which leads to improvements in the extraction efficiency of chemicals and, therefore, the saccharification of cellulose material (Zhang et al. 2013). Moreover, ultrasound pretreatment improves the extraction and fractionation of components such as phenolics, sugars, and other valuable products from raw materials (Ashokkumar 2015).

It seems that ultrasound technology can be a promising technique for the fractionation of plant biomass. Many studies have been carried out on biomass fractionation using ultrasonic technology and are summarized in Table 5.1. García et al. studied the fractionation of lignocellulosic material from olive tree pruning residues using ultrasonic-assisted technology. The treatment conditions were as follows: acetic acid 60% v/v, sodium hydroxide 7.5% w/w at low temperature, and different sonication parameters (Table 5.1). The results demonstrated that ultrasound technology improved the yield and selectivity of lignin as a target compound, and the use of infrared attenuated total reflection spectroscopy (IR-ATR), a non-destructive method frequently used in mineral identification, proved that this method did not change the physicochemical characters of lignin (García et al. 2011). Fermentable sugars (FSs) are other products derived from plant materials obtained after fractionation by ultrasound-assisted technology. Germec et al. (2016) investigated the ultrasound-assisted dilute acid hydrolysis (UADA) of tea processing waste (TPW) to produce *FS*. The authors optimized the *UADA* conditions using response surface methodology (RSM) under the conditions of ultrasound mentioned in Table 5.1. The results showed that although using *UADA* reduced the yield of *FSs* in comparison to the conventional method, this procedure requires lower temperatures and less liquid consumption, which makes it more economic. It was concluded that ultrasound-assisted technology is a promising method to hydrolyze biomass and produce biofuels, organic acids, and polysaccharides (Germec et al. 2016). Another study investigated the obtaining of fermentable sugars from cassava waste by using ultrasound-assisted technology. The effect of solid-to-liquid ratio, α-amylase, and amyloglucosidase concentration on *FSs* release was evaluated with and without ultrasound irradiation. The results illustrated that FSs release increased by about 40% when using sonication. Besides, specific total reducing sugars (TRS), expressed as g of TRS per g of cassava waste, varied from 0.18 to 0.27 for conventional hydrolysis and from 0.26 to 0.36 $g g^{-1}$ for ultrasound-assisted method. The authors concluded that ultrasound can have a positive effect on sugar yield production (Leaes et al. 2013). Lunelli et al. (2014) investigated the use of ultrasound-assisted technology to produce FSs from sugarcane bagasse (SCB). In this study, the effects of temperature, enzyme concentration, and moisture content were assessed in the presence and absence of sonication. The results showed that the yield of hydrolysis increased using ultrasound technology. Also, the amount of FSs increased by about two times in the presence of ultrasound compared to that without it. In another study, the use of ultrasound-assisted ammonia pretreatment (UAAP) for the production of FS from *SCB* was investigated. Cellulose recovery and delignification were evaluated based on the particle size of *SCB*, the concentration of ammonia, sonication time, temperature, and the liquid-to-solid ratio. At optimum conditions, the upper limit of cellulose recovery and delignification were 95.78 and 58.14%, respectively. Also, the use of this pretreatment resulted in the production of 16.58 $g·l^{-1}$ glucose, 8.21 $g·l^{-1}$ xylose, 2.78 $g·l^{-1}$ arabinose, 0.81 $g·l^{-1}$ furfural, and 1.79 $g·l^{-1}$ acetic acid. Moreover, the authors demonstrated that this method generated less inhibitors compared to the conventional method of FSs production (Ramadoss and Muthukumar 2014). In another interesting study, the use of ultrasound combined with enzymatic hydrolysis was investigated to produce FS from newspaper waste (Subhedar et al. 2015). They investigated the effect of different conditions such as substrate loading, enzyme loading, temperature, ultrasonic power,

Table 5.1 Example of applications and experimental conditions of ultrasound for biomass fractionation.

Biomass	Matrix	Treatment condition	Result	References
Lignocellulosic materials (lignin)	Olive tree pruning Residues	Ultrasonic bath: 30, 60, and 120 min, 420 W, 50/60 Hz	Improve the effectiveness of classic treatments, increase yield and selectivity, does not affect lignin characters	García et al. (2011)
Fermentable sugars	Tea processing Waste	Ultrasonic bath: 2 h, 900 W, 40 kHz	Ultrasound is a promising technology to hydrolysis of TPW compared with conventional hydrolysis method.	Germec et al. (2016)
Glucose	Corn slurry	20 and 40 s, 2.2 kW maximum power (at three different levels: low, medium, and high), 20 kHz	20-folds reduction in corn particle size, improved glucose release by about 30%, decrease of the fractionation time by about 60%, improved ethanol yield and improved profitability.	Khanal et al. (2007)
Fermentable sugars	Cassava waste	Ultrasonic bath: 8 h, 132 W, 40 kHz	Fermentable sugar release increases from 32.8 to 116.1 $g \cdot l^{-1}$.	Leaes et al. (2013)
Fermentable sugars	Sugarcane bagasse	Ultrasonic bath: 0, 60, 120, 180, and 240 min, 132 W, 40 kHz	Fermentable sugar release increases about twofold.	Lunelli et al. (2014)
Fermentable sugars	Sugarcane bagasse	Ultrasound probe type: 15, 30, and 45 min, 400 W, 24 kHz	Improved cellulose recovery and delignification, maximum cellulose recovery, useful for the production of fermentable sugars.	Ramadoss and Muthukumar (2014)
Lignocellulosic biomass	Areca nut husk (*Areca catechu*), bon bogori (*Ziziphus rugosa*), and moj (*Albizia lucida*)	Ultrasound probe type: 60–180 min, 100 W, 30 kHz	Higher yield and selectivity, reduced cellulose degradation, higher hemicellulose and lignin, no modification on lignin.	Sasmal et al. (2012)
Fermentable sugars	Waste newspaper	Ultrasound probe type: 6.5 h, 60 W, 20 kHz	Better yield in reducing sugars about 2.4 times, reduce hydrolysis time, and loaded enzymes.	Subhedar et al. (2015)

Table 5.1 (Continued)

Biomass	Matrix	Treatment condition	Result	References
Lignocellulosic biomass (fermentable sugars)	Groundnut (*Arachis hypogaea*) shells, coconut (*Cocos nucifera*) coir, and pistachio (*Pistacia vera*) shells	Probe-type ultrasonic: 7 h, 80 W, 20 kHz	1.7–2.0-folds enhancement in the delignification extent, high reducing sugar yields, decrease in the required time (almost 10 times) compared to the conventional methods.	Subhedar et al. (2018)
Fermentable sugars	Sugarcane bagasse	Probe-type ultrasonic: 5–50 min, 400 W, 24 kHz	Better reducing sugar yield, reduced time and temperature of operation with improved efficiency.	Velmurugan and Muthukumar (2012)
Fermentable sugars	Palm leaves	Ultrasonic bath: 0, 15, 30, 60, 120, 180, 240, and 300 min, 132 W, 40 kHz	83% of total fermentable sugars were released in the period of constant reaction rate. The amount of fermentable sugars released ranged from 0.15 to 0.59 g sugar g^{-1} dry material.	Werle et al. (2013)

and a duty cycle on the fractionation of newspaper waste. The optimum operating conditions were as follows: enzyme loading of 0.8% (w/v), sonication power of 60 w, duty cycle of 70%, and hydrolysis time of 6.5 hours, which resulted in 27.6 $g \cdot l^{-1}$ of reducing sugars. This amount was 2.4-folds higher than that obtained by hydrolysis without using ultrasound. The authors reported that the combination of enzymatic hydrolysis with ultrasound technology decreased the diffusion-limiting barrier to enzyme/substrate binding and has a synergistic effect on the yield of FS obtained from newspaper waste. Subhedar et al. (2018) studied the effect of enzymatic hydrolysis combined with ultrasounds on the yield of reducing sugars from biomass such as groundnut shells, coconut coir, and pistachio shell. The results showed that using ultrasound increased the delignification by ≈80–100%, and the reducing sugars increased from 10.2, 12.1, and 8.1 $g \cdot l^{-1}$ to 21.3, 23.9, and 18.4 $g \cdot l^{-1}$ for groundnut shells, coconut coir, and pistachio shells, respectively. To produce bioethanol from lignocellulosic biomass, Sasmal et al. (2012) applied ultrasound-assisted technology. They used Areca nut husk (*Areca catechu*), bon bogori (*Ziziphus rugosa*), and moj (*Albizia lucida*) as the source of lignocellulosic biomass. The results illustrated the removal of lignin by 68, 65, and 64% for Areca nut husk, bon bogori, and moj, respectively. Also, good recovery of total solid and FS was shown, which resulted in 0.32–0.43 yield of ethanol production.

To summarize, the use of ultrasound-assisted technology for the fractionation of biomass has many advantages including reducing the processing time, the use of lower temperatures, reducing the added substances amount (e.g. enzymes, chemicals, etc.), the increase of productivity, and the easy implementation at industrial scale.

5.3 Microwave Application for Biomass Fractionation

The underlying principle of microwave-assisted extraction (MAE) is the electromagnetic irradiation of the polar solvent and materials resulting in the superheating of the plant cells and their disruption (Deng et al. 2015). Microwave spectrum is ranging between far infrared light and radio frequencies, from 300 MHz to 300 GHz. The most used frequencies in scientific and industrial applications are 915 ± 25, 2450 ± 13, 5800 ± 75, and 22.125 ± 125 MHz (Lebovka et al. 2016). Dipole rotation and ionic conduction are the corresponding processes resulting in heating up of the molecules by microwaves. The resistance of the medium to ion flow causes friction, and the heating of the sample occurs with trace amount of moisture (Chan et al. 2011). This procedure creates huge pressure on the cell wall until rupture and the release of active compounds from the intracellular compartments. The mechanism of extraction by microwave has been described in three steps, namely (i) separation of solutes from the sample matrix under escalated temperature and pressure, (ii) the solvent diffusion across the matrix of the sample, and (iii) the release of solutes from the matrix of the sample to the solvent (Alupului et al. 2012).

The microwave-assisted technique has some advantages, including faster heating to extract biomass from plant materials, reduced thermal gradients, reduced equipment size, and improved extract productivity. Also, it is a selective procedure to extract organic and inorganic compounds (Alupului et al. 2012). Moreover, MAE reduces energy consumption and allows uniform processing and the immediate ability to start and stop the process (Grant and Halstead 1998). Many studies have revealed that changes in cellulose structure, hemicellulose, and lignin degradation by microwave irradiation in lignocellulosic substances occur (Xiong et al. 2000). It was reported that microwave-assisted technology enhances the enzymatic hydrolysis of lignocellulosic materials (Azuma et al. 1984). It could be also used for the acid-catalyzed auto-hydrolysis of lignocellulosic substances, as usually microwave-assisted technology is applied at increased temperature (>160°C), which leads to increased acidity during microwave irradiation (Intanakul et al. 2003). Microwave irradiation can be used easily in combination with other methods like enzymatic, acid, and alkali methods to fractionate biomass and especially lignocellulosic substances. It seems that microwave-assisted technology can be a promising technique for the fractionation of plant biomass (Table 5.2). Many studies have been carried out on biomass fractionation by this technology and are summarized in Table 5.2. Lü and Zhou (2011) optimized microwave-assisted $FeCl_3$ pretreatment conditions to produce reducing sugars from rice straw. According to them, the optimum conditions were $1.14\,mol{\cdot}l^{-1}$ of $FeCl_3$ concentration, 160°C of temperature, 19 minutes of irradiation time, and $109\,g{\cdot}l^{-1}$ of substrate concentration. The productivity of reducing sugars was 2.9 times ($6.62\,g{\cdot}l^{-1}$) more than that of the untreated samples. The silicified waxy surface of rice straw was damaged by applying the pretreatment, and also the ether bonds between lignin and carbohydrate were broken up, which resulted in removing lignin. In another study, the effect of microwave-assisted pretreatment was evaluated to produce FS from SCB. The authors illustrated that microwave technology can be a good alternative for conventional pretreatments (acid and alkali), which decreases the time of operation at high temperatures. Three different combined methods, such as microwave-acid, microwave-alkali, and microwave-acid-alkali, followed by enzymatic hydrolysis were assessed (microwave conditions are shown in Table 5.2). The yield of reducing sugars for

Table 5.2 Examples of applications and experimental conditions of microwave for biomass fractionation.

Biomass	Matrix	Treatment condition	Result	References
Reducing sugars	Rice straw	800 W, 140–180°C, 10–20 min	Increase of the production of reducing sugars by about 2.9 times.	Lü and Zhou (2011)
Fermentable sugars	Sugarcane bagasse	Frequency of 2450 MHz, power of 100–850 W, 1–30 min	Reducing sugar yields with microwave-acid, microwave-alkali, and microwave-alkali-acid were $0.665\,g{\cdot}g^{-1}$, $0.246\,g{\cdot}g^{-1}$, and $0.83\,g{\cdot}g^{-1}$ dry biomass, respectively.	Binod et al. (2012)
Reducing sugars	Rice straw	Frequency of 2.45 GHz, power of 100–1000 W, 140°C (working pressure: 0.2–0.6 MPa), 6–72 h	The maximum reducing sugar yield of $69.3\,g{\cdot}100\,g^{-1}$. Total volatile solid (TVS) and hydrogen yield of $155\,mlg^{-1}$ TVS are obtained in dark fermentation. On combination of dark- and photofermentation, the maximum hydrogen yield of $463\,ml{\cdot}g^{-1}$ TVS is obtained, which is 43.2% of the theoretical hydrogen yield.	Cheng et al. (2011)
Fermentable sugars	Lignocellulosic biomass (corn straw and rice husk)	2450 MHz, 1300 W, 180°C, 2 min	Improved enzyme hydrolysis of corn straw and rice husk, more efficient hydrolysis of corn straw, more significant modification of biomass structure in accordance with FTIR (Fourier Transform Infrared Spectroscopy), TG (Thermogravimetric), and SEM (Scanning Electron Microscopy) analysis.	Diaz et al. (2015)
Sugars	Palm kernel cake	120–200°C, 0–20 min, 1–3 MPa	Results in maximum yields, 77.11 and 92.11% of total monosaccharide and mannose, respectively. Effective method for monosaccharides production.	Fan et al. (2014)

(Continued)

Table 5.2 (Continued)

Biomass	Matrix	Treatment condition	Result	References
Fermentable sugars	Agricultural bioresources (barley husk, oat husk, wheat bran, and rye bran)	Frequency of 2450 MHz, power of 460, 600, and 700 W, 2–7 min	Maximum fermentable sugar was 49.65 $g \cdot l^{-1}$ (0.83 $g \cdot g^{-1}$) from wheat bran. The results showed that a microwave system was a promising technology for fermentable sugar from agricultural bioresource materials.	Germec et al. (2017)
Fermentable sugars	Oil palm empty fruit bunch fiber	180 W, time of 3, 6, 9, 12, 15, 18, and 21 min	Reducing sugars production had shown a reaction rate increase of 2.3-folds and with less enzyme requirement.	Nomanbhay et al. (2013)
Fucoidan, alginate sodium, sugars, and biochar	*Ascophyllum nodosum*	600 W, 24 min, 80–94°C,	Sugars obtained contained large proportion of fermentable sugars such as glucose, xylose, etc.	Yuan and Macquarrie (2015)

microwave-alkali and microwave-acid-alkali pretreatments were 0.665 and $0.83\,g{\cdot}g^{-1}$ dry biomass, respectively. Moreover, about 90% of lignin bagasse was eliminated by microwave-alkali pretreatment (450W, 5 minutes). Finally, it was concluded that microwave-acid-alkali method can improve the productivity of FS from SCB (Binod et al. 2012). Cheng et al. (2011) studied microwave-assisted alkali pretreatment to enhance enzymatic hydrolysis and hydrogen production from rice straw in dark and photofermentation. Using microwave heating at 140°C for 15 minutes in 0.5% sodium hydroxide solution followed by enzymatic hydrolysis for 96 hours resulted in the maximum reducing sugars productivity ($69.3\,g.100\,g^{-1}$ total volatile solid). The productivity of sugars from palm kernel cake affected by microwave-assisted dilute acid hydrolysis was studied. In optimum conditions including 148°C, 0.75N H_2SO_4, 10 minutes 31 seconds, and substrate-to-solvent ratio of 1:49.69 (w/v), the maximum mannose was achieved. Also, at 170°C, 0.181N H_2SO_4, 6 minutes 6 seconds, and a substrate-to-solvent ratio of 1:40 (w/v), a 77.11% productivity of FS was obtained. The authors concluded that microwave-assisted dilute acid hydrolysis is a promising method to produce sugars and the fractionation of palm kernel cake as biomass (Fan et al. 2014). Germec et al. (2017) studied the FS production from some lignocellulose sources by microwave-assisted dilute acid pretreatment. The optimum conditions of the pretreatment were determined (power level, time, solid-to-liquid ratio, and concentration of dilute acid) for barley husk (700W, 6.92 minutes, 1:15.36 w/v, and 3.67%, respectively), oat husk (600W, 6.96 minutes, 1:17.22 w/v, and 3.47%, respectively), wheat bran (600W, 6.62 minutes, 1:16.69 w/v, and 1.85%, respectively), and rye bran (460W, 6.15 minutes, 1:14.14 w/v, and 2.72%, respectively). The results demonstrated that the yields of FS of 0.68, 0.67, 0.83, and $0.62\,g{\cdot}g^{-1}$ for barley husk, oat husk, wheat bran, and rye bran, respectively, were achieved. The authors concluded that MAE is a promising technique for the fractionation and hydrolysis of these lignocellulosic materials. In another study, the combination of microwave with alkaline pretreatment and enzymatic saccharification to obtain FS from oil palm empty fruit bunch fiber was assessed (Nomanbhay et al. 2013). Treating the samples with microwave coupled with alkali pretreatment led to the removal of lignin and hemicellulose and improved accessibility to hydrolytic enzymes, which resulted in better saccharification. It was reported that the optimized conditions of microwave-alkali pretreatment were 3% w/v of NaOH, 180W of power, and 12 minutes of processing time, which resulted in 74 and 24.5% of lignin and hemicellulose removal, respectively. The later enzymatic saccharification resulted in 411 mg of reducing sugars per gram of sample, which showed improved sugar productivity by about 5.8 times compared to the untreated samples. The study showed thereby that microwave-assisted technology could well control the fractionation process of the lignocellulosic materials.

In general, microwave irradiation can cause the removal of the external fibers and reduce the biomass particle size, which results in increasing the specific surface of biomass. Also, the surface of biomass becomes disordered, and the disruption happens by applying microwave irradiation. These phenomena lead to enhance the accessibility of enzymes or chemicals to biomass to get better fractionation. Moreover, microwave irradiation improves the saponification of intermolecular ester bonds cross-linking the hemicelluloses, which results in decreasing the O—H bond intensity and thereby increasing the accessibility of enzymes and chemicals to biomass. Finally, microwave technology accelerates the rate of reactions and therefore decreases the reaction time and the energy consumption, which makes it a promising emerging technology for fractionation of biomass.

5.4 PEF Application for Biomass Fractionation

One of the other promising techniques for the recovery of valuable compounds from by-products and food waste is PEF. The principle consists of placing materials between two electrodes and applying a PEF treatment, with pulses having an intensity ranging from 100–300 V·cm^{-1} to 20–80 kV·cm^{-1} (Koubaa et al. 2015). Applying PEF treatment to plant cells can damage their cell membrane, which causes the creation of reversible or irreversible pores in a phenomenon called electroporation. When an intact cell is placed in an electric field, electric potential transits through the membrane of the cell and segregates molecules based on the dipole nature of membrane molecules and according to their charge in the cell membrane. At around a critical value of 1 V of transmembrane potential, repulsion takes place among the charge-carrying molecules, which creates pores in the weak areas of the membrane and results in extreme spread of permeability (Bryant and Wolfe 1987). Generally, in PEF treatment of plant biomass or materials, a simple circuit with pulses, which decay exponentially, is applied. The PEF chamber is designed to allow working in batch or continuous modes (Puértolas et al. 2010). The efficiency of PEF treatment depends on the process parameters including the field strength, the specific energy input, the pulse number, the temperature of treatment, and the characteristics of the treated material (Heinz et al. 2003). Mild PEF treatment (500 and 1000 V·cm^{-1} for 10^{-4}–10^{-2} s) could damage the plant cell membrane along with a slight increase of temperature, which reduces the degradation of the heat-sensitive molecules (Lebovka et al. 2002). Some examples of biomass fractionation by the application of PEF technology are summarized in Table 5.3.

López et al. (2009) studied the extraction of sucrose from sugar beet by PEF at different electric field strengths, frequency, time, and temperatures. Increasing the strength and time of pulses along with temperature resulted in increasing the sucrose extraction yield. Moreover, it was concluded that the pulse intensity has a higher effect on sucrose productivity than the temperature. Using PEF at 7 kV·cm^{-1} (3.9 kJ·kg^{-1}) with 20 pulses increased the sucrose yield by 7- and 1.6-folds, compared to the conventional extraction method at 20 and 40°C, respectively. Finally, the authors provided the optimum conditions of 80% sucrose yield after 60 minutes at 40°C (50% reduction in thermal energy consumption), which were 20 pulses of 7 kV·cm^{-1} intensity (López et al. 2009). Another study investigated the use of PEF as a novel technology for the generation of FS from sugar beet tails, a sustainable feedstock, to produce bioethanol. The effects of PEF strength and time on solute productivity were investigated. The optimum conditions of PEF were a field strength of 450 V·cm^{-1} and a time of 10 ms. Under these conditions, the solute productivity increased from 16.8 to 79.85%. Also, the concentrations of solute and sucrose content increased. The results showed that higher FS, obtained by the PEF method, resulted in higher ethanol content (6.1 vs. 2.95%) compared to the conventional method (Almohammed et al. 2016). In a case study, PEF technology was applied for bioactive extraction from thinned peach by-products. It was reported that unripe fruits contain a sufficient amount of antioxidant compounds. Here, the authors used an RSM to determine the optimum conditions of PEF-assisted extraction for the recovery of bioactive compounds. The optimum conditions were provided to allow obtaining increased amounts of chlorogenic acid, coumaric acid, and neochlorogenic acid. It was, therefore,

Table 5.3 Examples of applications and experimental conditions of pulsed electric fields (PEF) for biomass fractionation.

Biomass	Matrix	Treatment condition	Result	References
Sucrose	Sugar beet (*Beta vulgaris*)	Distance between the electrodes of 1 cm and an area of 4.9 cm^2; pulse widths of 2–5 ms; frequencies from 1 to 10 Hz; and field strengths from 1 to 7 $kV \cdot cm^{-1}$; corresponding to specific energies from 0.006 to 0.19 $kJ \cdot kg^{-1}$ per pulse	Decrease of the extraction temperature, time, and energy requirements compared to traditional thermal extraction. Applied electric field strength, energy input, temperature, and time of the extraction affected the yield of sucrose.	López et al. (2009)
Fermentable sugars (bioethanol)	Sugar beet tails	Pulse generator of 5 kV and 1 kA, surface of 14–24 cm^2, electric field from 0 to 600 $V \cdot cm^{-1}$, pulse repetition time 100 μs and 100 ms; the number of pulses ranged from 0 to 150 pulses	The yield was increased by about 3.75 times and twice more concentrated in sucrose as compared to the untreated one. More ethanol concentration was obtained from PEF treatment.	Almohammed et al. (2016)
Sugars	Sugar beet tissue	Various intensities (E = 270, 400, 540, 670, and 800 $V \cdot cm^{-1}$), various numbers of rectangular monopolar pulses (N = 50, 250, 500, and 1000), pulse duration t_i = 100 μs and frequency f = 1000 Hz	Enhanced sugar extraction, increased diffusion coefficient at moderate temperature of the process.	El-Belghiti et al. (2005)
Bioactive compounds	Thinned peach by-products	PEF treatments ranging from 10 to 50 pulses of 3 μs (30–150 μs), electric field strengths ranging from 0 to 5 $kV \cdot cm^{-1}$, specific energy from 0.61 to 9.98 $kJ \cdot kg^{-1}$, pulse frequency of 1 Hz	Electric field strength and the temperature of the process affected significantly the yield of bioactives. PEF could be used as an alternative method for bioactive extraction compared to the traditional methods. Reduced need in organic solvents, and shorter processing time.	Redondo et al. (2018)

concluded that PEF-assisted technology could be a promising alternative method to extract bioactive compounds in comparison to conventional methods, such as maceration (Redondo et al. 2018). Kumar et al. (2011) applied PEF treatment on switchgrass and wood chip species to produce biofuel. They designed a PEF system in which they were able to treat switchgrass and wood chip samples from biomass to fuel conversion. Different conditions of PEF (1000–5000 pulses, 1 and 10 $kV \cdot cm^{-1}$ intensity, a constant width pulse of 100 μs, and 3 Hz frequency) were investigated. The results showed that the use of a PEF intensity of more than 8 kV·cm^{-1} increased the porosity of the samples. It was then concluded that PEF could be applied to enhance cellulose fractionation to sugar for biofuels production.

In summary, the PEF-assisted technique for the fractionation of biomass can be applied at ambient conditions, and it also requires a low amount of energy as the pulse duration is short (about 100 μs).

5.5 Enzyme-Assisted Fractionation of Biomass

Enzyme-based extraction of bioactive compounds from plants is a safe and proper alternative to conventional solvent-based extraction methods (Gardossi et al. 2010). Different extraction strategies have been used to maximize the efficiency of the extraction of bioactive macromolecules from different parts of plants (Acosta-Estrada et al. 2014; Ajila et al. 2011). Enzymes are perfect catalysts to assist the extraction, modification, or synthesis of complex bioactive compounds of natural origin (Gardossi et al. 2010). Recently, several types of enzymes, particularly carbohydrate-hydrolyzing enzymes, such as pectinase, cellulase, hemicellulase, and glucanase, have been used for the extraction of polyphenols (Hansen et al. 2005; Landbo and Meyer 2001; Reclaru and Meyer 1998). These enzymes may decompose the plant cell wall matrix and facilitate polyphenol extraction (Le Bourvellec and Renard 2012; Pinelo et al. 2006; Stalikas 2007).

The cell wall complex polyphenols can be released using these enzymes and play a vital role in disintegrating the plant cell wall matrix and thus increasing the extraction of polyphenols (Le Bourvellec and Renard 2012). Enzyme-assisted extraction is based on the intrinsic ability of enzymes to catalyze reactions under mild processing conditions in aqueous solutions (Gardossi et al. 2010). Enzyme-assisted extraction methods are gaining more and more consideration because of the need for eco-friendly extraction technologies (Fu et al. 2005). A particularly useful application of enzymes includes increasing the solvent accessibility, and either decreasing the amount of solvent needed for the extraction or increasing the yield of extractable compounds.

Enzymes such as pectinases, cellulases, and hemicellulases are mostly used in juice processing and beer clarification to decompose the cell walls and improve the juice's extractability. The destruction of the cell wall matrix also releases molecules such as phenolic compounds into the juice, thereby improving the product quality. Enzyme-assisted extraction technique has been shown to reach high extraction yields of compounds such as polysaccharides, oils, natural pigments, flavors, and medicinal compounds (Barzana et al. 2002; Passos et al. 2009; Sowbhagya and Chitra 2010; Wu et al. 2007; Yang et al. 2010). Recent research on enzyme-assisted extraction showed faster extraction kinetic, higher output, reduced solvent usage, and lower energy consumption when compared to nonenzymatic methods.

Some examples of biomass fractionation for bioactive compounds recovery by enzyme-assisted extraction are shown in Table 5.4. The effective application of enzymes for the extraction of other products, including polyphenols (Yang et al. 2010), carotenoids from marigold flower (Barzana et al. 2002; Dehghan-Shoar et al. 2011; Ramadan et al. 2009; Yang et al. 2010), vanillin from vanilla green pods (Ruiz-Terán et al. 2001), polysaccharides

Table 5.4 Some examples of bioactive compounds fractionated by enzyme-assisted method.

Product type	Product	Source	Enzyme used	Maximum yield (%)
Oils and carotenoids	Oil	Grape seed	Cellulase, protease, xylase, and pectinase	17.5
	Carotenoids	Marigold flower	Viscozyme, Pectinex, Neutrase, Corolase and HT-proteolytic	97
	Volatile oil	Mandarin peel	Xylan-degrading enzymes	15
	Carotene	Carrot pomace	Pectinex Ultra SP-L	0.0064
	Lycopene	Tomato	Pancreatin	2.5-folds
		Tomato	Cellulase and pectinase	206
	Capsaicin	Chili	Cellulase, hemicellulase and pectinase	n.d.
	Colorant	Pitaya	Pectinolytic, hemicellulolytic, and cellulolytic enzymes	83.5
	Anthocyanin	Grape skin	Pectinex BE3-L	n.d.
Glycosides	Sugar	Grapefruit peel waste	Cellulase and pectinase	0.6377
	Oligosaccharide	Rice bran	Cellulase	39.9
	Inulin	Jerusalem artichoke	Inulinase	n.d.
	Starch	Cassava	Pectinase enzyme	45.6
	Pectin	Pumpkin	Xylase, cellulose, b-glucosidase, endopolygalacturonase, and pectinesterase	14.0

(*Continued*)

Table 5.4 (Continued)

Product type	Product	Source	Enzyme used	Maximum yield (%)
Others	Vanillin	Vanilla green pods	β-glucosidase and pectinase	14–21
	Flavonoid (naringin)	Kinnow peel	Recombinant rhamnosidase	n.d.
	Phenolics	Citrus peel	Celluzyme MX	65.5
	Proteins	Lentils and white beans	Glucoamylases	50.3
	Polyphenols	Grape pomace	Pectinolytic	98.1
	Catechins	Tea beverage	Pepsin	80
	Lignans	Flax	Cellulase and glycosidase	$40.75\,mg{\cdot}g^{-1}$
	Soluble fiber	Carrot pomace	Cellulase-rich crude preparation	77.3

Source: Puri et al. (2012).

from sterculia (Yang et al. 2010), and oil from grape seeds (Passos et al. 2009), shows that enzymes can also be useful for the extraction of bioactives from other plant sources. Enzyme-based extractions are the subject of ongoing research and have the potential to be commercially interesting. Enzymes can be obtained from bacteria, fungi, animal organs, or vegetable/fruit extracts. To use enzymes most efficiently for extraction applications, it is important to know their catalytic feature and mode of action, desirable operational conditions, and the type of enzyme that is suitable for the plant matter selected.

Different enzymes such as pectinases, hemicellulases, and cellulases hydrolyze the cell wall compounds, thereby increasing the cell wall penetrance and resulting in higher extraction products of bioactives (Stafford 2002). As an example, cellulose, pectin, and hemicellulose in grapefruit peel waste can be hydrolyzed by cellulase and pectinase enzymes into monomer sugars, which can then be used by microorganisms to produce ethanol and other fermentation products (Wilkins et al. 2007). For instance, cellulase was used at the liquefaction phase for the saccharification process during the treatment of SCB to produce bioethanol. The cellulases improved the saccharification process ($81\,g{\cdot}l^{-1}$ of total sugars), which remarkably increased ethanol production (Geddes et al. 2011). Some other researchers have shown that enzymes can also be used to disrupt the pectin–cellulose complex in citrus peels and enhance the flavonoid production (Puri et al. 2011).

Other investigations showed an enhancement in the phenol composition (90–39.72%) and sugars ($12–14\,g{\cdot}l^{-1}$) after enzyme-assisted extraction from grape pomace and citrus peels (Bildstein et al. 2008; Kammerer et al. 2005; Li et al. 2006).

Indeed, enzyme usage improved the extraction of total phenolic content from 32.33 to 61.90% (Stafford 2002) and provided a 59.4% production improvement in comparison with a nonenzymatic oil extraction process (Tobar et al. 2005). The extraction product under optimized conditions was $28.3\,mg{\cdot}g^{-1}$ dry weight, which was 102% more than the amount of extraction in nonenzymatic conditions (Chen et al. 2011).

Traditional extraction methods suffer from time inefficiency and lower extract quality due to unavoidable existence of traces of organic solvents in them and the usual long exposure to high extraction temperatures. Replacing these technologies by emerging ones, such as enzyme-assisted extraction, leads to reduced extraction time, minimal use of solvents, and enhanced product yields and quality (Meyer 2010; Sowbhagya and Chitra 2010). Complete knowledge of the cell wall composition of the raw materials helps in the selection of the appropriate enzyme for the pretreatment. Reduced solvent use during extraction is especially important for both regulatory and environmental reasons, providing a "greener" option than the traditional nonenzymatic extraction (Puri et al. 2012).

Unlike other nonthermal techniques, such as supercritical carbon dioxide (SC-CO2), high electric field pulses (HELP), high pressure (HP), and compressed carbon dioxide (cCO2), enzyme-assisted extraction can easily be done at the laboratory scale (Figure 5.1). Enzymes can be selected based on specific properties such as temperature and concentration. Enzymes normally operate at an optimal temperature; they can still be used over a range of temperatures, providing flexibility for both cost and yield quality. Enzyme-assisted extraction procedure from different plant sources is summarized in Figure 5.1. It should be noted that the parameters impacting enzyme-assisted extraction include pH, time, temperature, and enzyme concentration (Concha et al. 2004).

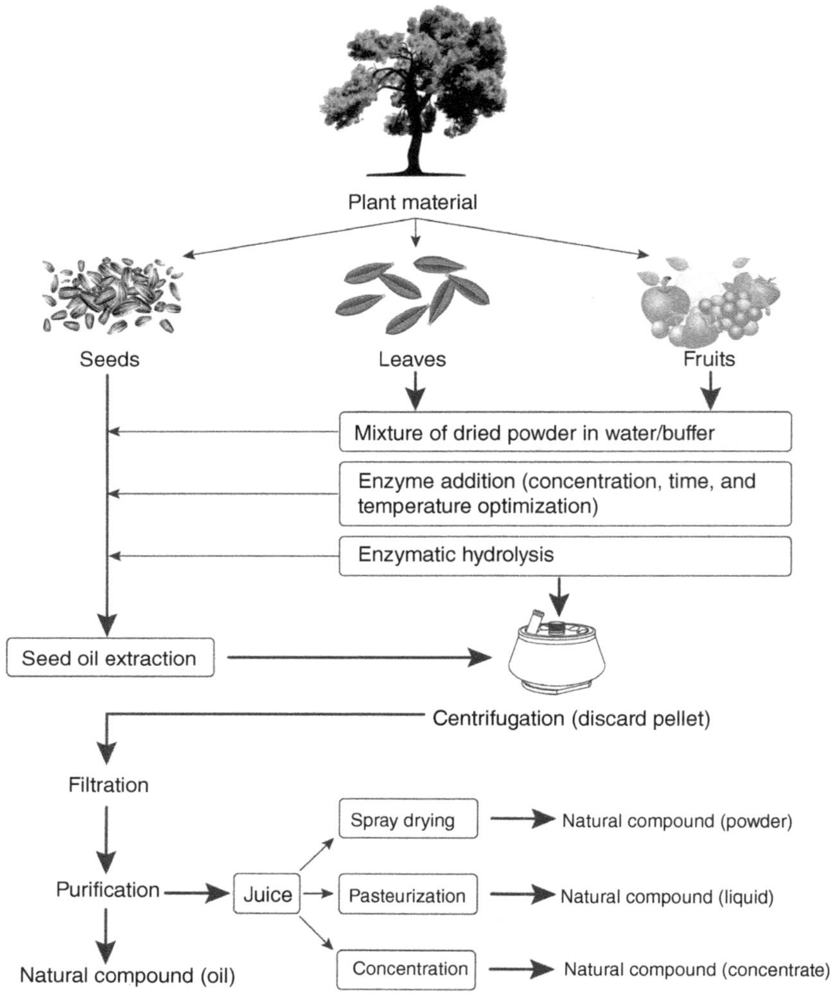

Figure 5.1 Enzyme-assisted extraction of compounds from plant materials. *Source:* Puri et al. (2012).

5.6 SCF Fractionation of Biomass

The supercritical fluid extraction (SFE) technique is another efficient extraction method (Kumarasamy et al. 2005). The critical point of a pure matter is defined as the highest pressure and temperature at which the material can exist in a vapor–liquid balance. Indeed, at pressures and temperatures above the critical point, a homogeneous fluid known as SCF is formed, which is heavy as a liquid with a permeation power of gas, and this makes SCF an efficient and selective solvent. SCF can be produced by heating a gas above the critical temperature or by compressing a liquid above the critical pressure (Gurdial et al. 1992). The application of SFE with SC-CO_2 was introduced for the decaffeination of coffee beans by Zosel (Zosel 1978). Extraction by SCFs relates to the selective extraction of desired materials or removal of undesired materials from natural materials (Marcus 2005). The most commonly used SCF is SC-CO_2. Other SCFs including trifluoromethane, butane, pentane, ethane nitrous oxide, supercritical ammonia, and water are also used (Gurdial et al. 1992).

SFE provides several effective privileges over traditional extraction techniques since it uses supercritical solvents, with different physicochemical properties such as density, diffusivity, viscosity, and dielectric constant. Because of their low viscosity and relatively high emission, SCFs have elevated transport features than liquids, can diffuse easily through solid matter, and can, therefore, recover the molecules from the plant material. One of the great characteristics of a SCF is the possibility of changing the density of the fluid by changing its pressure and/or temperature. As the density is related to solvability, by altering the extraction compression, the solvent power of the fluid can be changed (Herrero et al. 2006). Other advantages, compared to other extraction methods, are the use of solvents generally recognized as safe, the higher efficiency of the extraction process in terms of increasing products and lower extraction times, and the possibility of direct coupling with analytical chromatographic techniques such as gas chromatography or supercritical fluid chromatography (Herrero et al. 2006).

Water, methanol, CO_2, ethanol, and also their mixtures are among the SCFs that have so far been used for the extraction (Marcus 2005). It is clear that much milder positions are involved with the use of SC-CO2 than for the other SCFs, but it is necessary to mention that each fluid has its particular advantages (Barbosa et al. 2014; Marcus 2005). CO_2 is the most commonly used solvent for several reasons: firstly, it is harmless to human health and environment; secondly, it has medium critical temperature (31.2°C), which is a key issue for the protection of bioactive compounds in extracts; finally, the extract is preserved from contact with air and light, contributing to oxidation reactions (Barbosa et al. 2014). As CO_2 is a gas at room temperature, it is eliminated from the sample after system decompression, producing a solvent-free extract. At the industrial scale, when CO_2 utilization is high, the operation can be controlled to recycle it. However, because of its low polarity, CO_2 is less efficient in extracting most polar compounds from a natural substance, and modifiers (also called cosolvents) are commonly used (Solana et al. 2014; Zougagh et al. 2004).

Supercritical extraction originally occurs in two steps: the solubilization of the chemical compounds present in the solid matrix and its segregation into the supercritical solvent. During the extraction process, the solvent flows via the packed bed, solubilizing the compounds available in the matrix. Then, the solvent present in the extractor will carry out the solubilized compounds, and with a decrease in pressure and/or an increase in temperature, the extract becomes solvent-free (Brunner 1994).

The interesting versatility of the SFE method renders it a great tool to get substances of therapeutic nature, allowing the isolation and processing of drugs into new medicines for special drug delivery. SCF extracts have also produced promising and positive results in applications in the food, pharmaceutical, and cosmetic industries with the production of flavor enhancers, anti-inflammatory, and pain killer drugs as well as in the search for treatments for drastic diseases such as cancer, stroke, and Alzheimer (Sovilj et al. 2011).

Table 5.5 shows a selection of different bioactive compounds extracted by SFE methodology from distinct sample sources where the main bioactivities are antioxidant (41%), antitumor (18%), and antibacterial activity (10%), followed by antimicrobial, antiviral, anti-inflammatory, and anticholinesterase (total of 5%).

Table 5.5 Selected examples of bioactive compounds extracted from several sources using supercritical fluid extraction.

	SFE conditions									
Sample source	Solvent	Temp (°C)	P (bar)	Flow rate	Time (min)	Cosolvent	Bioactivity	Target/main compounds	SFE results	References
Bidens pilosa	CO_2	40	250	$15\,g{\cdot}min^{-1}$	240	—	Antitumor (MCF-7 breast cancer cells)	Polyacetylenes	ηtotal = 2.00%	Kviecinski et al. (2011)
Apium graveolens	CO_2	40	100	$0.30\,kg{\cdot}h^{-1}$	—	—	Antimicrobial (*Staphylococcus aureus*, *Listeria monocytogenes*, and *Listeria ivanovii* strains)	sedanenolide, sedanolide, 3-*n*-butylphthalide	ηtotal = 2.74%	Mišić et al. (2008)
Ceratonia siliqua	CO_2	40	220	$0.29\,kg{\cdot}h^{-1}$	450	10% (ethanol/water 8:2)	Antitumor (rat N1E-115 neuroblastoma cells, human HeLa cervical cancer cell, MCF-7 breast cancer cells)	Cinnamic acid, chrysoeriol, tricetin-3′,5′-dimethyl ether, naringenin and ferulic acid	ηtotal = 0.33% AA = 0.130 mmol TEAC·g^{-1} dry	Kviecinski et al. (2011)
Citrus sinensis	CO_2	40	200	$17\,g{\cdot}min^{-1}$	300	—	Antimicrobial (*S. aureus*)	Octane, 2,3,7-trimethyl, Decane and N-(2-Cyano-ethyl)-Benzenesulfonamide	ηtotal = 1.2% MIC (*S. aureus*) = 500 μg.ml^{-1}	Benelli et al. (2010)
Bidens pilosa nes	CO_2	40	250	$15\,g{\cdot}min^{-1}$	240	—	Antitumor (MCF-7 breast cancer cells)	Polyacetyle	ηtotal = 2.00%	Kviecinski et al. (2011)
Vaccinium macrocarpon Ait.	CO_2	40	655–621	$125\,g{\cdot}min^{-1}$	—	0–30% (ethanol)	Bacterial agglutination (ExPEC agglutination) and bacterial invasion (ExPEC enterocyte invasion inhibition)	Proanthocyanidin	—	Feliciano et al. (2014)

SFE = supercritical fluid extraction; CO_2 = carbon dioxide; ExPEC = extraintestinal pathogenic *Escherichia coli*; MIC = minimum inhibitory concentration.

5.7 Conclusions

The discussion reported in this chapter has illustrated the superiority of emerging technologies, such as ultrasonic, microwave, PEFs, enzyme-assisted, and SCFs, compared to traditional methods for biomass fractionation from agricultural and food residues. The ultrasound-assisted method showed advantages including lower operation time and temperature, higher productivity, easy combination with conventional and nonconventional methods, and eco-friendly industrial implementation. Moreover, MAE revealed lower energy and time consumption as compared to traditional methods. Also, in contrast to conventional methods for biomass fractionation, PEF could be carried out at an ambient temperature, which resulted in lower energy consumption. Besides, both enzyme-assisted and SCF methods for biomass fractionation not only showed benefits of emerging techniques that are mentioned above but also illustrated that they are more friendly environmental technologies for biomass fractionation as compared to traditional ones. Finally, it can be concluded that the use of emerging technologies can be a promising way for low-energy biomass fractionation.

References

Acosta-Estrada, B.A., Gutiérrez-Uribe, J.A., and Serna-Saldívar, S.O. (2014). Bound phenolics in foods, a review. *Food Chem.* 152: 46–55.

Ajila, C.M., Brar, S.K., Verma, M. et al. (2011). Extraction and analysis of polyphenols: recent trends. *Crit. Rev. Biotechnol.* 31: 227–249.

Almohammed, F., Mhemdi, H., and Vorobiev, E. (2016). Pulsed electric field treatment of sugar beet tails as a sustainable feedstock for bioethanol production. *Appl. Energy* 162: 49–57.

Alupului, A., Calinescu, I., and Lavric, V. (2012). Microwave extraction of active principles from medicinal plants. *UPB Sci. Bull. Ser. B* 74: 1454–2331.

Ashokkumar, M. (2015). Applications of ultrasound in food and bioprocessing. *Ultrason. Sonochem.* 25: 17–23.

Azuma, J., Tanaka, F., and Koshijima, T. (1984). Enhancement of enzymatic susceptibility of lignocellulosic wastes by microwave irradiation. *J. Ferment. Technol.* 62: 377–384.

Barba, F.J., Putnik, P., Bursać Kovačević, D. et al. (2017). Impact of conventional and non-conventional processing on prickly pear (*Opuntia* spp.) and their derived products: from preservation of beverages to valorization of by-products. *Trends Food Sci. Technol.* 67: 260–270.

Barbosa, H.M.A., de Melo, M.M.R., Coimbra, M.A. et al. (2014). Optimization of the supercritical fluid coextraction of oil and diterpenes from spent coffee grounds using experimental design and response surface methodology. *J. Supercrit. Fluids* 85: 165–172.

Barzana, E., Rubio, D., Santamaria, R.I. et al. (2002). Enzyme-mediated solvent extraction of carotenoids from marigold flower (*Tagetes erecta*). *J. Agric. Food Chem.* 50: 4491–4496.

Benelli, P., Riehl, C.A.S., Smânia, A. et al. (2010). Bioactive extracts of orange (*Citrus sinensis* L. Osbeck) pomace obtained by SFE and low pressure techniques: mathematical modeling and extract composition. *J. Supercrit. Fluids* 55: 132–141.

Bildstein, M., Lohmann, M., Hennigs, C. et al. (2008). An enzyme-based extraction process for the purification and enrichment of vegetable proteins to be applied in bakery products. *Eur. Food Res. Technol.* 228: 177–186.

Binod, P., Satyanagalakshmi, K., Sindhu, R. et al. (2012). Short duration microwave assisted pretreatment enhances the enzymatic saccharification and fermentable sugar yield from sugarcane bagasse. *Renew. Energy* 37: 109–116.

Brunner, G. (1994). Gas Extraction: An Introduction to Fundamentals of Supercritical Fluids and the Application to Separation Processes, Topics in Physical Chemistry. Heidelberg: Steinkopff-Verlag.

Bryant, G. and Wolfe, J. (1987). Electromechanical stresses produced in the plasma membranes of suspended cells by applied electric fields. *J. Membr. Biol.* 96: 129–139.

de Camargo, A.C., Regitano-d'Arce, M.A.B., Biasoto, A.C.T., and Shahidi, F. (2016). Enzyme-assisted extraction of phenolics from winemaking by-products: antioxidant potential and inhibition of alpha-glucosidase and lipase activities. *Food Chem.* 212: 395–402.

Cardenas-Toro, F.P., Forster-Carneiro, T., Rostagno, M.A. et al. (2014). Integrated supercritical fluid extraction and subcritical water hydrolysis for the recovery of bioactive compounds from pressed palm fiber. *J. Supercrit. Fluids* 93: 42–48.

Chan, C.-H., Yusoff, R., Ngoh, G.-C., and Kung, F.W.-L. (2011). Microwave-assisted extractions of active ingredients from plants. *J. Chromatogr. A* 1218: 6213–6225.

Chemat, F., Tomao, V., and Virot, M. (2008). Ultrasound-assisted extraction in food analysis. *Handb. Food Anal. Instrum.*: 85–103.

Chemat, F., Rombaut, N., Meullemiestre, A. et al. (2017). Review of green food processing techniques. Preservation, transformation, and extraction. *Innov. Food Sci. Emerg. Technol.* 41: 357–377.

Chen, S., Xing, X.-H., Huang, J.-J., and Xu, M.-S. (2011). Enzyme-assisted extraction of flavonoids from *Ginkgo biloba* leaves: improvement effect of flavonol transglycosylation catalyzed by *Penicillium decumbens* cellulase. *Enzym. Microb. Technol.* 48: 100–105.

Cheng, J., Su, H., Zhou, J. et al. (2011). Microwave-assisted alkali pretreatment of rice straw to promote enzymatic hydrolysis and hydrogen production in dark- and photo-fermentation. *Int. J. Hydrog. Energy* 36: 2093–2101.

Concha, J., Soto, C., Chamy, R., and Zuniga, M.E. (2004). Enzymatic pretreatment on rose-hip oil extraction: hydrolysis and pressing conditions. *J. Am. Oil Chem. Soc.* 81: 549–552.

Dehghan-Shoar, Z., Hardacre, A.K., Meerdink, G., and Brennan, C.S. (2011). Lycopene extraction from extruded products containing tomato skin. *Int. J. Food Sci. Technol.* 46: 365–371.

Deng, Q., Zinoviadou, K.G., Galanakis, C.M. et al. (2015). The effects of conventional and non-conventional processing on glucosinolates and its derived forms, isothiocyanates: extraction, degradation, and applications. *Food Eng. Rev.* 7: 357–381.

Diaz, A.B., Moretti, M.M.d.S., Bezerra-Bussoli, C. et al. (2015). Evaluation of microwave-assisted pretreatment of lignocellulosic biomass immersed in alkaline glycerol for fermentable sugars production. *Bioresour. Technol.* 185: 316–323.

El-Belghiti, K., Rabhi, Z., and Vorobiev, E. (2005). Kinetic model of sugar diffusion from sugar beet tissue treated by pulsed electric field. *J. Sci. Food Agric.* 85: 213–218.

Fan, S.-P., Jiang, L.-Q., Chia, C.-H. et al. (2014). High yield production of sugars from deproteinated palm kernel cake under microwave irradiation via dilute sulfuric acid hydrolysis. *Bioresour. Technol.* 153: 69–78.

Feliciano, R.P., Meudt, J.J., Shanmuganayagam, D. et al. (2014). Supercritical fluid extraction (SFE) of cranberries does not extract oligomeric proanthocyanidins (PAC) but does alter the chromatography and bioactivity of PAC fractions extracted from SFE residues. *J. Agric. Food Chem.* 62: 7730–7737.

Fu, G.Z., Chan, A., and Minns, D. (2005). Preliminary assessment of the environmental benefits of enzyme bleaching for pulp and paper making (7 pp). *Int. J. Life Cycle Assess.* 10: 136–142.

García, A., Alriols, M.G., Llano-Ponte, R., and Labidi, J. (2011). Ultrasound-assisted fractionation of the lignocellulosic material. *Bioresour. Technol.* 102: 6326–6330.

Gardossi, L., Poulsen, P.B., Ballesteros, A. et al. (2010). Guidelines for reporting of biocatalytic reactions. *Trends Biotechnol.* 28: 171–180.

Geddes, C.C., Mullinnix, M.T., Nieves, I.U. et al. (2011). Simplified process for ethanol production from sugarcane bagasse using hydrolysate-resistant *Escherichia coli* strain MM160. *Bioresour. Technol.* 102: 2702–2711.

Germec, M., Tarhan, K., Yatmaz, E. et al. (2016). Ultrasound-assisted dilute acid hydrolysis of tea processing waste for production of fermentable sugar. *Biotechnol. Prog.* 32: 393–403.

Germec, M., Demirel, F., Tas, N. et al. (2017). Microwave-assisted dilute acid pretreatment of different agricultural bioresources for fermentable sugar production. *Cellulose* 24: 4337–4353.

Grant, E. and Halstead, B.J. (1998). Dielectric parameters relevant to microwave dielectric heating. *Chem. Soc. Rev.* 27: 213–224.

Gurdial, G.S., Foster, N.R., Yun, S.L.J., and Tilly, K.D. (1992). Phase behavior of supercritical fluid – Entrainer systems. In: Supercritical Fluid Engineering Science: Fundamentals and Applications (eds. E. Kiran and J. Brennecke), 34–45. ACS Symposium Series: American Chemical Society.

Hansen, M.H., Hansen, A., Larsen, T.J. et al. (2005). Control Design for a Pitch-Regulated, Variable Speed Wind Turbine. Denmark: Risø National Lab. ed. Roskilde (DK). Wind Energy Department.

Heinz, V., Toepfl, S., and Knorr, D. (2003). Impact of temperature on lethality and energy efficiency of apple juice pasteurization by pulsed electric fields treatment. *Innov. Food Sci. Emerg. Technol.* 4: 167–175.

Herrera, M.C. and De Castro, M.D.L. (2005). Ultrasound-assisted extraction of phenolic compounds from strawberries prior to liquid chromatographic separation and photodiode array ultraviolet detection. *J. Chromatogr. A* 1100: 1–7.

Herrero, M., Cifuentes, A., and Ibañez, E. (2006). Sub-and supercritical fluid extraction of functional ingredients from different natural sources: plants, food-by-products, algae and microalgae: a review. *Food Chem.* 98: 136–148.

Huynh, N.T., Smagghe, G., Gonzales, G.B. et al. (2014). Enzyme-assisted extraction enhancing the phenolic release from cauliflower (*Brassica oleracea* L. var. botrytis) outer leaves. *J. Agric. Food Chem.* 62: 7468–7476.

Intanakul, P., Krairiksh, M., and Kitchaiya, P. (2003). Enhancement of enzymatic hydrolysis of lignocellulosic wastes by microwave pretreatment under atmospheric pressure. *J. Wood Chem. Technol.* 23: 217–225.

Kammerer, D., Claus, A., Schieber, A., and Carle, R. (2005). A novel process for the recovery of polyphenols from grape (*Vitis vinifera* L.) pomace. *J. Food Sci.* 70: C157–C163.

Khan, M.K., Abert-Vian, M., Fabiano-Tixier, A.-S. et al. (2010). Ultrasound-assisted extraction of polyphenols (flavanone glycosides) from orange (*Citrus sinensis* L.) peel. *Food Chem.* 119: 851–858.

Khanal, S.K., Montalbo, M., van Leeuwen, J. et al. (2007). Ultrasound enhanced glucose release from corn in ethanol plants. *Biotechnol. Bioeng.* 98: 978–985.

Koubaa, M., Roselló-Soto, E., Šic Žlabur, J. et al. (2015). Current and new insights in the sustainable and green recovery of nutritionally valuable compounds from *Stevia rebaudiana* Bertoni. *J. Agric. Food Chem.* 63: 6835–6846.

Koubaa, M., Mhemdi, H., Barba, F.J. et al. (2016). Oilseed treatment by ultrasounds and microwaves to improve oil yield and quality: an overview. *Food Res. Int.* 85: 59–66.

Koubaa, M., Mhemdi, H., and Fages, J. (2018). Recovery of valuable components and inactivating microorganisms in the agro-food industry with ultrasound-assisted supercritical fluid technology. *J. Supercrit. Fluids* 134: 71–79.

Kumar, P., Barrett, D.M., Delwiche, M.J., and Stroeve, P. (2011). Pulsed electric field pretreatment of switchgrass and wood chip species for biofuel production. *Ind. Eng. Chem. Res.* 50: 10996–11001.

Kumarasamy, Y., Nahar, L., Byres, M. et al. (2005). The assessment of biological activities associated with the major constituents of the methanol extract of "wild carrot" (*Daucus carota* L) seeds. *J. Herb. Pharmacother.* 5: 61–72.

Kviecinski, M.R., Benelli, P., Felipe, K.B. et al. (2011). SFE from *Bidens pilosa* Linné to obtain extracts rich in cytotoxic polyacetylenes with antitumor activity. *J. Supercrit. Fluids* 56: 243–248.

Landbo, A.-K. and Meyer, A.S. (2001). Enzyme-assisted extraction of antioxidative phenols from black currant juice press residues (*Ribes nigrum*). *J. Agric. Food Chem.* 49: 3169–3177.

Le Bourvellec, C. and Renard, C. (2012). Interactions between polyphenols and macromolecules: quantification methods and mechanisms. *Crit. Rev. Food Sci. Nutr.* 52: 213–248.

Leaes, E.X., Zimmermann, E., Souza, M. et al. (2013). Ultrasound-assisted enzymatic hydrolysis of cassava waste to obtain fermentable sugars. *Biosyst. Eng.* 115: 1–6.

Lebovka, N.I., Bazhal, M.I., and Vorobiev, E. (2002). Estimation of characteristic damage time of food materials in pulsed-electric fields. *J. Food Eng.* 54: 337–346.

Lebovka, N., Vorobiev, E., and Chemat, F. (2016). Enhancing Extraction Processes in the Food Industry. CRC Press.

Li, B.B., Smith, B., and Hossain, M.M. (2006). Extraction of phenolics from citrus peels: II. Enzyme-assisted extraction method. *Sep. Purif. Technol.* 48: 189–196.

López, N., Puértolas, E., Condón, S. et al. (2009). Enhancement of the solid-liquid extraction of sucrose from sugar beet (*Beta vulgaris*) by pulsed electric fields. *LWT Food Sci. Technol.* 42: 1674–1680.

Lü, J. and Zhou, P. (2011). Optimization of microwave-assisted $FeCl_3$ pretreatment conditions of rice straw and utilization of *Trichoderma viride* and *Bacillus pumilus* for production of reducing sugars. *Bioresour. Technol.* 102: 6966–6971.

Lunelli, F.C., Sfalcin, P., Souza, M. et al. (2014). Ultrasound-assisted enzymatic hydrolysis of sugarcane bagasse for the production of fermentable sugars. *Biosyst. Eng.* 124: 24–28.

Mai, N.L., Ha, S.H., and Koo, Y.-M. (2014). Efficient pretreatment of lignocellulose in ionic liquids/co-solvent for enzymatic hydrolysis enhancement into fermentable sugars. *Process Biochem.* 49: 1144–1151.

Marcus, Y. (2005). Solvatochromic probes in supercritical fluids. *J. Phys. Org. Chem.* 18: 373–384.

Mason, T.J., Paniwnyk, L., and Lorimer, J.P. (1996). The uses of ultrasound in food technology. *Ultrason. Sonochem.* 3: S253–S260.

Meyer, A.S. (2010). Enzyme technology for precision functional food ingredient processes. *Ann. N. Y. Acad. Sci.* 1190: 126–132.

Mišić, D., Zizovic, I., Stamenić, M. et al. (2008). Antimicrobial activity of celery fruit isolates and SFE process modeling. *Biochem. Eng. J.* 42: 148–152.

Nomanbhay, S.M., Hussain, R., and Palanisamy, K. (2013). Microwave-assisted alkaline pretreatment and microwave assisted enzymatic saccharification of oil palm empty fruit bunch fiber for enhanced fermentable sugar yield. *J. Sustain. Bioenergy Syst.* 3: 7–17.

Passos, C.P., Yilmaz, S., Silva, C.M., and Coimbra, M.A. (2009). Enhancement of grape seed oil extraction using a cell wall degrading enzyme cocktail. *Food Chem.* 115: 48–53.

Pinelo, M., Arnous, A., and Meyer, A.S. (2006). Upgrading of grape skins: significance of plant cell-wall structural components and extraction techniques for phenol release. *Trends Food Sci. Technol.* 17: 579–590.

Pothula, A.K., Igathinathane, C., Faller, T., and Whittaker, R. (2014). Novel front end processing method of industrial beet juice extraction for biofuels and bioproducts industries. *Biomass Bioenergy* 68: 161–174.

Prado, J.M., Forster-Carneiro, T., Rostagno, M.A. et al. (2014). Obtaining sugars from coconut husk, defatted grape seed, and pressed palm fiber by hydrolysis with subcritical water. *J. Supercrit. Fluids* 89: 89–98.

Prado, J.M., Lachos-Perez, D., Forster-Carneiro, T., and Rostagno, M.A. (2016). Sub- and supercritical water hydrolysis of agricultural and food industry residues for the production of fermentable sugars: a review. *Food Bioprod. Process.* 98: 95–123.

Puértolas, E., López, N., Saldaña, G. et al. (2010). Evaluation of phenolic extraction during fermentation of red grapes treated by a continuous pulsed electric fields process at pilot-plant scale. *J. Food Eng.* 98: 120–125.

Puri, M., Kaur, A., Schwarz, W.H. et al. (2011). Molecular characterization and enzymatic hydrolysis of naringin extracted from kinnow peel waste. *Int. J. Biol. Macromol.* 48: 58–62.

Puri, M., Sharma, D., and Barrow, C.J. (2012). Enzyme-assisted extraction of bioactives from plants. *Trends Biotechnol.* 30: 37–44.

Putnik, P., Bursać Kovačević, D., Režek Jambrak, A. et al. (2017). Innovative "green" and novel strategies for the extraction of bioactive added value compounds from citrus wastes – a review. *Molecules* 22: 680.

Ramadan, M.F., Wahdan, K.M.M., Hefnawy, H.T.M. et al. (2009). Chromatographic analysis for fatty acids and lipid-soluble bioactives of Derris indica crude seed oil. *Chromatographia* 70: 103–108.

Ramadoss, G. and Muthukumar, K. (2014). Ultrasound assisted ammonia pretreatment of sugarcane bagasse for fermentable sugar production. *Biochem. Eng. J.* 83: 33–41.

Reclaru, L. and Meyer, J.-M. (1998). Effects of fluorides on titanium and other dental alloys in dentistry. *Biomaterials* 19: 85–92.

Redondo, D., Venturini, M.E., Luengo, E. et al. (2018). Pulsed electric fields as a green technology for the extraction of bioactive compounds from thinned peach by-products. *Innov. Food Sci. Emerg. Technol.* 45: 335–343.

Ruiz-Terán, F., Perez-Amador, I., and López-Munguia, A. (2001). Enzymatic extraction and transformation of glucovanillin to vanillin from vanilla green pods. *J. Agric. Food Chem.* 49: 5207–5209.

Sasmal, S., Goud, V.V., and Mohanty, K. (2012). Ultrasound assisted lime pretreatment of lignocellulosic biomass toward bioethanol production. *Energy Fuel* 26: 3777–3784.

Solana, M., Boschiero, I., Dall'Acqua, S., and Bertucco, A. (2014). Extraction of bioactive enriched fractions from *Eruca sativa* leaves by supercritical CO_2 technology using different co-solvents. *J. Supercrit. Fluids* 94: 245–251.

Sovilj, M.N., Nikolovski, B.G., and Spasojević, M.Đ. (2011). Critical review of supercritical fluid extraction of selected spice plant materials. *Maced. J. Chem. Chem. Eng.* 30: 197–220.

Sowbhagya, H.B. and Chitra, V.N. (2010). Enzyme-assisted extraction of flavorings and colorants from plant materials. *Crit. Rev. Food Sci. Nutr.* 50: 146–161.

Stafford, A.M. (2002). Plant cell cultures as a source of bioactive small molecules. *Curr. Opin. Drug Discov. Devel.* 5: 296–303.

Stalikas, C.D. (2007). Extraction, separation, and detection methods for phenolic acids and flavonoids. *J. Sep. Sci.* 30: 3268–3295.

Subhedar, P.B., Babu, N.R., and Gogate, P.R. (2015). Intensification of enzymatic hydrolysis of waste newspaper using ultrasound for fermentable sugar production. *Ultrason. Sonochem.* 22: 326–332.

Subhedar, P.B., Ray, P., and Gogate, P.R. (2018). Intensification of delignification and subsequent hydrolysis for the fermentable sugar production from lignocellulosic biomass using ultrasonic irradiation. *Ultrason. Sonochem.* 40: 140–150.

Suslick, K.S. and Doktycz, S.J. (1990). The Effects of Ultrasound on Solids in Advances in Sonochemistry (ed. T.J. Mason). JAI Press Inc.

Tobar, P., Moure, A., Soto, C. et al. (2005). Winery solid residue revalorization into oil and antioxidant with nutraceutical properties by an enzyme assisted process. *Water Sci. Technol.* 51: 47–52.

Velmurugan, R. and Muthukumar, K. (2012). Ultrasound-assisted alkaline pretreatment of sugarcane bagasse for fermentable sugar production: optimization through response surface methodology. *Bioresour. Technol.* 112: 293–299.

Vorobiev, E. and Lebovka, N. (2016). Applications of pulsed electric energy for biomass pretreatment in biorefinery. In: Biomass Fractionation Technologies for a Lignocellulosic Feedstock Based Biorefinery (ed. S.I. Mussatto), 151–168. Amsterdam: Elsevier.

Werle, L.B., Garcia, J.C., Kuhn, R.C. et al. (2013). Ultrasound-assisted acid hydrolysis of palm leaves (*Roystonea oleracea*) for production of fermentable sugars. *Ind. Crop. Prod.* 45: 128–132.

Wilkins, M.R., Widmer, W.W., Grohmann, K., and Cameron, R.G. (2007). Hydrolysis of grapefruit peel waste with cellulase and pectinase enzymes. *Bioresour. Technol.* 98: 1596–1601.

Wu, Y., Cui, S.W., Tang, J., and Gu, X. (2007). Optimization of extraction process of crude polysaccharides from boat-fruited sterculia seeds by response surface methodology. *Food Chem.* 105: 1599–1605.

Xiong, J., Ye, J., Liang, W.Z., and Fan, P.M. (2000). Influence of microwave on the ultrastructure of cellulose I. *J. South China Univ. Technol.* 28: 84–89.

Yang, Y.-C., Li, J., Zu, Y.-G. et al. (2010). Optimisation of microwave-assisted enzymatic extraction of corilagin and geraniin from *Geranium sibiricum* Linne and evaluation of antioxidant activity. *Food Chem.* 122: 373–380.

Yuan, Y. and Macquarrie, D.J. (2015). Microwave assisted step-by-step process for the production of fucoidan, alginate sodium, sugars and biochar from *Ascophyllum nodosum* through a biorefinery concept. *Bioresour. Technol.* 198: 819–827.

Yunus, R., Salleh, S.F., Abdullah, N., and Biak, D.R.A. (2010). Effect of ultrasonic pre-treatment on low temperature acid hydrolysis of oil palm empty fruit bunch. *Bioresour. Technol.* 101: 9792–9796.

Zhang, Q., Benoit, M., Vigier, K.D.O. et al. (2013). Pretreatment of microcrystalline cellulose by ultrasounds: effect of particle size in the heterogeneously-catalyzed hydrolysis of cellulose to glucose. *Green Chem.* 15: 963–969.

Zosel, K. (1978). Praktische Anwendungen der Stofftrennung mit überkritischen Gasen. *Angew. Chem.* 90: 748–755.

Zougagh, M., Valcárcel, M., and Rıos, A. (2004). Supercritical fluid extraction: a critical review of its analytical usefulness. *TrAC Trends Anal. Chem.* 23: 399–405.

6

Enhancing Microbial Growth Using Emerging Technologies

Maria J. Mota[1], Rita P. Lopes[1], Ana M. Gomes[2], Ivonne Delgadillo[1], and Jorge A. Saraiva[1]

[1]*LAQV-REQUIMTE, Department of Chemistry, University of Aveiro, Aveiro, Portugal*
[2]*Escola Superior de Biotecnologia, Universidade Católica Portuguesa, Porto, Portugal*

6.1 Introduction

Stress is one of the major driving forces of microbial evolution and adaptation, bringing forward new strains (Serrazanetti et al. 2009). Microorganisms may show different reactions when exposed to stress: at more extreme conditions, cells are unable to withstand and adapt, leading to microbial destruction; at milder stress conditions, microorganisms may be able to survive and grow, due to activation of general and specific stress response mechanisms (Huang et al. 2014; Lado and Yousef 2002). These stress responses rely on the coordinated expression of genes that alter different cellular processes (e.g. cell division, DNA metabolism, housekeeping, membrane composition, transport, etc.) to improve the stress tolerance (Storz and Hengge 2000; van de Guchte et al. 2002). Some of the stress-induced genes seem to be genuinely specific, while others correspond to general stress response genes that can be induced by a wide variety of stresses (De Angelis et al. 2001; Serrazanetti et al. 2009).

The exposure of microbial cells to stressful conditions during growth and fermentation involves a complex network of response mechanisms, with several metabolic activities that will reflect upon the metabolome of the fermentative microorganisms and thus on the bioproducts and on the bioprocess itself (Serrazanetti et al. 2009). Some of the changes promoted by these stress responses may have a positive outcome, such as enhanced cell growth, improved fermentation rates and yields, or even changes in metabolic selectivity. Therefore, the concept of performing fermentations using stressful "nonconventional" conditions is arising, as recently reviewed by Mota et al. (2018) and Galván-D'Alessandro and Carciochi (2018). This approach is based on the use of emerging processing technologies (typically applied for food pasteurization) at sublethal levels in order to affect microbial growth and fermentation but without causing microbial inactivation. The most studied technologies in this context include electric fields (EFs), ultrasound (US), and high

Fermentation Processes: Emerging and Conventional Technologies, First Edition.
Edited by Mohamed Koubaa, Francisco J. Barba, and Shahin Roohinejad.

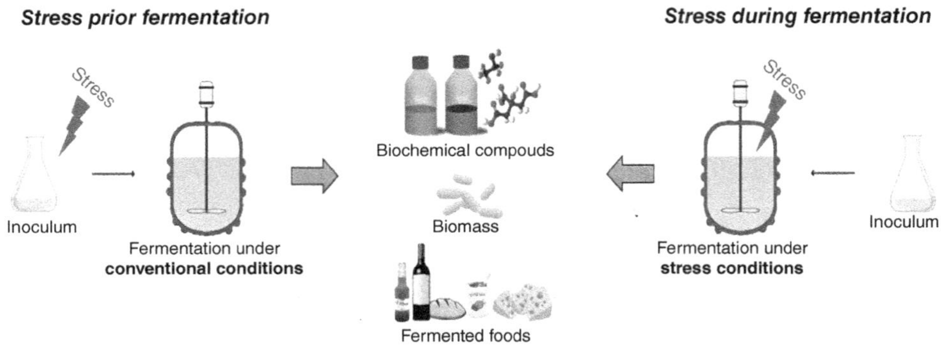

Figure 6.1 The two main approaches presented in the literature for the use of emerging technologies (electric fields, ultrasound, or high pressure) to stimulate cell growth and fermentation: (a) application of the stress to the inoculum, with fermentation subsequently taking place at normal conditions and (b) application of the stress during fermentation, whether continuously during the entire process or intermittently with fixed duration pulses.

pressure (HP). There are two distinct approaches for the use of these emerging technologies (Figure 6.1), and both will be addressed in the course of this chapter: the first one is the application of stress (EF, US, or HP) to the inoculum, with fermentation subsequently taking place at normal conditions, while the second corresponds to the application of stress during fermentation, whether continuously during the entire process or intermittently with fixed duration pulses.

At the moment, several microorganisms were tested under these nonconventional conditions, aiming not only the stimulation of cell growth but also the improvement of relevant fermentation processes. These industrially relevant processes include not only food fermentations (e.g. for the production of dairy products, alcoholic beverages, and others) but also the production of commodity biochemicals (e.g. acetic acid, citric acid, and ethanol) and high-value bioproducts (e.g. vitamins, antibiotics, and biopolymers).

6.2 Microbial Stimulation Using EFs

The application of EFs to fermentation comprises the application of direct current electric pulses on the fermentative sample, which is placed between two electrodes in a treatment chamber. When cell suspensions are exposed to EF, the ions inside the cell move along the field until there is an accumulation of the free charges at both membrane surfaces (Panagopoulos et al. 2002). When the intensity of an applied EF increases, the potential difference across a cell membrane also increases. If this transmembrane potential exceeds a stated threshold value (typically 0.2–1 V), a temporary loss of membrane semi-permeability occurs. This phenomenon of cell damage is called electroporation and, depending on the given EF, leads to the formation of temporary (reversible) or permanent (irreversible) pores (Galván-D'Alessandro and Carciochi 2018).

Microbial growth could be enhanced or inhibited by electrical current depending on the process parameters and the properties of the treated matrix. Regarding the process parameters, the most important factors include EF intensity, the number of pulses, pulse width and waveform, treatment time, and total specific energy conveyed to the cells. Regarding the properties of the fermentation medium, the key factors are related to the electrical properties of samples, such as their electrical conductivity and their electric strength. Other important aspects are related to the morphological and physiological characteristics of the treated microorganisms, such as cell type, size and shape, cell density, and cell growth stage of the microorganisms. For instance, Gram-positive cells are more resistant to EF than Gram-negative cells, possibly due to the thick and more robust cell wall. In addition, large cells are usually more sensitive to EF because of the larger surface of contact (Ben Ammar et al. 2011). Yeasts are more sensitive to EF than Gram-positive bacteria, but they may be more resistant than Gram-negative bacteria. In this case, S—S bonds in the yeast walls seem to stabilize EF-treated yeast cells (Galván-D'Alessandro and Carciochi 2018).

Stone (cited by Rowley (1972)) was the first author to report the stimulation of microbial growth by electrical current through electrochemical reactions, which resulted in bacterial counts 100-fold greater than controls. Nowadays, two different approaches using EF are being tested in this context in biotechnological processes for microbial stimulation: (i) pulsed electric fields (PEF) and (ii) moderated electric fields (MEF).

For the purpose of cell growth stimulation, PEF stresses usually involve the application of low-voltage pulses (lower than $10\,kV\,cm^{-1}$) for less than 1 second in order to trigger a microbial stress response without causing microbial inactivation and destruction. There are two mechanisms of action that can explain the positive effects of PEF-mediated processes on microbial growth and fermentation. One is the stimulation of specific metabolic pathways by EF (Hunt et al. 2009), and the other one is the formation of pores in the cell membrane due to the externally applied EF (phenomenon called electroporation) (Shil et al. 2008). Electroporation, as a physical process, increases the cell permeability and diffusion of ions and molecules across the cell membrane, enhancing the nutrient uptake and, consequently, promoting the cell growth and/or fermentative processes. In 1999, Velizarov (1999) reviewed the PEF-mediated processes tested to date, and both stimulation and inhibition of microbial growth, as well as changes in the bioprocesses productivity were observed. However, we will focus only on microbial stimulation by PEF, whether enhancement of microbial growth or productivity. The main findings about this subject are presented in Table 6.1, and they are briefly described below.

The first study with a PEF-mediated process was published by Grosse et al. (1988), where both stimulation of *Streptomyces noursei* growth and production of antibiotic (nourseothricin) were observed after the application of pulses of $1.5\,mV\,cm^{-1}$. Similar effects were obtained for *Trichoderma reesei*, *Kluyveromyces marxianus*, and *Saccharomyces cerevisiae* cultures (Fologea et al. 1998; Kerns et al. 1993; McCabe et al. 1995). In addition, PEF also caused a metabolic shift on kimchi-making cultures when applied during the fermentative process (Joo et al. 2013). On one hand, electric pulses of $0.6\,V\,cm^{-1}$ decreased the production of lactic acid and ethanol, and on the other hand, it increased acetic acid production. The authors explained this difference with the variation of redox potential, since lactic acid and ethanol are produced under low redox potential (i.e. low PEF intensity), while acetic acid is produced under high redox potential (i.e. high PEF intensity). Therefore, the application of PEF on kimchi-making cultures affected the sensorial and nutritional properties of the kimchi obtained.

Table 6.1 General effects of sublethal electric fields on microbial cell growth and fermentation.

Microorganism	Main effects	References
Streptomyces noursei	• Stimulation of cell growth • Increased antibiotic (nourseothricin) production	Grosse et al. (1988)
Trichoderma reesei	• Increased membrane permeability • Enhanced cellulase activity (≈60%) and secretion (≈80%)	Kerns et al. (1993)
Kluyveromyces marxianus	• Increased bioconversion of cellobiose into ethanol (40%)	McCabe et al. (1995)
Saccharomyces cerevisiae	• Stimulation of yeast growth • Modification of cell cycle • Increased sugar consumption • Increased ethanol production • Induction of the expression of oxidation stress response genes and not heat stress genes	Araújo et al. (2007), Fologea et al. (1998), Mattar et al. (2015), Nakanishi et al. (1998), Tanino et al. (2012)
Recombinant *Saccharomyces cerevisiae* (able to grow on lactose)	• Increased lag phase (21–53%) • Increased β-galactosidase yield • Lower ethanol yield	Castro et al. (2012)
Kimchi-making cultures	• Decreased production of lactic acid and ethanol • Increased production of acetic acid	Joo et al. (2013)
Lactococcus cremoris	• Increased exopolysaccharide (EPS) yield • EPS produced with lower molecular weight	Ohba et al. (2016)
Aspergillus brasiliensis	• Higher citric acid yield (≈1.4-fold) • Increased consumption of dissolved oxygen	Fiedurek (1999)
Lactobacilli strains	• Stimulation of cell growth (>1-log) • Increased membrane permeability • Increased nutrient uptake • Improved cholesterol assimilation by incorporation (50.0–59.6%) into lactobacilli cell membrane	Lye et al. (2011)
Bifidobacteria and Lactobacilli strains	• Increased intracellular and extracellular β-glucosidase activities • Increased bioconversion of isoflavones into bioactive aglycones	Ewe and Alias (2012), Yeo and Liong (2013)
Streptomyces avermitilis	• Stimulation of cell growth • Higher avermectin production (42%) • Enhanced expression of positive regulators for avermectin biosynthesis	Guo et al. (2016)

Table 6.1 (Continued)

Microorganism	Main effects	References
Lactobacillus acidophilus	• Increased bacteriocin activity when combining EF fermentation with conventional fermentation • Decreased bacteriocin activity when applying EF continuously • EF waveform and frequency affected lag phase and bacteriocin production • Cells at lag phase with the highest susceptibility to permeabilization • No electroporation during the stationary phase	Loghavi et al. (2007, 2008, 2009)

EF = Electric field.

Lactococcus lactis ssp. *cremoris* is a bacterial strain able to produce an exopolysaccharide (EPS) that confers unique rheological properties as well as several health-promoting benefits to fermented milk products (Ohba et al. 2016). PEF treatments can be a potential engineering technique to enhance EPS production by *L. lactis* ssp. *cremoris*. The application of PEF treatments to this fermentative process was found to increase the EPS yield up to 94%, compared to nontreated fermentation. However, the size of EPS produced with PEF treatments was 1.4 million Da smaller than the EPS obtained by nontreated fermentation, but both EPS showed almost identical chemical composition.

In addition to the application of PEF to fermentation, several studies were performed where PEF was applied to the microbial culture prior to inoculation in order to possibly improve the subsequent fermentative process. For instance, PEF-treated *Aspergillus brasiliensis* spores enhanced the citric acid yield by approximately 1.4-fold, due to a higher consumption of dissolved oxygen (Fiedurek 1999). *S. cerevisiae* cells were also successfully stimulated by PEF (Araújo et al. 2007; Mattar et al. 2015; Nakanishi et al. 1998), which was reflected in fermentation rate enhancement, mainly during the initial phase of fermentation (lag phase). This effect of PEF on alcoholic fermentation is still not clearly understood. While Tanino et al. (2012) presumed that the difference in expression of oxidation stress response genes during PEF treatment is the cause of the differences in the *S. cerevisiae* fermentation, Fologea et al. (1998) assumed that they are related to changes in the activity of some membrane proteins. Even so, and given the practical importance of *S. cerevisiae*, the increasing performance by PEF may prove the great potential of the technology when applied to fermentation processes.

Some lactobacilli strains were also electrically stimulated prior to fermentation, which was reflected in a higher cell growth and cholesterol assimilation and incorporation into cell membrane (50.0–59.6%) (Lye et al. 2011). The authors assume that electroporation is the main cause of the differences in this process. On one hand, the increase in membrane permeability due to electroporation causes the increase in cell growth and cholesterol assimilation by the cells (Wu and Nyborg 2008), and on the other hand, oxidative deterioration of membrane phospholipids during electroporation allows the cholesterol incorporation

(Tang et al. 2008). Therefore, PEF-treated lactobacilli cells can be used as cholesterol-lowering adjuncts in dairy foods (Lye et al. 2011).

Fermentation of soymilk by lactobacilli and bifidobacteria enhances the bioconversion of isoflavones to bioactive aglycones, due to the action of bacterial β-glucosidase (Ewe et al. 2011; Setchell et al. 2002; Yeo and Liong 2010). However, the semi-permeable nature of the bacterial cell membrane hinders the bioconversion rate at conventional conditions. Thus, Ewe et al. (2012) and Yeo and Liong (2013) applied PEF treatments (2.5–7.5 $kV\,cm^{-1}$ for 3.0–4.0 ms) prior to fermentation, which successfully increased the bioconversion process, possibly due to the increase in intracellular and extracellular β-glucosidase activities.

Avermectins are a group of anthelmintic and insecticidal agents produced by *Streptomyces avermitilis* during fermentation. Application of PEF to *S. avermitilis* inoculum prior to fermentation was performed by Guo et al. (2016), and both cell growth and avermectin production were stimulated after 20 pulses at 10–20 $kV\,cm^{-1}$. The authors reported that this increase was closely associated with a decrease in oxidation–reduction potential and an increase in temperature, in addition to the expression enhancement of aveR and malE genes in *S. avermitilis* during fermentation, which are positive regulators for avermectin biosynthesis.

With regard to MEF, EFs are applied continuously at moderate levels (usually lower than 1000 $V\,cm^{-1}$) with temperature control to avoid the occurrence of heating during the process (Sastry et al. 2011). Despite the few studies that are published regarding MEF-assisted processes, the results obtained seem to be very promising (Castro et al. 2012). Most studies relating to this subject focus on lactic acid fermentation processes performed by *Lactobacillus acidophilus*. Loghavi et al. (2007) and Loghavi et al. (2008) tested the bacteriocin activity after *L. acidophilus* fermentation, and the highest activity was obtained when MEF were firstly applied during the initial stage of fermentation (5 hours), followed by conventional fermentation for the remaining time (35 hours). This positive effect on bacteriocin activity could be related to the stress caused by EF, since stress conditions usually promote the production of defensive molecules, such as bacteriocins (Verluyten et al. 2003). However, when the EF was applied during the entire growth time (40 hours), the bacteriocin activity decreased, which may be caused by the transport of prebacteriocins due to temporary electroporation, thereby bypassing the critical translocation process by adenosine triphosphate (ATP)-binding cassette (ABC) transporter proteins, which is necessary for activation of bacteriocin. Therefore, in this case, some of the bacteriocin molecules may be released in the environment without getting activated (Loghavi et al. 2008). In consequence, Loghavi et al. (2009) evaluated the cell membrane permeabilization caused by the MEF treatment and observed that the stage of microbial growth influences the level of permeabilization. In fact, cells at lag and exponential phases showed the greatest susceptibility to permeabilization, while no evidence of electroporation was observed during the stationary phase. These results are consistent with the findings of Zimmermann et al. (1974), who showed that the critical dielectric cell membrane breakdown voltage increased about 30% at the stationary phase, possibly due to differences in the dielectric properties of the cell membrane caused by modification of protein and lipid content. An alternative explanation could be the pH decline during cell growth, which may affect the cell membrane permeabilization.

Similar results were observed in the fermentation of a recombinant *S. cerevisiae* strain growing on lactose to produce ethanol and β-galactosidase (Castro et al. 2012). In this case,

MEF was applied during the entire fermentation process, and the early stages of fermentation were positively affected, since a decrease in lag phase and an increase in biomass yield were observed at $2\,V\,cm^{-1}$. Regarding the substrate conversion, a higher β-galactosidase yield was observed, while the ethanol yield was lower. This difference may be due to a shift toward pure respiratory metabolism (where biomass build-up is preferred) instead of ethanol production.

Overall, fermentation processes may be improved by the use of EF, mainly due to increased membrane permeability (and consequent improved cell transport), as well as possible changes in genetic expression and increase in the activity of some enzymes. However, when EFs are applied continuously or even intermittently during the whole fermentation process, cell viability may be seriously compromised due to the occurrence of heating and consequently severe cell damage. Therefore, the temperature increase should be controlled by adapted cooling systems during the entire process, leading to higher energetic costs of the process. Thus, most of the studies presented in this section did not apply the treatment during the fermentation process, since PEF was applied shortly prior to inoculation and MEF only in the early stages of fermentation. Therefore, despite the possible improvement of fermentative processes by EF, this technology may lead to some drawbacks when applied during the whole fermentation process.

6.3 Stimulation Using US

US used at low intensities ($<1\,W\,cm^{-2}$) may result in intensification of several biotechnological processes without severe damages in the cells (Chisti 2003; Gogate and Kabadi 2009). However, the mechanisms behind these productivity improvements have not been clearly identified. The main physical effects of US are induced during cavitation, with the collapse of the bubbles generating shockwave-induced damage and microjet impacts (Esclapez et al. 2011). Other physical phenomena include agitation, turbulence, vibration, pressure, shear forces, and acoustic streaming. The application of US also promote different chemical effects, and one of the most important chemical effects is the sonolysis of water (the main component of the fermentative media), which leads to the production of free radicals (O'donnell et al. 2010). These radicals can react with each other to form other molecules, such as hydrogen peroxide, or even react with other components in the surrounding medium. According to Ashokkumar et al. (2010), the physical effects of US are usually dominant in the lower frequency range of 20–100 kHz with a higher level of transient cavitation. On the other hand, the chemical effects are dominant at higher frequencies (>200 kHz), with larger amounts of free radicals being produced due to the generation of a higher number of active bubbles (Galván-D'Alessandro and Carciochi 2018).

Several studies have already demonstrated the beneficial effects of sonication on bacteria (Bar 1988; Chu et al. 2000; Joyce et al. 2003; Runyan et al. 2006; Sakakibara et al. 1994; Wang and Sakakibara 1997; Wood et al. 1997; Wu et al. 2000; Zabaneh and Bar 1991), yeasts (Anderson 1953; Jomdecha and Prateepasen 2006; Lanchun et al. 2003; Matsuura et al. 1994), and filamentous fungi (Chuanyun et al. 2003, 2004; Herrán et al. 2008). In this section, several examples of the microbial stimulation by US are briefly described, and the main results are presented in Table 6.2.

Table 6.2 General effects of ultrasounds on microbial cell growth and fermentation.

Microorganism	Main effects	References
Saccharomyces cerevisiae	• Stimulation of the yeast growth • Increase in glucose consumption • Increase in ethanol yields and productivities • Decrease in the concentration of inhibitory dissolved carbon dioxide • Acceleration of fermentation (50–64%) for the production of wine, beer, and sake	Choi et al. (2015), Hu et al. (2011), Matsuura et al. (1994), Schläfer et al. (2000), Singh et al. (2015), Subhedar and Gogate (2015)
Kluyveromyces marxianus	• Stimulation of the yeast growth • Increase in ethanol productivity	Sulaiman et al. (2011)
Lactobacilli strains	• Leak of intracellular enzymes • Enhancement of β-galactosidase activity • Increase in lactose hydrolysis, depending on the cell strain and the cell growth phase • Acceleration of lactic acid production • Reduction of fermentation time by 0.5 h	Dakubu (1976), Mett et al. (1988), Sakakibara et al. (1994), Wang et al. (1996), Wang and Sakakibara (1997)
Bifidobacteria strains	• Strain-dependent resistance of bifidobacteria to US • Increased lactose consumption • Modification of acid profiles	Nguyen et al. (2009, 2012)
Penicillium brevicompactum	• Enhanced mycophenolic acid (up to 60%)	Zhao et al. (2012)
Ecemothecium ashbyii	• Stimulation of cell growth • Decrease in fermentation time (from 96 to 60 hours) • Increased production of riboflavin	Chuanyun et al. (2003, 2004)
Micromonospora spp.	• Increased release of intracellular gentamicin into the broth	(Chu et al. 2000)
Aspergillus terreus	• No effect on cell growth • Modification of cell morphology • Reduction of lovastatin production	Herrán et al. (2008)
Streptomyces venezuelae	• Stimulation of cell growth • Increased thrombinase activity	Naveena et al. (2012)

US = ultrasound.

There is a great variability regarding the resistance of microorganisms to US, with spores being more resistant than vegetative cells, fungi more resistant than bacteria, aerobes more resistant than anaerobes, and cocci more resistant than bacillus (Chandrapala et al. 2012). The findings concerning the effect of US on Gram-positive and Gram-negative bacteria are controversial: according to some studies, Gram-positive bacteria are more resistant than the Gram-negative ones, possibly because of their thicker cell wall, which provides a better protection against US effects (Drakopoulou et al. 2009; Villamiel and de Jong 2000), while others have reported no significant differences between these two groups (Scherba et al. 1991).

In fact, the effect of US on cells is highly dependent on the properties of the medium and the applied process parameters (e.g. US time function – continuous or intermittent). For instance, continuous sonication (1 MHz, 10.5 W cm^{-2}) inhibited *S. cerevisiae* fermentation, while intermittent sonication at the same intensity was less damaging (Anderson 1953). In another study using *S. cerevisiae*, intermittent sonication (25 kHz, 300 W m^{-3}) applied during the entire fermentation process more than doubled the ethanol yield, while this effect was not observed in continuously sonicated cultures (Schläfer et al. 2000).

S. cerevisiae is one of the most studied microorganisms relatively to this subject, and productivity enhancement by sonication has been claimed in several fermentation studies, whether in culture medium fermentations or food fermentations. A fourfold increase in ethanol productivity was observed by Singh et al. (2015) when applying US (35 kHz, 1.48 W cm^{-2}) at a 10% duty cycle (equivalent to sonication for 1 minute followed by a rest period of 9 minutes). In another study, ethanol productivity was improved by 1.8-fold when optimized conditions of US (25 kHz, 160 W, 10 minutes duration, 20% duty cycle) were applied at a 12-hour growth phase (Subhedar and Gogate 2015). Regarding food fermentations, continuous US treatments (43 kHz, 30 mW cm^{-2}) were applied during the production of wine, beer, and sake by immobilized cells of *S. cerevisiae* (Matsuura et al. 1994). The treatment reduced the fermentation time by 50–64%, possibly due to an increase in the yeast growth and a decrease in the concentration of inhibitory dissolved carbon dioxide. In addition, a US treatment of 40 kHz and 160 W was applied to beer production during four days of primary fermentation, and the ethanol productivity increased (13.18%), with the US-treated beer showing acceptable sensorial properties (Choi et al. 2015).

Hu et al. (2011) evaluated the effects of US at different frequencies (0.3–2.4 MHz) and voltages (3–24 V) on *S. cerevisiae* fermentation process. The authors observed that both cell growth phase and US intensity influence the US effects on *S. cerevisiae* cells. While US application during the exponential phase exhibited the strongest influence on growth, application during the stationary phase showed no visible effect, which may be explained by the different shapes, structures, and physiological conditions of *S. cerevisiae* at the different growth phases. Regarding US intensity, the lower the intensity applied, the more pronounced the effect on yeast growth is. Together with this, both glucose consumption and ethanol production also increased when sonication was applied, possibly due to mechanical stimulation of cells and the increased diffusion of nutrients and metabolites across the cell membrane. Several other studies have also demonstrated that low-intensity US promoted cell proliferation via activation of DNA and protein synthesis (Kobayashi et al. 2009; Saura et al. 2003; Takeuchi et al. 2008).

The effects of pulsed low-intensity US (20 kHz, 11.8 W cm^{-2}) were also evaluated in the yeast *K. marxianus*, which is able to produce bioethanol from lactose (Sulaiman et al. 2011). Different duty cycles were tested, being the duty cycles of 10 and 20% (i.e. 1 second sonication followed by a rest period of 10 seconds and 5 seconds, respectively) the most effective to stimulate both ethanol production and biomass growth. While the increase in biomass may be explained by the improved gas–liquid mass transfer of oxygen by US (Ashokkumar et al. 2007), it does not explain the increase in ethanol concentration, since ethanol is usually produced under low dissolved oxygen concentrations (Guimarães et al. 2010). However, the authors hypothesized that the improved gas–liquid mass transfer may have contributed to improved removal of dissolved carbon dioxide, which is produced in elevated concentrations during alcoholic fermentation and is known to inhibit the growth of *K. marxianus* (Guimarães et al. 2010).

The effectiveness of US on cell growth stimulation has been demonstrated in lactic acid fermentation, particularly in the production of yogurt or other fermented milk products. In these cases, fermentative activity of the microbial strains was also improved, reducing the production time (Nguyen et al. 2009). In fact, Wang et al. (1996) observed an increase in the fermentative rate when applying US treatments (20 kHz, 90, 225, or 450 W for 1, 6, or 10 minutes, respectively) on inoculated milk samples with *Lactobacillus bulgaricus*, which result in the reduction of fermentation time by 0.5 hour compared to control fermentation. The authors explained that this acceleration with the release of intracellular enzymes (such as β-galactosidase) from the cells is due to sonication, which may lead to increased lactic acid production.

Acceleration of lactose hydrolysis was also observed in some cases, causing a sweetening effect in yogurt without increasing its caloric content (Kreft and Jelen 2000; Masuzawa and Ohdaira 2002; Povey and Mason 1998; Sastry et al. 2011; Shimada et al. 2004; Toba et al. 1990; Wang and Sakakibara 1997; Wu et al. 2000). The enhancement of lactose hydrolysis may result from damage to the cells and leak of intracellular β-galactosidase caused by US, with consequent increase in β-galactosidase activity (Dakubu 1976; Mett et al. 1988; Wang and Sakakibara 1997). Wang and Sakakibara (1997) observed that the enzyme release occurs more easily in the exponential phase of cell growth because the cells are less resistant to US during this phase. Additionally, it was found that lactose hydrolysis induced by US was highly dependent on the cell strain. For instance, while higher lactose hydrolysis was observed after US-assisted fermentation with *L. bulgaricus* and *L. helveticus*, lower lactose hydrolysis was observed with *L. lactis* and *L. acidophilus*, when compared with the respective conventional fermentations. Therefore, depending on the selected cell strain and the cell growth phase, sonicated fermentation may represent a promising process for manufacturing of fermented milk products with low lactose content (Sakakibara et al. 1994).

The effects of US on milk fermentation by four *Bifidobacterium* strains (*Bifidobacterium breve*, *Bifidobacterium infantis*, *Bifidobacterium animalis* ssp. *lactis*, and *Bifidobacterium longum*) were evaluated by Nguyen et al. (2009), who observed that carbohydrate metabolism was affected by US during the fermentation of milk. Lactose consumption was accelerated for all strains during 24 hours of fermentation, and changes in the acid profiles of the strains were observed: *Bifidobacterium lactis* showed a decrease in the ratio of acetate to lactate, and *B. longum* showed a decrease in the ratio of acetate and propionate to lactate.

Low-intensity US was also applied to the production of compounds with pharmaceutical and/or biomedical interest. For instance, fermentation by *Penicillium brevicompactum* was studied in order to promote the stimulation of mycophenolic acid (MPA) production, an immunosuppressive agent used to prevent rejection after organ transplantation (Zhao et al. 2012). For that, the authors used low-intensity pulsed ultrasound (LIPUS), which is a type of pulsed US that uses intensities around ≈0.1–0.2 W cm^{-2}, and MPA production was enhanced by up to 60% when treated eight times a day for 10 minutes at an intensity of 0.2 W cm^{-2}. This increase could be explained by the observed significant impact of LIPUS on cell growth during and after the treatment.

In addition, riboflavin production (vitamin B2) by *Ecemothecium ashbyii* was also tested. This fermentative process is advantageous for a safe and nontoxic production of riboflavin, but *E. ashbyii* cell wall structure hinders its release to the culture medium, leading to high recovery losses. In order to simplify this process, Chuanyun et al. (2003) and Chuanyun et al. (2004) evaluated the application of US to *E. ashbyii* fermentation. In fact, a fivefold increase in riboflavin production, together with an increase in biomass concentration and a decrease in fermentation time (from 96 to 60 hours), were observed, relatively to the non-sonicated control cultures. The authors explained this increase with the enhancement of cell membrane permeability and the consequent release of riboflavin into the medium, which stimulates its biosynthesis by feedback mechanisms. Chu et al. (2000) reported a similar effect with the production and secretion of gentamicin by *Micromonospora* spp. The application of US treatments increased the release of intracellular gentamicin into the culture medium, which could alleviate the feedback regulation of gentamicin biosynthesis, thus improving its productivity.

The effects of continuous US on fermentation of *Aspergillus terreus* to produce lovastatin (drug for lowering blood cholesterol) was tested by Herrán et al. (2008). Despite the fact that the biomass growth profiles were not affected by US when compared to control, biomass morphology was substantially modified, with pelleted growth (characteristic of control fermentation) being entirely replaced by growth as dispersed hyphae.

Fibrinolytic enzymes are used as thrombolytic agents for the treatment of myocardial infarction and stroke (Holden 1990). However, production of new fibrinolytic agents is needed, since the thrombolytic agents used nowadays in clinical field have several limitations (Blann et al. 2002; Bode et al. 1996; Turpie et al. 2002). Therefore, Naveena et al. (2012) used low-intensity US to produce and isolate thrombinase (a potent fibrinolytic protease) from marine actinomycetes. Successful results were obtained, since higher biomass concentration and higher enzyme activity were obtained in US-treated samples, compared to the control. US stimulation improved the transport of substrates into the cell and increased the oxygen transfer rate, thus promoting cell growth and enzyme production.

Some patents have been submitted regarding microbial stimulation using US (e.g. CN104450517, CN102618440, US2011189748, US2013022957, WO2010022508, etc.), showing either the effect of US on cell growth or the conception of sonobioreactors.

In summary, suitably controlled US treatments have shown beneficial effects on cell metabolism, and, in consequence, this technology may be applied to improve/accelerate relevant fermentative processes, ranging from food fermentations to production of added-value compounds (Chisti 2003). Although US-based fermentation processes could have an additional requirement for energy to avoid the heating caused by US, this can be compensated by the increased productivity. Therefore, the potential benefits of US, along with its cost effectiveness for biotechnology industry, must be further investigated and evaluated (Sulaiman et al. 2011).

6.4 Microbial Stimulation Using HP

Similar to the other technologies mentioned in this chapter, HP may bring novel characteristics to fermentative processes when applied at sublethal levels. When mesophilic microorganisms are exposed to HP, cell morphology and metabolism are modified accordingly. The structures and functions of cell wall and cell membrane, biochemical reactions, and gene expression can all be affected by pressure (Coelho et al. 2004), subsequently leading to a remodeled cell metabolism through either directly targeting the chemical reactions or changing the expression of relevant genes involved in these reactions (Lee et al. 2006).

Huang et al. (2014) reviewed the response of microorganisms to pressure and concluded that microorganisms are more likely to be stressed or injured than killed under HP, particularly when lower intensity treatments are applied. In this case, microorganisms possess regulatory genes responsible for environmental adaptation, which are generally translated in heat-shock proteins (HSPs). Their accumulation within the cell enhances the cell resistance to multiple environmental stresses (Lou and Yousef 1997; Wemekamp-Kamphuis et al. 2004). For instance, Welch et al. (1993) reported that an HP treatment of 55 MPa induced a stress response on cells where cold-shock proteins, HSPs, and other protective proteins were accumulated. In addition, Hörmann et al. (2006) used a comparative proteome approach to characterize the HP effects on *Lactobacillus sanfranciscensis*, concluding that HP stress response uses subsets of other stress responses (such as cold and high salinity).

The effects of HP on microbial cells are influenced by several factors, which include the level and duration of the pressure treatments, the compression/decompression rates, and other process parameters (temperature, media composition, pH, etc.). In addition, each microbial strain has a specific degree of HP tolerance according to their intrinsic cellular characteristics. In general, prokaryotes are more HP resistant than eukaryotes, Gram-positive bacteria are more HP resistant than Gram-negative bacteria, and cocci are more HP resistant than bacilli (Huang et al. 2014). The cell growth stage was also found to affect the microbial tolerance to HP treatments, which is usually higher during the stationary phase than during the exponential phase.

In general, application of sublethal HP treatments to fermentative processes may lead to possible acquisition of new desirable characteristics, obtained by inhibition or even suppression of some metabolic pathways and/or utilization of new ones (Mota et al. 2013). This concept is gaining relevance over the last years, since piezo-tolerant strains may have numerous interesting applications in different fields (Aertsen et al. 2009; Hörmann et al. 2006). Some fermentative processes were already studied, and the obtained results are summarized in Table 6.3.

Most studies regarding the effects of HP on microbial growth and metabolism were performed using the yeast *S. cerevisiae*. Picard et al. (2007) studied the application of sublethal levels of HP during the alcoholic fermentation by *S. cerevisiae*. Fermentation performed at 5 and 10 MPa proceeded threefold faster than control fermentation, with the fermentation yield enhanced by 6 and 5%, respectively. However, at pressures above 20 MPa, fermentation was slowed down, and the authors estimated that alcoholic fermentation was interrupted at 87 ± 7 MPa. The authors attributed this increase in fermentative rate to an enhancement in the activity of one or more enzymes involved in the glycolytic pathway up to 10 MPa, and with the pressure increasing, they are progressively repressed. Therefore, these results reveal a great practical significance, mainly because of the widespread applicability of *S. cerevisiae*. Similarly, Bravim et al. (2013) observed that a pretreatment of *S. cerevisiae* with HP led to an increase in ethanol content upon fermentation. In parallel, these authors conducted a global transcriptional analysis and demonstrated the overexpression of several genes related to cell recovery and stress tolerance induced by HP. The most

Table 6.3 General effects of high pressure on microbial cell growth and fermentation.

Microorganism	Main effects	References
Saccharomyces cerevisiae	• Acceleration of alcoholic fermentation (up to threefold) • Increased ethanol yields (5–6%) • Overexpression of several genes related to cell recovery and stress tolerance, including the gene SYM1 • Increased glutathione concentration (58.7%) • Increased trehalose yield (82.9%)	Bravim et al. (2013), Dong et al. (2007), Picard et al. (2007), Qiao et al. (2006)
Clostridium thermocellum	• Metabolic shift, with increased ethanol production and decreased acetate (by-product) production • Higher ethanol: acetate ratio	Bothun et al. (2004)
Gluconacetobacter xylinus	• Decreased cell growth • Decreased production of bacterial cellulose • Cellulose ribbons with profound morphological differences	Kato et al. (2007)
Streptococcus thermophilus, *Lactobacillus bulgaricus*, and *Bifidobacterium lactis*	• Decreased fermentation rate • At higher pressures, microorganisms metabolically inhibited • Bacterial strains still viable at lower pressures, with ability to produce yogurt	Mota et al. (2015)
Oenococcus oeni	• Fermentation with *Oenococcus oeni* during and after HP stresses • Decreased concentrations of L-lactic acid • Increased concentrations of D-lactic acid	Neto et al. (2016)

HP = high pressure.

relevant case was the gene SYM1, whose overexpression results in enhanced ethanol production and stress tolerance upon fermentation.

Trehalose and glutathione are two major stress-induced metabolites with industrial value, which could be produced under HP conditions. *S. cerevisiae* CICC1339 growing at 0.5 MPa showed an increase in 58.7% in glutathione concentration in comparison with the control cells at atmospheric pressure (Qiao et al. 2006). Similarly, application of HP at 1 MPa on *S. cerevisiae* resulted in an increased yield of trehalose by 82.9% (Dong et al. 2007). Trehalose protects cells through a nonspecific mechanism under adverse conditions (Liang et al. 2013), while glutathione is involved in apoptosis (Kiriyama et al. 2012). These two products are normally present at very low concentrations in microorganisms (Bachhawat et al. 2013), but their production increases when cells undergo stress, possibly for protection (Dong and Jiang 2016).

Clostridium thermocellum is a thermophilic bacterium that efficiently converts cellobiose to bioethanol and other chemicals. However, several undesired coproducts are also produced during this process, such as organic acids (acetate, lactate) and gaseous end products (H_2, CO_2), which limit the commercialization of biofuel produced by this microorganism (Béguin and Aubert 1994; Bothun et al. 2004; Herrero et al. 1985; Wiegel 1980). Pressure was applied to this fermentative process in order to improve the biofuel production by this bacterium. In fact, pressures of 7 and 17 MPa redirected the fermentative metabolism of *C. thermocellum* from the production of organic acids (such as acetate) to ethanol, leading to a 60-fold increase in the ethanol to acetate ratio under HP compared to atmospheric pressure. These shifts could be explained by the higher concentration of dissolved gases at HP, which consequently changes the metabolic pathways involved in the production of gases (Bothun et al. 2004).

Since the application of HP to fermentative processes is relatively new, there are few studies where microbial stimulation was actually achieved. However, application of HP to some other fermentative processes was investigated, and promising results were obtained. For instance, Kato et al. (2007) observed that the bacterial cellulose (BC) produced by *Gluconacetobacter xylinus* under HP had morphological differences, even though the amounts of bacterial cells and BC produced were lower than control samples. In this case, the BC fibers produced under HP had significantly higher density compared with the BC produced at atmospheric pressure. Nevertheless, the result of these morphological changes on the functional properties of the polymer still needs further investigation (Nishi et al. 1990).

Another case was the study of the effect of HP (5–100 MPa) on lactic acid fermentation responsible for the probiotic yogurt production (Mota et al. 2015). A decrease in the fermentative rate under pressure was observed, until fermentation does not occur at 100 MPa. However, extension of the fermentation time at 5 MPa (used as a case study) yielded a typical pH for probiotic yogurt, showing that lactic acid fermentation can be performed at this pressure without compromising the viability of the bacterial strains used (*Streptococcus thermophilus*, *L. bulgaricus*, and *B. lactis*) and their ability to produce yogurt. Additionally, the authors found that microorganisms were metabolically inhibited (but not inactivated) at 100 MPa, since the samples subjected to 100 MPa revealed a normal metabolic activity when they were returned to atmospheric pressure. Taking this into account, the authors suggested the possible use of pressure as a variable to control the metabolic activity of microorganisms, similarly to how refrigeration is widely used today. In summary, despite lactic acid fermentation was not stimulated, these results were the starting point for the study of HP influence when applied to food fermentation.

Neto et al. (2016) applied different HP stresses (50 MPa, 8 hours; 100 MPa, 8 hours; 300 MPa, 0.5 hour) in the beginning of fermentation with *Oenococcus oeni*, a lactic acid bacterium employed by wine industry to perform malolactic fermentation. *O. oeni* was able to grow and ferment during and after HP stresses of 50 and 100 MPa, although with some metabolic changes. Particularly, the HP-stress of 100 MPa lead to lower concentrations of L-lactic acid and higher concentrations of D-lactic acid, compared to the control. In contrast, the HP-stress of 300 MPa, 0.5 hour resulted in complete inactivation of *O. oeni*, but malolactic fermentation was still observed at some extent, showing that malolactic enzyme maintained some residual activity at these conditions. Although the impact of these metabolic changes in wine malolactic fermentation is still not understood, it provides important insights on the possible effects of HP on biocatalysts, which in some cases can be more resistant to pressure than the analogous microbial strain itself.

HP application to fermentative processes may allow more novel possibilities than the technologies presented in the other sections. In fact, HP can be applied intermittently or continuously during the whole fermentation time, without serious cell loss (if adequate sublethal pressure is applied) and no heating effects. Therefore, refrigeration is not needed in this case, lowering the energetic costs of the fermentative process when compared with the technologies discussed above. Additionally, the energetic costs involved in the application of HP stress during the whole fermentation process are also lower, since HP only needs energy to generate the pressure, with no energy needed to maintain the pressure. Therefore, the whole fermentative processes can be easily performed under HP stress conditions, contrary to other technologies. This new feature might be an advantage for using HP as a nonconventional technology for fermentative processes, while the main drawback is still the design of more affordable high-pressure equipment. Moreover, HP can be rather easily combined with other nonconventional technologies such as US and EF.

6.5 Conclusions

Stimulation of microbial growth is a topic of major interest for the scientific community and also for food and biotechnological industries. Apart from the improvements obtained by optimizing the more obvious process parameters (substrate and air availability, agitation, temperature, pH), there is also the possibility of performing cell growth and fermentation under nonconventional conditions. This approach comprises the application of EF, US, and HP (or other still unexplored technologies) at sublethal levels, before or during cell growth, intermittently or continuously, in order to promote stress conditions able to cause substantial changes in the cells and in the process itself. The studies currently available in the literature show numerous improvements in microbial processes, including enhanced cell growth, higher fermentative rates and yields, lower accumulation of by-products, and/or production of different compounds. Although some differences were observed between the technologies that have been tested so far, all of them seem to be suitable for application during the microbial growth and fermentation, when applied at specific and adjusted conditions.

Application of EF under certain conditions was found to stimulate cell growth and also to promote beneficial effects to fermentation processes. However, a careful optimization of process parameters was always necessary in order to avoid inhibition of cell growth or even cell destruction, since the EF (particularly in the case of PEF) may be very aggressive to the microorganisms. The most relevant drawbacks of this technology include difficulties in the scaling up of systems, their high initial cost, and the reduced availability of commercial units. In contrast, US is a well-studied technology, with several applications in different fields and with lower equipment costs. A further benefit is that US is usually less aggressive to the cells and thus more suitable for stimulation of cell growth and fermentation. However, application of US on biological processes implies the use of a refrigeration system to control the temperature, which usually tends to rise during these treatments.

In consequence, the use of US may promote higher energy requirements and higher energetic costs. In contrast, HP can be applied continuously without inactivation problems or heat generation. Therefore, application of sublethal HP has higher versatility and showed some interesting distinctive features, as it can be applied as intermittent stress or continuously, without heating effects, leading to more possibilities to influence microbial processes. In addition, pressure is an important variable of life, with interesting effects on all living systems. Such an example is the microbial growth under deep sea (at HP conditions), which is increasingly attracting attention of several pharmaceutical companies wishing to obtain novel compounds with biological activity. Despite the strong interest of pressure application on biological processes, information in the literature is scarce, possibly due to the low availability of HP systems and the still high cost of equipment, making these two aspects the main challenges to a more widespread study of this technology.

In general, the information available so far regarding the use of these nonconventional conditions to enhance microbial growth and fermentation is still scarce. At this point, it is difficult to assess the economic feasibility of these approaches, since there are many factors affecting the processes that still need better understanding and optimization. In addition, since most studies were performed at laboratory scale, the scale-up of these processes will be an important requirement to their implementation. Therefore, further work conducted in this field should focus on addressing these issues, as well as on the evaluation of these technologies on other fermentative processes with industrial relevance.

Acknowledgments

FCT/MEC for the financial support to the LAQV-REQUIMTE and QOPNA research Units (UIDB/50006/2020 and FCT UID/QUI/00062/2019, respectively) and CBQF (project UID/Multi/50016/2013), through national funds and where applicable cofinanced by the FEDER, within the PT2020 Partnership Agreement. Authors Maria J. Mota and Rita P. Lopes thank FCT for the grants SFRH/BD/97061/2013 and SFRH/BD/97062/2013, respectively.

References

Aertsen, A., Meersman, F., Hendrickx, M.E.G. et al. (2009). Biotechnology under high pressure: applications and implications. *Trends Biotechnol.* 27: 434–441.

Anderson, J.M. (1953). Effects of ultrasonic radiation on growth and fermentation in the yeast, *Saccharomyces cerevisiae*. *Biochim. Biophys. Acta* 11: 122–137.

Araújo, O.Q.F., Oliveira, A.A.C., Torres, C.C.O. et al. (2007). Glucose uptake in electrically stimulated cultures of Saccharomyces cerevisiae, in: Proceedings of European Congress of Chemical Engineering (ECCE-6). Copenhagen, Denmark, pp. 1–9.

Ashokkumar, M., Lee, J., Kentish, S., and Grieser, F. (2007). Bubbles in an acoustic field: an overview. *Ultrason. Sonochem.* 14: 470–475.

Ashokkumar, M., Bhaskaracharya, R., Kentish, S. et al. (2010). The ultrasonic processing of dairy products – an overview. *Dairy Sci. Technol.* 90: 147–168.

Bachhawat, A.K., Thakur, A., Kaur, J., and Zulkifli, M. (2013). Glutathione transporters. *Biochim. Biophys. Acta – Gen. Subj.* 1830: 3154–3164.

Bar, R. (1988). Ultrasound enhanced bioprocesses: cholesterol oxidation by *Rhodococcus erythropolis*. *Biotechnol. Bioeng.* 32: 655–663.

Béguin, P. and Aubert, J.-P. (1994). The biological degradation of cellulose. *FEMS Microbiol. Rev.* 13: 25–58.

Ben Ammar, J., Lanoisellé, J., Lebovka, N.I. et al. (2011). Impact of a pulsed electric field on damage of plant tissues: effects of cell size and tissue electrical conductivity. *J. Food Sci.* 76.

Blann, A.D., Landray, M.J., and Lip, G.Y.H. (2002). An overview of antithrombotic therapy. *BMJ* 325: 762–765.

Bode, C., Runge, M.S., and Smalling, R.W. (1996). The future of thrombolysis in the treatment of acute myocardial infarction. *Eur. Heart J.* 17: 55–60.

Bothun, G.D., Knutson, B.L., Berberich, J.A. et al. (2004). Metabolic selectivity and growth of *Clostridium thermocellum* in continuous culture under elevated hydrostatic pressure. *Appl. Microbiol. Biotechnol.* 65: 149–157.

Bravim, F., Lippman, S.I., Da Silva, L.F. et al. (2013). High hydrostatic pressure activates gene expression that leads to ethanol production enhancement in a *Saccharomyces cerevisiae* distillery strain. *Appl. Microbiol. Biotechnol.* 97: 2093–2107.

Castro, I., Oliveira, C., Domingues, L. et al. (2012). The effect of the electric field on lag-phase, ethanol and β-galactosidase production of a recombinant *S. cerevisiae* growing on lactose. *Food Bioprocess Technol.* 5: 3014–3020.

Chandrapala, J., Oliver, C., Kentish, S., and Ashokkumar, M. (2012). Ultrasonics in food processing – food quality assurance and food safety. *Trends Food Sci. Technol.* 26: 88–98.

Chisti, Y. (2003). Sonobioreactors: using ultrasound for enhanced microbial productivity. *Trends Biotechnol.* 21: 89–93.

Choi, E.J., Ahn, H., Kim, M. et al. (2015). Effect of ultrasonication on fermentation kinetics of beer using six-row barley cultivated in Korea. *J. Inst. Brew.* 121: 510–517.

Chu, J., Li, B., Zhang, S., and Li, Y. (2000). On-line ultrasound stimulates the secretion and production of gentamicin by *Micromonospora echinospora*. *Process Biochem.* 35: 569–572.

Chuanyun, D., Bochu, W., Chuanren, D., and Sakanishi, A. (2003). Low ultrasonic stimulates fermentation of riboflavin producing strain *Ecemothecium ashbyii*. *Colloids Surf. B – Biointerfaces* 30: 37–41.

Chuanyun, D., Bochu, W., Huan, Z. et al. (2004). Effect of low frequency ultrasonic stimulation on the secretion of siboflavin produced by *Ecemothecium Ashbyii*. *Colloids Surf. B – Biointerfaces* 34: 7–11.

Coelho, M.A.Z., Belo, I., Pinheiro, R. et al. (2004). Effect of hyperbaric stress on yeast morphology: study by automated image analysis. *Appl. Microbiol. Biotechnol.* 66: 318–324.

Dakubu, S. (1976). Cell inactivation by ultrasound. *Biotechnol. Bioeng.* 18: 465–471.

De Angelis, M., Bini, L., Pallini, V. et al. (2001). The acid-stress response in *Lactobacillus sanfranciscensis* CB1. *Microbiology* 147: 1863–1873.

Dong, Y. and Jiang, H. (2016). Microbial production of metabolites and associated enzymatic reactions under high pressure. *World J. Microbiol. Biotechnol.* 32: 178.

Dong, Y., Yang, Q., Jia, S., and Qiao, C. (2007). Effects of high pressure on the accumulation of trehalose and glutathione in the *Saccharomyces cerevisiae* cells. *Biochem. Eng. J.* 37: 226–230. https://doi.org/10.1016/j.bej.2007.04.004.

Drakopoulou, S., Terzakis, S., Fountoulakis, M.S. et al. (2009). Ultrasound-induced inactivation of gram-negative and gram-positive bacteria in secondary treated municipal wastewater. *Ultrason. Sonochem.* 16: 629–634.

Esclapez, M.D., García-Pérez, J.V., Mulet, A., and Cárcel, J.A. (2011). Ultrasound-assisted extraction of natural products. *Food Eng. Rev.* 3: 108.

Ewe, J.A. and Alias, A.K. (2012). Bioconversion of isoflavones and the probiotic properties of the electroporated parent and subsequent three subcultures of Lactobacillus fermentum BT 8219 in biotin-soymilk. *J. Microbiol. Biotechnol.* 22: 947–959.

Ewe, J.-A., Wan-Abdullah, W.-N., Karim Alias, A. et al. (2011). ACE inhibitory activity and bioconversion of isoflavones by *Lactobacillus* in soymilk supplemented with B-vitamins. *Br. Food J.* 113: 1127–1146.

Ewe, J.-A., Wan-Abdullah, W.-N., Alias, A.K., and Liong, M.-T. (2012). Enhanced growth of lactobacilli and bioconversion of isoflavones in biotin-supplemented soymilk by electroporation. *Int. J. Food Sci. Nutr.* 63: 580–596.

Fiedurek, J. (1999). Influence of a pulsed electric field on the spores and oxygen consumption of *Aspergillus niger* and its citric acid production. *Acta Biotechnol.* 19: 179–186.

Fologea, D., Vassu-Dimov, T., Stoica, I. et al. (1998). Increase of *Saccharomyces cerevisiae* plating efficiency after treatment with bipolar electric pulses. *Bioelectrochem. Bioenerg.* 46: 285–287.

Galván-D'Alessandro, L. and Carciochi, R.A. (2018). Fermentation assisted by pulsed electric field and ultrasound: a review. *Fermentation* 4: 1.

Gogate, P.R. and Kabadi, A.M. (2009). A review of applications of cavitation in biochemical engineering/biotechnology. *Biochem. Eng. J.* 44: 60–72.

Grosse, H.-H., Bauer, E., and Berg, H. (1988). Electrostimulation during fermentation. *Bioelectrochem. Bioenerg.* 20: 279–285.

van de Guchte, M., Serror, P., Chervaux, C. et al. (2002). Stress responses in lactic acid bacteria. *Antonie Van Leeuwenhoek* 82: 187–216.

Guimarães, P.M.R., Teixeira, J.A., and Domingues, L. (2010). Fermentation of lactose to bio-ethanol by yeasts as part of integrated solutions for the valorisation of cheese whey. *Biotechnol. Adv.* 28: 375–384.

Guo, J., Ma, R., Su, B. et al. (2016). Raising the avermectins production in *Streptomyces avermitilis* by utilizing nanosecond pulsed electric fields (nsPEFs). *Sci. Rep.* 6: 25949.

Herrán, N.S., López, J.L.C., Pérez, J.A.S., and Chisti, Y. (2008). Effects of ultrasound on culture of *Aspergillus terreus*. *J. Chem. Technol. Biotechnol.* 83: 593–600.

Herrero, A.A., Gomez, R.F., and Roberts, M.F. (1985). 31P NMR studies of *Clostridium thermocellum*. Mechanism of end product inhibition by ethanol. *J. Biol. Chem.* 260: 7442–7451.

Holden, R.W. (1990). Plasminogen activators: pharmacology and therapy. *Radiology* 174: 993–1001.

Hörmann, S., Scheyhing, C., Behr, J. et al. (2006). Comparative proteome approach to characterize the high-pressure stress response of *Lactobacillus sanfranciscensis* DSM 20451T. *Proteomics* 6: 1878–1885.

Hu, H., Yonezawa, Y., Matsuda, A. et al. (2011). Influences of ultrasound and ohmic heating on growth of Sake yeast, in: 11th International Congress on Engineering and Food. Athens, Greece.

Huang, H.-W., Lung, H.-M., Yang, B.B., and Wang, C.-Y. (2014). Responses of microorganisms to high hydrostatic pressure processing. *Food Control* 40: 250–259.

Hunt, R.W., Zavalin, A., Bhatnagar, A. et al. (2009). Electromagnetic biostimulation of living cultures for biotechnology, biofuel and bioenergy applications. *Int. J. Mol. Sci.* 10: 4515–4558.

Jomdecha, C. and Prateepasen, A. (2006). The research of low-ultrasonic energy affects to yeast growthin fermentation process, in: 12th Asia-Pacific Conference on Nondestructive Testing. Auckland, New Zealand, pp. 5–10.

Joo, D.H., Jeon, B.Y., and Park, D.H. (2013). Effects of an electric pulse on variation of bacterial community and metabolite production in kimchi-making culture. *Biotechnol. Bioprocess Eng.* 18: 909–917.

Joyce, E., Phull, S.S., Lorimer, J.P., and Mason, T.J. (2003). The development and evaluation of ultrasound for the treatment of bacterial suspensions. A study of frequency, power and sonication time on cultured *Bacillus* species. *Ultrason. Sonochem.* 10: 315–318.

Kato, N., Sato, T., Kato, C. et al. (2007). Viability and cellulose synthesizing ability of *Gluconacetobacter xylinus* cells under high-hydrostatic pressure. *Extremophiles* 11: 693–698.

Kerns, G., Bauer, E., and Berg, H. (1993). Electrostimulation of cellulase fermentation by pulsatile electromagnetically induced currents. *Bioelectrochem. Bioenerg.* 32: 89–94.

Kiriyama, K., Hara, K.Y., and Kondo, A. (2012). Extracellular glutathione fermentation using engineered *Saccharomyces cerevisiae* expressing a novel glutathione exporter. *Appl. Microbiol. Biotechnol.* 96: 1021–1027.

Kobayashi, Y., Sakai, D., Iwashina, T. et al. (2009). Low-intensity pulsed ultrasound stimulates cell proliferation, proteoglycan synthesis and expression of growth factor-related genes in human nucleus pulposus cell line. *Eur. Cell. Mater.* 17: 15–22.

Kreft, M.E. and Jelen, P. (2000). Stability and activity of β-galactosidase in sonicated cultures of *Lactobacillus delbrueckii* ssp. *bulgaricus* 11842 as affected by temperature and ionic environments. *J. Food Sci.* 65: 1364–1368.

Lado, B.H. and Yousef, A.E. (2002). Alternative food-preservation technologies: efficacy and mechanisms. *Microbes Infect.* 4: 433–440.

Lanchun, S., Bochu, W., Zhiming, L. et al. (2003). The research into the influence of low-intensity ultrasonic on the growth of *S. cerevisiae*. *Colloids Surf. B – Biointerfaces* 30: 43–49.

Lee, J.M., Gianchandani, E.P., and Papin, J.A. (2006). Flux balance analysis in the era of metabolomics. *Brief. Bioinform.* 7: 140–150.

Liang, J., Huang, R., Huang, Y. et al. (2013). Cloning, expression, properties, and functional amino acid residues of new trehalose synthase from *Thermomonospora curvata* DSM 43183. *J. Mol. Catal. B Enzym.* 90: 26–32.

Loghavi, L., Sastry, S.K., and Yousef, A.E. (2007). Effect of moderate electric field on the metabolic activity and growth kinetics of *Lactobacillus acidophilus*. *Biotechnol. Bioeng.* 98: 872–881.

Loghavi, L., Sastry, S.K., and Yousef, A.E. (2008). Effect of moderate electric field frequency on growth kinetics and metabolic activity of *Lactobacillus acidophilus*. *Biotechnol. Prog.* 24: 148–153.

Loghavi, L., Sastry, S.K., and Yousef, A.E. (2009). Effect of moderate electric field frequency and growth stage on the cell membrane permeability of *Lactobacillus acidophilus*. *Biotechnol. Prog.* 25: 85–94.

Lou, Y. and Yousef, A.E. (1997). Adaptation to sublethal environmental stresses protects *Listeria monocytogenes* against lethal preservation factors. *Appl. Environ. Microbiol.* 63: 1252–1255.

Lye, H.S., Karim, A.A., Rusul, G., and Liong, M.T. (2011). Electroporation enhances the ability of lactobacilli to remove cholesterol. *J. Dairy Sci.* 94: 4820–4830.

Masuzawa, N. and Ohdaira, E. (2002). Attempts to shorten the time of lactic fermentation by ultrasonic irradiation. *Jpn. J. Appl. Phys.* 41: 3277–3278.

Matsuura, K., Hirotsune, M., Nunokawa, Y. et al. (1994). Acceleration of cell growth and ester formation by ultrasonic wave irradiation. *J. Ferment. Bioeng.* 77: 36–40.

Mattar, J.R., Turk, M.F., Nonus, M. et al. (2015). *S. cerevisiae* fermentation activity after moderate pulsed electric field pre-treatments. *Bioelectrochemistry* 103: 92–97.

McCabe, A., Barron, N., McHale, L., and McHale, A.P. (1995). Increased efficiency of substrate utilization by exposure of the thermotolerant yeast strain, *Kluyveromyces marxianus* IMB3 to electric-field stimulation. *Biotechnol. Tech.* 9: 133–136.

Mett, H., Schacher, B., and Wegmann, L. (1988). Ultrasonic disintegration of bacteria may lead to irreversible inactivation of β-lactamase. *J. Antimicrob. Chemother.* 22: 293–298.

Mota, M.J., Lopes, R.P., Delgadillo, I., and Saraiva, J.A. (2013). Microorganisms under high pressure - adaptation, growth and biotechnological potential. *Biotechnol. Adv.* 31: 1426–1434.

Mota, M.J., Lopes, R.P., Delgadillo, I., and Saraiva, J.A. (2015). Probiotic yogurt production under high pressure and the possible use of pressure as an on/off switch to stop/start fermentation. *Process Biochem.* 50: 906–911.

Mota, M.J., Lopes, R.P., Koubaa, M. et al. (2018). Fermentation at non-conventional conditions in food-and bio-sciences by the application of advanced processing technologies. *Crit. Rev. Biotechnol.* 38: 122–140.

Nakanishi, K., Tokuda, H., Soga, T. et al. (1998). Effect of electric current on growth and alcohol production by yeast cells. *J. Ferment. Bioeng.* 85: 250–253.

Naveena, B., Sakthiselvan, P., Elaiyaraju, P., and Partha, N. (2012). Ultrasound induced production of thrombinase by marine actinomycetes: kinetic and optimization studies. *Biochem. Eng. J.* 61: 34–42.

Neto, R., Mota, M.J., Lopes, R.P. et al. (2016). Growth and metabolism of *Oenococcus oeni* for malolactic fermentation under pressure. *Lett. Appl. Microbiol.* 63: 426–433.

Nguyen, T.M.P., Lee, Y.K., and Zhou, W. (2009). Stimulating fermentative activities of bifidobacteria in milk by high intensity ultrasound. *Int. Dairy J.* 19: 410–416.

Nguyen, T.M.P., Lee, Y.K., and Zhou, W. (2012). Effect of high intensity ultrasound on carbohydrate metabolism of bifidobacteria in milk fermentation. *Food Chem.* 130: 866–874.

Nishi, Y., Uryu, M., Yamanaka, S. et al. (1990). The structure and mechanical properties of sheets prepared from bacterial cellulose. *J. Mater. Sci.* 25: 2997–3001.

O'donnell, C.P., Tiwari, B.K., Bourke, P., and Cullen, P.J. (2010). Effect of ultrasonic processing on food enzymes of industrial importance. *Trends Food Sci. Technol.* 21: 358–367.

Ohba, T., Uemura, K., and Nabetani, H. (2016). Moderate pulsed electric field treatment enhances exopolysaccharide production by *Lactococcus lactis* subspecies *cremoris*. *Process Biochem.* 51: 1120–1128.

Panagopoulos, D.J., Karabarbounis, A., and Margaritis, L.H. (2002). Mechanism for action of electromagnetic fields on cells. *Biochem. Biophys. Res. Commun.* 298: 95–102.

Picard, A., Daniel, I., Montagnac, G., and Oger, P. (2007). In situ monitoring by quantitative raman spectroscopy of alcoholic fermentation by Saccharomyces cerevisiae under high pressure. *Extremophiles* 11: 445–452.

Povey, M.J.W. and Mason, T.J. (1998). Ultrasound in Food Processing. London, UK: Blackie Academic & Professional.

Qiao, C., Liu, B., Xu, X., and Jia, S. (2006). The changes of glutathione and ergosterol in *Saccharomyces cerevisiae* under high pressure. *J. Chin. Biotechnol.* 26: 59.

Rowley, B.A. (1972). Electrical current effects on *E. coli* growth rates. *Proc. Soc. Exp. Biol. Med.* 139: 929–934.

Runyan, C.M., Carmen, J.C., Beckstead, B.L. et al. (2006). Low-frequency ultrasound increases outer membrane permeability of *Pseudomonas aeruginosa*. *J. Gen. Appl. Microbiol.* 52: 295–301.

Sakakibara, M., Wang, D., Ikeda, K., and Suzuki, K. (1994). Effect of ultrasonic irradiation on production of fermented milk with *Lactobacillus delbrueckii*. *Ultrason. Sonochem.* 1: S107–S110.

Sastry, S.K., Shynkaryk, M., and Somavat, R. (2011). Ohmic and moderate electric field processing: developments and new applications, in: 11th International Congress on Engineering and Food. Athens, Greece.

Saura, R., Terashi, H., Lee, S. et al. (2003). Low intensity pulsed ultrasound exposure increase prostaglandin E2 release in human dermal fibroblasts. *Bull. Health Sci. Kobe* 19: 121–128.

Scherba, G., Weigel, R.M., and O'brien, W.D. (1991). Quantitative assessment of the germicidal efficacy of ultrasonic energy. *Appl. Environ. Microbiol.* 57: 2079–2084.

Schläfer, O., Sievers, M., Klotzbücher, H., and Onyeche, T.I. (2000). Improvement of biological activity by low energy ultrasound assisted bioreactors. *Ultrasonics* 38: 711–716.

Serrazanetti, D.I., Guerzoni, M.E., Corsetti, A., and Vogel, R. (2009). Metabolic impact and potential exploitation of the stress reactions in lactobacilli. *Food Microbiol.* 26: 700–711.

Setchell, K.D.R., Brown, N.M., Zimmer-Nechemias, L. et al. (2002). Evidence for lack of absorption of soy isoflavone glycosides in humans, supporting the crucial role of intestinal metabolism for bioavailability. *Am. J. Clin. Nutr.* 76: 447–453.

Shil, P., Bidaye, S., and Vidyasagar, P.B. (2008). Analysing the effects of surface distribution of pores in cell electroporation for a cell membrane containing cholesterol. *J. Phys. Appl. Phys.* 41: 55502.

Shimada, T., Ohdaira, E., and Masuzawa, N. (2004). Effect of ultrasonic frequency on lactic acid fermentation promotion by ultrasonic irradiation. *Jpn. J. Appl. Phys.* 43: 2831–2832.

Singh, S., Agarwal, M., Sarma, S. et al. (2015). Mechanistic insight into ultrasound induced enhancement of simultaneous saccharification and fermentation of *Parthenium hysterophorus* for ethanol production. *Ultrason. Sonochem.* 26: 249–256. https://doi.org/10.1016/j.ultsonch.2015.02.011.

Storz, G. and Hengge, R. (2000). Bacterial Stress Responses. Washington DC, US: American Society for Microbiology Press.

Subhedar, P.B. and Gogate, P.R. (2015). Ultrasound-assisted bioethanol production from waste newspaper. *Ultrason. Sonochem.* 27: 37–45.

Sulaiman, A.Z., Ajit, A., Yunus, R.M., and Chisti, Y. (2011). Ultrasound-assisted fermentation enhances bioethanol productivity. *Biochem. Eng. J.* 54: 141–150.

Takeuchi, R., Ryo, A., Komitsu, N. et al. (2008). Low-intensity pulsed ultrasound activates the phosphatidylinositol 3 kinase/Akt pathway and stimulates the growth of chondrocytes in three-dimensional cultures: a basic science study. *Arthritis Res. Ther.* 10: R77.

Tang, W., Liu, Q., Wang, X. et al. (2008). Membrane fluidity altering and enzyme inactivating in sarcoma 180 cells post the exposure to sonoactivated hematoporphyrin *in vitro*. *Ultrasonics* 48: 66–73.

Tanino, T., Sato, S., Oshige, M., and Ohshima, T. (2012). Analysis of the stress response of yeast *Saccharomyces cerevisiae* toward pulsed electric field. *J. Electrost.* 70: 212–216.

Toba, T., Hayasaka, I., Taguchi, S., and Adachi, S. (1990). A new method for manufacture of lactose-hydrolysed fermented milk. *J. Sci. Food Agric.* 52: 403–407.

Turpie, A.G.G., Chin, B.S.P., and Lip, G.Y.H. (2002). Venous thromboembolism: treatment strategies. *BMJ* 325: 948–950.

Velizarov, S. (1999). Electric and magnetic fields in microbial biotechnology: possibilities, limitations, and perspectives. *Electro- Magnetobiol.* 18: 185–212.

Verluyten, J., Messens, W., and De Vuyst, L. (2003). The curing agent sodium nitrite, used in the production of fermented sausages, is less inhibiting to the bacteriocin-producing meat starter culture *Lactobacillus curvatus* LTH 1174 under anaerobic conditions. *Appl. Environ. Microbiol.* 69: 3833–3839.

Villamiel, M. and de Jong, P. (2000). Inactivation of *Pseudomonas fluorescens* and *Streptococcus thermophilus* in trypticase soy broth and total bacteria in milk by continuous-flow ultrasonic treatment and conventional heating. *J. Food Eng.* 45: 171–179.

Wang, D. and Sakakibara, M. (1997). Lactose hydrolysis and β-galactosidase activity in sonicated fermentation with *Lactobacillus* strains. *Ultrason. Sonochem.* 4: 255–261.

Wang, D., Sakakibara, M., Kondoh, N., and Suzuki, K. (1996). Ultrasound-enhanced lactose hydrolysis in milk fermentation with *Lactobacillus bulgaricus*. *J. Chem. Technol. Biotechnol.* 65: 86–92.

Welch, T.J., Farewell, A., Neidhardt, F.C., and Bartlett, D.H. (1993). Stress response of *Escherichia coli* to elevated hydrostatic pressure. *J. Bacteriol.* 175: 7170–7177.

Wemekamp-Kamphuis, H.H., Wouters, J.A., de Leeuw, P.P.L.A. et al. (2004). Identification of sigma factor σB-controlled genes and their impact on acid stress, high hydrostatic pressure, and freeze survival in *Listeria monocytogenes* EGD-e. *Appl. Environ. Microbiol.* 70: 3457–3466.

Wiegel, J. (1980). Formation of ethanol by bacteria. A pledge for the use of extreme thermophilic anaerobic bacteria in industrial ethanol fermentation processes. *Experientia* 36: 1434–1446.

Wood, B.E., Aldrich, H.C., and Ingram, L.O. (1997). Ultrasound stimulates ethanol production during the simultaneous saccharification and fermentation of mixed waste office paper. *Biotechnol. Prog.* 13: 232–237.

Wu, J. and Nyborg, W.L. (2008). Ultrasound, cavitation bubbles and their interaction with cells. *Adv. Drug Deliv. Rev.* 60: 1103–1116.

Wu, H., Hulbert, G.J., and Mount, J.R. (2000). Effects of ultrasound on milk homogenization and fermentation with yogurt starter. *Innov. Food Sci. Emerg. Technol.* 1: 211–218.

Yeo, S.-K. and Liong, M.-T. (2010). Effect of prebiotics on viability and growth characteristics of probiotics in soymilk. *J. Sci. Food Agric.* 90: 267–275.

Yeo, S.-K. and Liong, M.-T. (2013). Effect of electroporation on viability and bioconversion of isoflavones in mannitol-soymilk fermented by lactobacilli and bifidobacteria. *J. Sci. Food Agric.* 93: 396–409.

Zabaneh, M. and Bar, R. (1991). Ultrasound-enhanced bioprocess. II: dehydrogenation of hydrocortisone by *Arthrobacter simplex*. *Biotechnol. Bioeng.* 37: 998–1003.

Zhao, Y., Ang, W.T., Xing, J. et al. (2012). Applications of ultrasound to enhance mycophenolic acid production. *Ultrasound Med. Biol.* 38: 1582–1588.

Zimmermann, U., Pilwat, G., and Riemann, F. (1974). Dielectric breakdown of cell membranes. *Biophys. J.* 14: 881–899.

7

Application of Fermentation to Recover High-Added Value Compounds from Food By-Products

Antifungals and Antioxidants

Sucheta Khubber[1], *Francisco J. Marti-Quijal*[2], *Igor Tomasevic*[3], *Fabienne Remize*[4], *and Francisco J. Barba*[2]

[1]*Food Engineering and Nutrition, Center of Innovative and Applied Bioprocessing, Mohali, India*
[2]*Faculty of Pharmacy, Preventive Medicine and Public Health, Food Sciences, Toxicology and Forensic Medicine Department, Universitat de València, València, Spain*
[3]*Faculty of Agriculture, University of Belgrade, Belgrade, Serbia*
[4]*UMR QualiSud, Université de La Réunion, CIRAD, Université Montpellier, Institut Agro, Université d'Avignon, Sainte Clotilde, France*

7.1 Introduction

The world's rising population and modern lifestyle have stimulated the demand for direct consumable products, which have led to the growth and diversification of industrial processing of food raw materials. The extensive food processing generates large quantity of food waste from manufacturing industries every year, which count to 88 million tons (MT) with a cost estimated at €143 billion (FUSIONS 2016). The food waste or by-products are remnants of processed food and include peel, leaf, stem, pomace, bran, broken kernels, seed, oil cakes, etc. (Barba et al. 2020). Food sources including cereals, pulses or oilseeds, dairy, fruits or vegetables, and meat products generate different kinds of leftovers varying in fibers, proteins, phytochemicals, and micronutrients, which determine their potential to be utilized for value-added products. The by-products may be either converted into number of useful compounds through chemical conversions or extracted for enzymes, phytochemicals, fibers, and micronutrients (Chemat et al. 2020; Kamani et al. 2019; Ravindran and Jaiswal 2016). However, microbial fermentation, an emerging sustainable approach, finds possibilities for bioconversion of various food by-products into value-added bioproducts (e.g. biofuel, organic acids, pigments, phenolics, sugars, etc.) of enhanced functionality (Martí-Quijal et al. 2020a; Ng et al. 2020; Ravindran and Jaiswal 2016; Sindhu et al. 2020). It is an expedient process as it results in the production or extraction of high-quality bioactives without any risk of toxicity as perceived in solvent extraction or chemical conversion processes. Solid-state fermentation (SSF) or submerged fermentation (SmF) is widely employed to utilize the bioactives from food by-products such as shrimp, rice bran, etc. (Mapelli-Brahm et al. 2020; Roselló-Soto et al. 2019; Shin et al. 2019). SSF is a low-cost process using low water quantity and energy giving minimal waste streams. A number of newly isolated

Fermentation Processes: Emerging and Conventional Technologies, First Edition.
Edited by Mohamed Koubaa, Francisco J. Barba, and Shahin Roohinejad.

bacterial (e.g. lactic acid bacteria (LAB), *Propionibacteria*, *Chromobacterium*, *Acetobacter*, etc.) and fungal strains (e.g. *Aspergillus*, *Saccharomyces*, *Rhodosporidium*, etc.) have been demonstrated to grow on a wide variety of food by-products as substrate (Sindhu et al. 2020). The principle of using fermentation for value addition to by-products is based on the ability of microbial endogenous enzymes such as hydrolases, esterases, dehydrogenases, isomerases, decarboxylases, or lyases to catalyze the conversion of biopolymers or polyphenolics in food by-products via oxidation–reduction pathways into sugars, alcohol, phenolic acid derivatives, fatty acids, bioactive peptides, and release of certain vitamins (Marti-Quijal et al. 2020). The released microbial metabolites can exhibit strong antioxidant activity or broad-range antifungal activity. The richness of food by-products in bioactives or biopolymers and further enhancement of the contents of such compounds through microbial fermentation to contribute toward antifungal or antioxidant potential has been reported previously in several studies (de Francisco et al. 2018; Martins et al. 2011; Martí-Quijal et al. 2020a, b; Smuda et al. 2018). The utilization of food by-products can generate incomes to the food industries, and fermentation, being an eco-friendly process, underpins a commercial potential to recover antifungals and antioxidants. Overall, fermentation acts as an eco-friendly economical alternative for the recovery of antifungals and antioxidants from the food industry by-products (Sadh et al. 2018). This chapter discusses an overview of the food by-products as a source of antifungals and antioxidants and fermentation as a strategy to utilize by-products and enhance the recovery of antifungals and antioxidants. The commercial value of antifungal and antioxidant compounds recovered using fermentation and technical hurdles associated with it is also discussed.

7.2 Food Industry By-Products and Global Estimates

A number of food sources are partially processed or processed on an industrial scale for further treatment or direct human consumption, thus resulting in by-products generation in large quantities. The varied food by-products are commercially preferred for animal feed due to unrealized bioactivity or nutritional benefits to humans. Diverse categories of food by-products from cereals, fruits, vegetables, meat, poultry, etc. (Figure 7.1) comprise dietary fibers, proteins, phenolics, and polysaccharides, opening the possibility to utilize them for nutraceuticals, extruded food products, or as food ingredients (Barba et al. 2020).

Bran is one of the largely produced cereal by-products, which is a rich source of phenolics, proteins, or peptides. Spent malt grains from barley or distiller's grains from corn or cornmeal comprises fibers, fatty acids, and phytochemicals such as carotenoids (Klempová et al. 2020). About 4.24 MT dry distiller's grains were produced around the European Union in 2019 (Statista 2020). Fruits and vegetables show great diversity in by-products with various kinds and quantities of phytochemicals. Peel, pomace, seeds, or husk are the major ones generated in large quantities (<0.5 billion tons) worldwide (Coman et al. 2020). Most of the by-products from meat and poultry industry are generated during slaughtering and includes blood, skin, organs, or visceral biomass. Blood, edible fats, proteins, and collagen are potentially used as edible by-products after cooking, braising, or boiling. Egg white flavoprotein is also a useful food ingredient. The

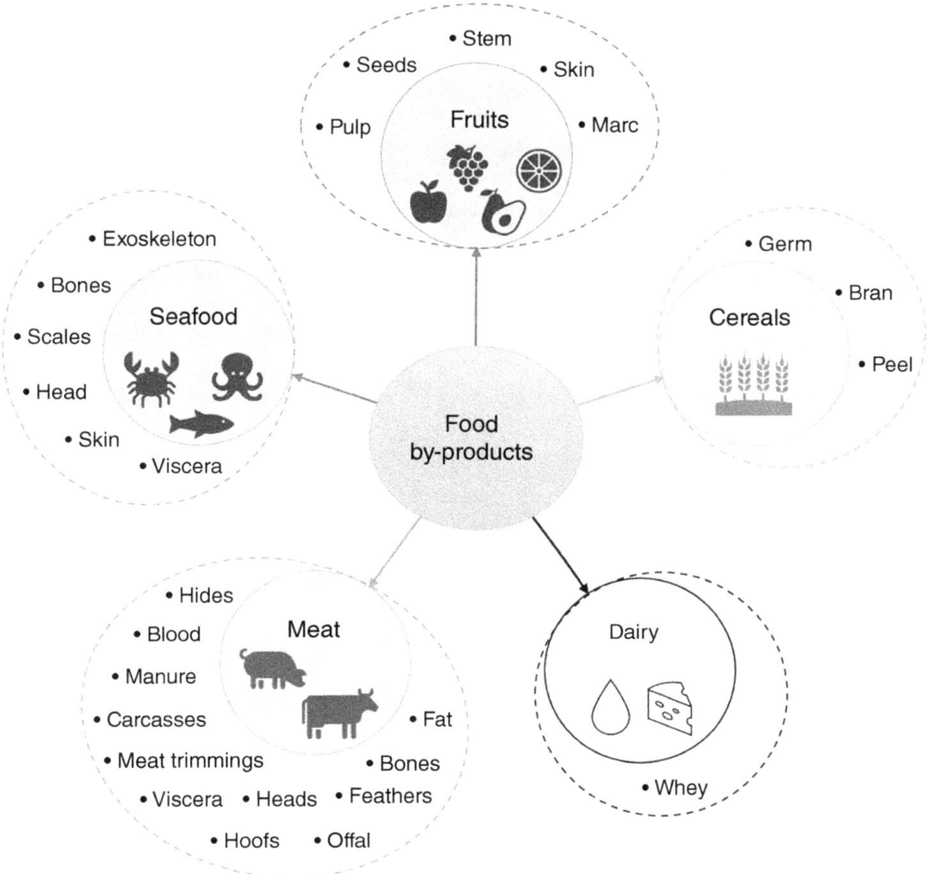

Figure 7.1 Main by-products generated from meat, seafood, dairy, fruit, and cereal processing.

world production of edible by-products from pigs in 2004 was 625 MT (Jayathilakan et al. 2012). Fish by-products constitute about 60% of the total fish weight and can serve as a good source of proteins, enzymes, oil, or gelatin for human consumption. About 1.5 MT of fish waste is generated in the United States every year and is discarded (Marti-Quijal et al. 2020). Seafood by-products or aquatic processing waste can be completely valorized for value-added compounds such as chitin, carotenoids, enzymes, or minerals (Bruno et al. 2019). The volume of seafood by-products accounts for about 30–70% of whole seafood, which is produced in approximately 40–80 MT (FAO 2018). Whey is a major by-product of dairy industry and is produced in large quantities worldwide (≈40 MT in 2012), and 10% is processed to protein concentrate. The whey permeate can lead to environmental pollution owing to its high biological oxygen demand. Therefore, recent efforts were focused on its bioconversion into value-added compounds through microbial fermentation (Pescuma et al. 2015).

7.3 Food By-Products as Sources of Antifungals or Antioxidants

A number of food sources are fermented for sensory and health benefits and increase in shelf life. The by-products are also composed of similar macro-, micronutrients, and phytochemicals, which enable a number of microbial species to thrive and release bioactive metabolites (Figure 7.2). Such compounds show considerable antifungal or antioxidant potential, which could be attributing shelf life effects in food upon fermentation. The food by-products may be categorized into plant-derived, dairy, meat, fish, poultry, and seafood for their understanding as a source to recover antifungals or antioxidants using fermentation and are detailed below.

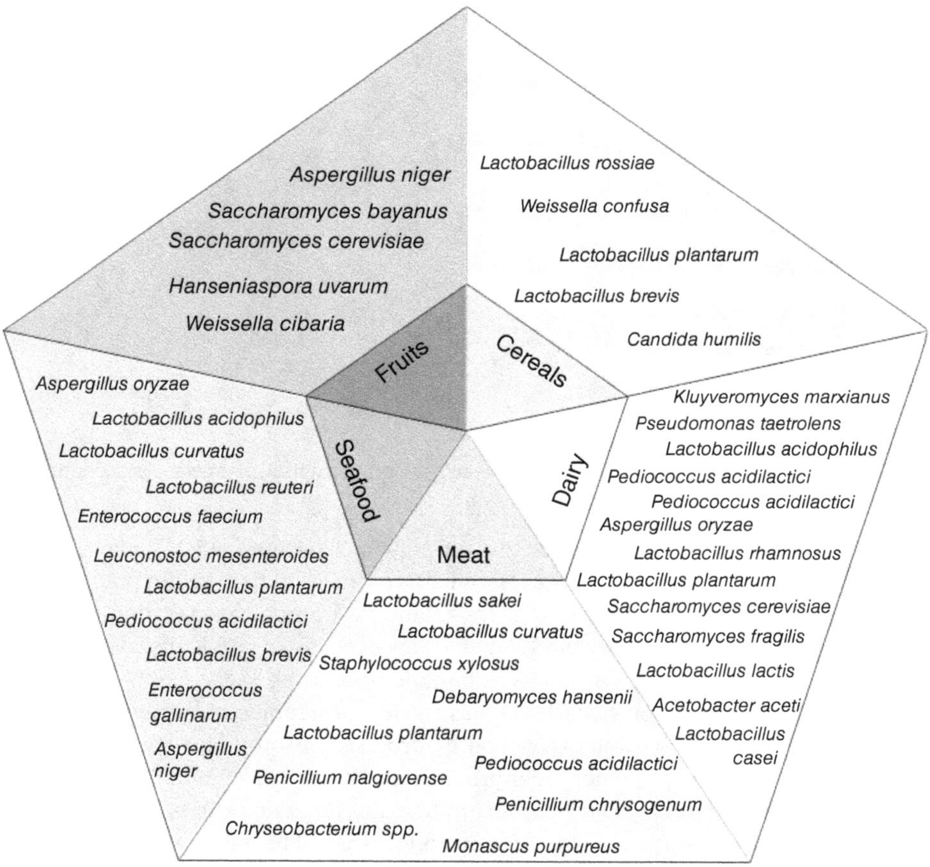

Figure 7.2 Microbial species thrive and release bioactive metabolites from meat, seafood, dairy, fruit, and cereals. *Source:* Tlais et al. (2020), Zotta et al. (2020), Marti-Quijal et al. (2020), Toldrá (2017), and Borrajo et al. (2019).

7.3.1 Fruit

Approximately 30% of the global fruit agricultural production is considered as waste and by-products (Ajila et al. 2012; Barba et al. 2020). These by-products are different according to the target matrix (Puértolas et al. 2016; Roselló-Soto et al. 2015a). For example, in the case of fruits, the main by-products are pulp, seeds, skins, stems and pomace. Among the fruits that generate most of the by-products, we can highlight apple, pomegranate, grape, avocado, and citrus (Tlais et al. 2020). Apple, for example, is mainly intended for the manufacture of juices and derivatives such as vinegar, thereby its main by-product is the pulp and the skin, followed by the seeds and stems (Gunes et al. 2019). These by-products contain significant amounts of bioactive compounds, especially polyphenols, being the predominant chlorogenic acid, quercetin glycosides, and floretine glycosides (Călinoiu et al. 2017). This composition so rich in polyphenols gives a great beneficial value to their extracts for health, mainly due to its antioxidant ability, although others have also been described due to their antidiabetic or antitumor roles (Rabetafika et al. 2014).

On the other hand, pomegranate is another fruit widely processed in the industry to obtain juice and other derivatives such as jam or syrup (Putnik et al. 2019). Therefore, after industrial processing, the main by-products obtained are hull, bagasse, and pulp. These by-products, especially peel, have been found to be rich in polyphenols, in particular flavonoids, anthocyanins, and ellagitannins, which have, among other characteristics, a high antioxidant activity (Ismail et al. 2016; Kowalska et al. 2017; Lorenzo et al. 2019). Extracts obtained from pomegranate by-products have also been shown to have antifungal activity. Specifically, Tehranifar et al. (2011) studied the antifungal activity of methanolic and aqueous extracts of different by-products of the pomegranate against *Penicillium italicum*, *Rhizopus stolonifer*, and *Botrytis cinerea*. They observed that the highest antifungal activity was obtained with methanolic extracts, being similar at concentrations of 500, 1000, or 1500 ppm. These authors also reported that the antioxidant activity of the husk extract was 54% higher than that of the seeds, although the content of polyphenols was similar (302 versus 325 mg/g, respectively). Tannin fractions are well-known for their antimicrobial activity (Reddy et al. 2007).

The grape by-products come mainly from its use in winemaking and juice production (Barba et al. 2015, 2016). After this process, grape marc is obtained as the main by-product, representing around 20% of the total weight of the grape. This grape pomace is made up of skins, stems, residual pulp, and seeds (Barba et al. 2016; Bordiga et al. 2019). These by-products are known to have high polyphenol content. The main ones are gallic acid, syringic acid, procyanidins B1 and B2, catechin, epicatechin, epicatechin gallate, quercetin 3-β-d-glucoside, delphinidin 3-glucoside, peonidin 3-O-glucoside, and malvidin 3-glucoside (Melo et al. 2015). However, they are not evenly distributed. Anthocyanins are abundant in the skin, while flavanols are predominant in the seeds (Bordiga et al. 2019). This high concentration of polyphenols in these by-products is correlated with a high antioxidant capacity, which has been shown to, among other health benefits, counteract the effects produced by a high intake of mycotoxins (Moccia et al. 2020; Taranu et al. 2019). Extracts obtained from these by-products have also been shown to have antifungal activity against *B. cinerea*, one of the main grape phytopathogen (Mendoza et al. 2013). Not only has antifungal activity been seen against pathogenic fungi of the vineyard, it has also been described against fungi harmful to human health, specifically against *Candida albicans*, in addition to having other properties such as antimicrobial or antiviral activity (Friedman 2014).

Citrus fruit is another group of fruit widely consumed worldwide. The main product obtained from them is juice, and therefore by-products such as seeds or peels are generated (Putnik et al. 2017). From these, compounds such as ascorbic acid and polyphenols (flavonoids, flavonols, and flavanones) can be obtained, all of them with a high antioxidant capacity (İnan et al. 2018). In addition, antifungal properties of extracts of peels, including essential oils, of different citrus fruits have also been described by inhibiting the *in vitro* growth of the fungus *Fusarium oxysporum* (Okwu et al. 2007).

Finally, avocado is another fruit widely cultivated and industrially processed in order to prepare mainly guacamole, although other derivatives such as oil are also obtained. After its industrial processing, the main by-products generated are the skin and the stone, which have a high content of polyphenols, specifically hydroxybenzoic and hydroxycinnamic acid, procyanidins, and flavonols (Coman et al. 2020). Tremocoldi et al. (2018) studied the antioxidant capacity of hydroalcoholic extracts obtained from these by-products. They reported great antioxidant capacity, as well as good anti-inflammatory activity, especially in the peel extracts of the "Fuerte" variety.

7.3.2 Cereals

In the cereal industry, the main by-products obtained after processing are the germ, the peel, and the bran (Anal et al. 2013; Rocchetti et al. 2019; Nikmaram et al. 2017). Among the main industrially processed cereals, we can find rice, wheat, and corn.

Wheat is processed globally, generally for flour, so its main by-product is wheat bran (Viuda-Martos et al. 2019). This by-product is rich in polyphenols, the most abundant being those derived from benzoic acid and cinnamic acid, ferulic acid standing out as the most abundant polyphenol (Viuda-Martos et al. 2019). Due to its high content of polyphenols, it has an *in vitro* high antioxidant activity (Janiak et al. 2019). In addition, wheat bran-fortified biscuits have also been shown to have a mild antifungal effect against mycotoxin-producing fungi (*Aspergillus flavus*, *Aspergillus niger*, *Fusarium solani*, and *Penicillium notatum*), in addition to clearly reducing the production of aflatoxin, a carcinogenic mycotoxin, produced by fungi of the genus *Aspergillus* (Hussien et al. 2019).

Corn is another kind of cereal most consumed by the world population, especially in Latin America and South of Africa, where it represents the third of its caloric intake (Díaz-Gómez et al. 2017). At an industrial level, its processing consists of dry milling to clean the grain, generating corn bran as the main by-product (Viuda-Martos et al. 2019). Corn bran is an important source of polyphenols, more specifically ferulic acid, which has great antioxidant capacity (Dapčević-Hadnadev et al. 2018). It can also be processed following a wet milling process, obtaining in this case fiber as the main by-product, which has also been described as having antioxidant activity (Viuda-Martos et al. 2019).

Finally, rice is the third most cultivated cereal in the world, after wheat and corn. After industrial processing to obtain the edible product from raw rice, by-products such as husk, bran, or germ are generated (Al-Doury et al. 2018; Xia et al. 2019, 2020). These by-products have a high content of polyphenols, highlighting vanillic and p-coumaric acid in the shell, and gallic acid, p-hydroxybenzoic acid, chlorogenic acid, vanillic acid, caffeic acid, ferulic acid, syringic acid, and p-coumaric acid in bran and germ (Peanparkdee and Iwamoto 2019). All of these compounds confer important antioxidant properties to these by-products (Peanparkdee and Iwamoto 2019).

7.3.3 Dairy Products

The main by-product of the dairy industry is whey, which is obtained mainly during cheesemaking (Marti-Quijal et al. 2019; Musina et al. 2018). Whey represents approximately 80–90% of the total volume of milk used to make cheese and is a source of different nutrients and bioactive peptides (Ajila et al. 2012; Carvalho et al. 2013). Proteins present in whey have been shown to enhance the activity of glutathione peroxidase, a key enzyme in the fight against oxidative stress (Yu et al. 2020). One of the components of whey is lactoferrin, which represents between 1 and 2% of the total proteins present in whey (Echegaray et al. 2019). Lactoferrin has been shown to have antioxidant capacity (Khan et al. 2019). This is due to its ability to scavenge free radicals due to its sulfur-containing amino acids in its structure and the chelation of transition metals (McCarthy et al. 2001). Furthermore, lactoferrin has also been shown to have antifungal activity (Daley 2009). On the other hand, antifungal properties have also been described for fatty acids present in whey. Specifically, after bovine serum saponification, the free fatty acids obtained have been found to inhibit the growth of *Aspergillus fumigatus* and *C. albicans*, with γ-linolenic acid being the most potent antifungal compound (Clément et al. 2008). Finally, during the elaboration of the yogurt, bioactive peptides are released due to the fermentation process. These peptides, present in the serum, have demonstrated *in vitro* antioxidant activity, in addition to possessing other biologically interesting characteristics such as antihypertensive activity (Sultan et al. 2017).

7.3.4 Meat

Meat, one of the most consumed foods worldwide, is an important source of essential amino acids, in addition to providing other nutrients such as iron, zinc, or proteins (Pateiro et al. 2019). However, many parts cannot be destined for human consumption, being discarded as by-products during processing. In general, these by-products are carcasses, hides, feathers, manure, offal, hoofs, heads, bones, fat, viscera, meat trimmings, and blood (Bharat Helkar and Sahoo 2016). From these by-products, bioactive peptides can be obtained by hydrolysis or fermentation (Bharat Helkar and Sahoo 2016; Borrajo et al. 2019; Lorenzo et al. 2018). To hydrolyze proteins, two different treatments can be applied: chemical or enzymatic hydrolysis. Chemical hydrolysis uses an acid or a base to hydrolyze proteins and generate peptides, whereas enzymatic hydrolysis uses endogenous or exogenous enzymes (Pateiro et al. 2019). Enzymatic hydrolysis is the most widely used because higher quality peptides are obtained, with better qualities and without residual contamination of toxic chemical compounds (Lafarga and Hayes 2014; Pateiro et al. 2019). Bioactive peptides are composed of between 2 and 20 amino acids, and their biological activity is defined by the composition of these amino acids and their location (Lorenzo et al. 2018). In addition, the length of the peptide is also important since the activity decreases with the increasing peptide length (Vermeirssen et al. 2004). Regarding this, different peptides with antioxidant and even antifungal activity have been obtained by hydrolysis of meat by-products. Przybylski et al. (2016) evaluated the antioxidant capacity of isolated bovine hemoglobin peptides. The results obtained showed that the hydrolysate inhibited lipid oxidation in a way comparable to a synthetic antioxidant in minced meat. Saiga et al. (2003) obtained antioxidant peptides by enzymatic hydrolysis of the myofibrillar proteins of the pig using papain. These peptides had the amino acid sequences DSGVT, IEAEGE, EELDNALN, VPSIDDQEELM, and DAQEKLE, with the latter peptide showing the highest antioxidant activity. Di Bernardini et al. (2012) obtained antioxidant peptides by hydrolysis of bovine brisket muscles with papain. On the other hand, four antioxidant peptides were obtained from hydrolyzed collagen from pig skin (Li et al. 2007). Antioxidant peptides have also been isolated from water buffalo horn, with the sequences QYDQGV, YEDCTDCGN, and AADNANELFPPN, confirming *in vitro* their protective antioxidant activity (Liu et al. 2010). Finally, Hu et al. (2011) obtained a peptide with the sequence VNFKLLSHSLLVTLASHL from bovine hemoglobin with antifungal activity against *C. albicans*.

7.3.5 Seafood

One of the most important components of the diet is fish and other foods from the sea such as cephalopods or crustaceans. The main by-products of fish are head, viscera, skin, bones, and scales, while the main by-product of shellfish after industrial processing is the exoskeleton (Al Khawli et al. 2020). Different biologically interesting compounds can be recovered from fish by-products, but the main compounds with antioxidant activity are bioactive peptides. Various studies have been carried out to reuse fish by-products in order to obtain these antioxidant compounds. Lima et al. (2019) studied how to recover antioxidant peptides from the industrial by-products of the Stripped weakfish by enzymatic hydrolysis process, using Alcalase® and Protamex® as enzymes. Analysis of the obtained antioxidant peptides revealed that the amino acids Phe, Leu, and Trp as well as the WDDMEK sequence have an important role in the antioxidant capacity of these peptides. On the other hand, Zamora-Sillero et al. (2018a) used these same enzymes to obtain antioxidant peptides from common carp by-products. After the hydrolysis process, peptides were obtained, and their antioxidant activity was confirmed in mouse hippocampal cell line (HT-22 cells). The search for bioactive peptides from fish by-products is a subject of great interest, so there are currently several bibliographic reviews on the subject (Hou et al. 2017; Sila and Bougatef 2016; Venkatesan et al. 2017; Zamora-Sillero et al. 2018b).

Apart from fish, by-products from other marine animals such as crustaceans or mollusks can also be used. From these by-products, the compounds obtained with antioxidant or antifungal capacity are mainly polysaccharides derived from chitosan. In this sense, chitosan has been shown to possess antioxidant capacity, which depends on its molecular weight and degree of acetylation (Anraku et al. 2018). Chitosan has also been shown to possess both antibacterial and antifungal activity (Verlee et al. 2017). Younes et al. (2014) obtained chitin and chitosan through enzymatic deproteinization of the shrimp processing by-products. The chitin and chitosan obtained had great antioxidant capacity, in addition to antifungal activity against the fungi *F. oxysporum*, *F. solani*, and *Fusarium* sp. Bioactive peptides with antioxidant activity can also be obtained from the by-products of shellfish. Vieira et al. (2016) obtained antioxidant peptides by enzymatic hydrolysis from shrimp by-products. On the other hand, Jiang et al. (2017) also obtained antioxidant peptides after an enzymatic hydrolysis process of crab by-products, specifically crab shells. Another interesting source of bioactive peptides is the squid by-products, mainly the viscera. Song et al. (2016) used autolysis to obtain antioxidant bioactive peptides from North Pacific squid viscera. Finally, Jiang et al. (2018) used pepsin to obtain hydrolysates with antioxidant activity from squid viscera and ink sacs.

7.4 Fermentation as a Strategy for Food By-Product Valorization

Fermentation is a metabolic process by which energy is obtained from organic compounds without using any external oxidizing agent such as oxygen (König and Fröhlich 2017). The fermentative metabolic pathways vary with the type of food sources and its nutritional value, containing carbon, fatty acids, nitrogen, or phenolic compounds, and is catalyzed by certain enzymes (e.g. hydrolase, esterase, dehydrogenase, isomerase, decarboxylase, lyase, and more). For instance, in lactic acid fermentation, the sugars (e.g. glucose, galactose, fructose, pentoses) are generally converted through Embden-Meyerhof or phosphoketolase pathway to pyruvate which is then oxidized to lactate, or a mix with acetate, acetaldehyde, ethanol, carbon dioxide, acetoin, diacetyl, and 2,3-butanediol in hetero lactic fermentation. Certain enzymes such as lipases bring out reduction of fats to fatty acids, which further oxidize, and the end products contribute to flavor. The phenolics of fruits and vegetables or cereals are de-esterified, degraded, or converted to other individual phenolic acids through fermentation. Several enzymes such as glycosyl hydrolases, decarboxylases, reductases, and esterases are involved in the formation of bioactive metabolites from phenolic acids through microbial metabolism. The fermentation process can be carried out in two different ways:

- *SSF*: In this procedure, the microorganisms, primarily fungi, grow on a solid substrate, without water in a liquid state, and it is possible because this substrate has enough humidity to support microbiological growth (Thomas et al. 2013). With this process, it is possible to use as much as possible the nutrients that the substrate contains. In turn, the SSF can be divided into two subtypes depending on whether the solid is the source of nutrients (for example, in the case of food by-products) or if, on the contrary, the solid is inert and is impregnated with a nutrient solution to microbiological growth. In turn, it is also possible to differentiate between whether the culture is pure (only one strain is used) or mixed (a combination of microorganisms is used) (Sadh et al. 2018). The SSF has drawn attention in recent years as a promising ecological technology that allows obtaining bioproducts of industrial interest using solid biomass (waste and by-products). The main advantages over conventional SmF are based on the lower water and energy requirements, which generate minimal waste streams. Among its main advantages, we can highlight the low consumption of both water and energy, as well as the low generation of waste. That is why, SSF has gained prominence in recent years as an ecological and promising process (Cerda et al. 2019).

- *SmF*: In this case, water is added to a solid substrate, or the solid is liquefied, to obtain a liquid medium (Sadh et al. 2018). It has some disadvantages compared to SSF such as the higher consumption of water and energy or the need for a large physical space. Despite this, this fermentation technique is widely used in industry for its low cost, high performance, and low contamination when processed discontinuously (Sadh et al. 2018).

Several studies have focused on the use of these processes to recover bioactive compounds. In this sense, Schmidt et al. (2014) used the SSF to recover polyphenols from rice bran by fermenting with *Rhizopus oryzae*. Shin et al. (2019) also used SSF for improving the production of phenolic compounds by *Aspergillus awamori* using black rice bran as a substrate. On the other hand, Zhang et al. (2017) used SmF to obtain chitin and chitosan using shrimp by-products as substrate.

According to the European Union, around 90 MT of food by-products are generated each year (European Commission 2010). These by-products are an ecological issue, so it is necessary to look for alternatives to reuse and revalue these by-products (Zhu et al. 2020). On the other hand, in the case of food additives, consumers currently prefer compounds of natural origin to those of synthetic origin. Therefore, the industry must be able to obtain compounds of natural origin that are capable of replacing those of synthetic origin (Granato et al. 2017, 2020; Putnik et al. 2018; Zhu et al. 2020). There are various naturally occurring antioxidant compounds that have been investigated as substitutes for others of synthetic origin such as butylhydroxytoluene (BHT) and butylhydroxyanisole (BHA) (Abdel-Wahab et al. 2020; Cilli et al. 2020; Drinić et al. 2020; Franco-Arnedo et al. 2020). Another industry with high economic value is the nutraceutical industry. Nutraceuticals are products of natural origin with beneficial biological properties for health (Corzo et al. 2020). Currently, obtaining bioactive compounds from food by-products and their use as nutraceuticals has been widely studied (Ashraf et al. 2020; Kumar et al. 2017; Liu et al. 2016; Mahato et al. 2018; Roselló-Soto et al. 2015b).

7.5 Recovery of High-Added Value Compounds from Food By-Products

7.5.1 Plant-Derived

The selection of starter is an imperative criterion for fermentation of food by-products to utilize them for functional compound production. It is generally considered as an important part of the common sourdough bread processing, and most techno-functional properties are analyzed. Antifungal potential against common food pathogenic strains and radical antioxidant assays for newly isolated strains form bases for quality improvements in fermented foods. SSF is mostly employed for providing higher amount of phenolic acids or other bioactives from agro-industry by-products. The hydrolysis of cell wall components through action of microbial enzymes stimulates the release of bound phenolics along with other metabolites such as organic acids, vitamins, and peptides (Arte et al. 2015). Higher bioactives released on fermentation provides enhanced antifungal activity against pathogenic fungi (e.g. *Aspergillus*, *Penicillium*, *Fusarium*, etc.) commonly found in foods (Denardi-Souza et al. 2018). Phenolic compounds (e.g. ferulic acid, chlorogenic acid, hydroxybenzoic acid, etc.) mostly contribute to antifungal potential as observed in rice bran fermented with *R. oryzae* (Schmidt et al. 2014). For instance, ferulic acid, being a strong antioxidant, and α-amylase inhibition activity in chlorogenic acid prevents the growth of microorganisms in fermented foods (Christ-Ribeiro et al. 2019). Apart from phenolics, SSF also lead to enhanced production of mucilage and bioactive peptides as reported in roasted wheat bran fermented with *Bacillus subtilis* (Lim et al. 2017). *Mucor* sp. strains have been identified as producers of γ-linolenic acid and β-carotene in SmF on the substrate wheat bran and cornmeal (Klempová et al. 2020). *Pediococcus acidilactici* strain LUHS29 led to the production of various antioxidant and antimicrobial bioactives (e.g. fatty acids, free amino acids, phenolic compounds, biogenic amines, lignans, alkylresorcinols, etc.) in barley by-products through SSF as well as SmF (Bartkiene et al. 2020). LAB are known to produce several bioactive antioxidant peptides from protein-rich cereal by-products. Newer or unexplored protein sources such as rice starch extraction by-product, i.e. hydrolyzed proteins showed an immense potential of fermentation with *Bacillus* to produce antioxidant bioactive peptides or amino acids in comparison to commercial enzyme hydrolysis (Dei Piu et al. 2014). Similar to cereals, the pulse milling by-products including broken grains, germ, and husk are rich in fibers and proteins, which indicates the microbial growth opportunities over these ingredients as substrate. For instance, soybean meal, a rich source of dietary proteins, is used mostly as feedstuff. Recent studies have reported higher bioactivities and metal chelating activity upon fermentation. Moreover, fermentation of soybean meal with *B. subtilis* enhanced the peptide and soluble protein fraction due to enhanced action of bacterial enzymes over proteolysis when subjected to sonication (Bi et al. 2015; Ruan et al. 2020). Peanut meal has also been fermented with *B. subtilis* for antioxidant peptides giving 63.28% antioxidant activity in comparison to 99.16% for BHT (Zhang et al. 2014). However, the utilization of pulse-fermented by-products for antifungals and antioxidants still needs to be extensively explored. Furthermore, oilseed cakes are also a major source of proteins in animal feed and may act as a substrate for SSF to make them usable as functional ingredients. Generally recognized as safe (GRAS) fungal strains, e.g. *A. awamori* grow well on protein-rich peanut press cake, which increases the phenolic and flavonoids content, and, therefore, enhances the antioxidant capacity upon fermentation (Sadh et al. 2018).

Fruit or vegetable by-products (e.g. peel, husk, pomace, etc.) contain dietary fibers and are rich in bioactive compounds including pigments (e.g. carotenoids, anthocyanins, betalains, lycopene, etc.), phenolic acids, flavonoids, and certain water-soluble vitamins. The presence of carbon makes a wide variety of microflora inhabitable and fresh-produced by-products being moist are favorable for SSF as reported in several findings. Like cereals, pulses, and oilseeds, the microbial enzymes (e.g. cellulases, pectinases, etc.) convert biopolymers in fruit and vegetable by-products and release antioxidant or antifungal compounds (McCue and Shetty 2003). The extent of release or bioconversion varies with the species and strains of microorganism used for fermentation. Certain by-products also act as support and as carbon source for fungal growth in SSF. The fungi with carbohydrase enhance the release of bound phenolic acids, and thereby bioavailability of phenolics in the fermented by-products (Zambrano et al. 2018). Moreover, phenolics and flavonoids play a key role in imparting enhanced bioactivity to fermented by-products as compared to peptides in other plant or animal foods. For instance, SSF of black grape pomace, apple, and yellow pitahaya peel, core, and peduncle mixtures with *Rhizomucor miehei* increased the gallic acid, 4-hydroxy benzoic acid, vanillic acid, catechin, and epicatechin concentrations (Zambrano et al. 2018). In another finding, an increase in quercetin-3-glucoside, a major phenolic acid and the total phenolic content of plum pomaces from juice industry subjected to fermentation with GRAS fungi (*A. niger* and *Rhizopus oligosporus*), has been elucidated (Dulf et al. 2016). The higher quantities of naringin and its derivatives were observed in phenolic spectra of fermented pomelo peel liquid, resulting in significant increase of antioxidant potential (Tian et al. 2018). Apart from the release of microbial metabolites, the growth pattern of bacteria or fungi also affects the stimulation of bioactivity in fermented by-products. In a study, the enhanced fungal growth released higher levels of metabolites, and thereby, higher antioxidant compounds may be recovered (Larios et al. 2017). Certain fungal species, such as *A. niger* and *R. oligosporus*, could also enhance the sterol ester and polar lipid content which contribute to increased recovery of antioxidants as reported in fermented plum by-products (Dulf et al. 2016). Above findings indicate that the incorporation of fermented by-products might enhance the antifungal or antioxidant activity in fortified foods. It was reported in a recent study that fermented apple by-products led to increase in the total dietary fibers and delayed mold contamination in fortified wheat breads (Cantatore et al. 2019). Overall, SSF seems to be a great alternative for value addition of low-cost fruit or vegetable by-products, thus indicating possible economic gains.

7.5.2 Dairy Foods

A large-scale production of processed foods by dairy industries includes cheese and yogurt which the postfermentation releases whey as by-product. Whey is a source of lactose, lactic acid, proteins, and minerals, which enable fermentation through a number of microbial species. It is directly utilized as a food ingredient or concentrated for protein powders. Whey acts as support for microbial growth, for instance, improved production of microalgae biomass which could result in higher carotenoid or chlorophyll content (Lappa et al. 2019). The fermentative processing of whey is targeted for biofuels and bioplastics and involves inoculation with specific species or combinations of microorganisms. However, whey proteins of high nutritional value are an important source of antifungal or antioxidant peptides. Certain common bacterial, yeast, or fungal species, however, produce individual compounds (e.g. ethanol, lactic acid, ethyl lactate, exopolysaccharides, etc.) of significant antifungal or antioxidant potential (Pescuma et al. 2015). Dairy fermented products contain numerous antifungal compounds, among which organic acids and fatty acids were identified. Fermented whey finds uses in fortified food or feed products for enhanced antifungal activity (Garnier et al. 2020). Generally, whey is fermented with LAB and leads to significant shelf life improvement of fortified foods. The poultry feed supplemented with fermented whey enhanced the antifungal potential two to four times than the control feed (Londero et al. 2014). The fortification of *Lactobacillus plantarum*-fermented whey pita bread inhibited *Penicillium expansum* and *Penicillium brevicompactum* strains and increased the shelf life (Izzo et al. 2020). In another example, lactic acid-fermented whey powder led to slight inactivation (0.5–0.6 log cfu/g) of *P. expansum*, a toxigenic fungus in loaf bread (Luz et al. 2020). Bioconversion of whey proteins into bioactive peptides and free amino acids enhances the antioxidant potential upon fermentation. Furthermore, the enhancement of antioxidant activity in fermented whey depends on the microbial strain. Few selected strains grow fast on whey as a medium and exhibit higher antioxidant activity. LAB mostly carry out proteolysis in whey, and higher degree of proteolysis relates to the DPPH or $ABTS^+$ antioxidant potential (Virtanen et al. 2007). Therefore, fermentation of whey provides immense potential of waste valorization to generate bioactives and proficient alternative for shelf life enhancement of fortified food products.

7.5.3 Animal Foods

Seafood by-products are rich sources of collagen, gelatin, protein, peptides, vitamins, minerals, and pigments. They constitute about 80% of inedible portion of seafoods. Recently, fermentation of seafood by-products has been applied for the production of bioactive peptides from collagen or gelatin as a result of hydrolysis by microbial endogenous or exogenous enzymes. The enzymes mostly include hydrolases (e.g. chitinase, protease, lipase) catalyzing the hydrolysis of chitin, protein, and lipids, respectively (Bruno et al. 2019). The peptides or fatty acids possess antioxidant potential as mentioned in several studies. Similar to plant and dairy foods, LAB are the most extensively explored for fermentation of seafood by-products. However, the recovery of antifungals or antioxidant bioactives depends on the strain used for fermentation. Crustacean waste, on the other hand, is also a rich source of chitin, protein, and calcium carbonate. The bioconversion of chitin demonstrates a straightforward waste disposal, and broad antifungal metabolites may be generated from bacterial species such as *Bacillus cereus* (Chang et al. 2007).

Eggshells and eggshell membranes constitute 11% of the egg weight and are produced in large quantities during processing of eggs. Being rich source of proteins, bioactive protein hydrolysates could be obtained through bacterial fermentation. For instance, chicken eggshell membrane fermented with *L. plantarum* resulted in 25.1% hydrolysis of proteins into hydrolysates with 70.5% DPPH radical scavenging activity (Jain and Anal 2017). The eggshells are also found to condition the fermentation medium for food waste, thus leading to higher volatile acids production (e.g. acetic acid). The addition of eggshells to food waste reactors increased the abundance of hydrolases (protease and α-glucosidase) and of *Bifidobacterium*, *Clostridium*, *Eubacterium*, *Butyriolbrio*, and *Megasphaera* responsible for acetic and butyric acid production (Luo et al. 2020). Also, naturally obtained peptides have received considerable attention these days owing to the absence of any side effects on health. Eggshells might also act as immobilization support for SmF such as in the case of brewery wastewater to yield fumaric acid, a food additive, widely used as broad-spectrum antimicrobial agent (Das et al. 2015). To conclude on the basis of the above findings, the strain dependency for eggshell or membranes fermentation and the recovery of antifungals or antioxidants need to be further elucidated.

As in plant or dairy foods, the fermentation of fish by-products preserves them and intensifies antioxidant activity due to lactic acid, phenolic acids, and peptides produced by microbial metabolism. LAB are the most investigated microorganisms for fermentation of fish and meat foods. However, mostly bioactive peptides, which are short sequences of amino acids, impart antioxidant activity. Fish by-products such as meat remain rich in proteins and lead to the production of bioactive peptides. Other metabolites such as lactic acid, acetic acid, succinic acid, and bacteriocins of LAB enhance the antimicrobial activity. For instance, fish sauce by-product is rich in peptides of high antioxidant activity, which result from protein conversion upon fish sauce processing and fermentation. The 10% protein content of fermented by-product and its profile was similar to that in the product indicating high potential of the by-product for bioactive peptide recovery (Choksawangkarn et al. 2018). In another study, fermentation of discarded turbot skin with *Aspergillus oryzae* generated bioactive peptides with significant antioxidant activity (Fang et al. 2017). Ruthu et al. (2014) also demonstrated antimicrobial and antifungal activity of obtained peptides after the fermentation of fish heads (from rohu and catla carps) with LAB. Nile tilapia fish head hydrolysate fermented by nonhalophilic LAB yielded high levels of volatiles (alcohols or esters), free amino acids, organic acids, and thus higher antioxidant activity (Gao et al. 2020). Overall, ample amount of nutrients in fish by-products underpins exploration possibilities for recovery of antifungal and antioxidant compounds upon fermentation by varied kinds of microorganisms.

Meat food by-products constitute a large proportion of the live animal weight. To ensure profits to meat industry, by-product valorization is imperative, and fermentation has been stated as an economic processing method. Acidity increase and release of extracellular metabolites such as lactic acid, acetic acid, succinic acid, and bacteriocins occur in LAB fermentation of chicken sausages (Yadav 2017). Moreover, bacteriocins mostly impart antimicrobial activity in fermented meat foods. However, extent of proteolysis differs in the peptides and free amino acids generated with the action of endogenous proteases (Takeda et al. 2017). A number of factors like pH, carbohydrates, nitrogen, or salt content in meat products also affect the antimicrobial activity of bacteriocins produced by LAB (Castilho et al. 2019). In an example, the presence of dextrose and starch enhanced the radical scavenging activity by *L. plantarum* in fermented chicken sausages. In the end, utilization of fermentation for bioactive peptide recovery from meat by-products has not been much explored, though extracted peptides have high antioxidant activity (Borrajo et al. 2019).

7.6 Technical and Economical Hurdles in Fermentation Assisted Recovery

The recovery of antifungals and antioxidants from fermentation process requires further investigations, as many technical hurdles exist:

- Upgradation for large-scale sustainable processing
- Increased research efforts for targeted culturing of raw material and generation of data bank for microbial species specificity
- Efficiency of bioconversion in association with raw material composition, especially the presence of antimicrobials, and environmental factors
- Cost-effective purification methods for recovered antifungals or antioxidants

The SSF has been found to be more advantageous than the SmF due to lower production costs, simpler process, and better yield (Abu Yazid et al. 2017). That is why, SSF becomes a good option for the valorization of food by-products through fermentation. However, most of the studies carried out so far are at a laboratory scale, so it is necessary to go deeper into the scaling process industrial scaling-up this process and applying it in a generalized way (Abu Yazid et al. 2017). The scaling-up procedure can be defined as the process in which different mechanisms (e.g. trial-errors, mathematical models, etc.) define a large system from a small one, developing rules and criteria to achieve it (de la Fuente et al. 2020). Also, the SSF lacks the complicated instrumental and control systems as compared to submerged one. However, the control of growth kinetics of microbes and medium requirements in SSF is a major drawback. Furthermore, the efficiency of metabolites production in response to raw materials, strain type, and environmental conditions needs to be elucidated through research efforts. The presence of antimicrobial compounds in by-products, naturally present in foods (cf. Sections 7.3.1, 7.3.2, and 7.3.5), has to be taken into account. Considering extraction efficiency, it has to be economical for the recovery of antifungal or antioxidant compounds through fermentation of food by-products.

7.7 Conclusions and Future Outlook

Food waste and by-products valorization through fermentation seems to be an economical eco-friendly alternative for the recovery of antifungals and antioxidant compounds. The reusability of agro-residues is being suggested as a new revenue source for targeted food industries. Strain dependency for certain metabolites would provide newer insights to fermentation application for food by-products. Microbial metabolic pathways for major food components, however, provide a prediction for strain type to be used for certain food categories. To conclude, the process optimization and cost analysis would possibly bring valorization of by-products from fermentation to industrial scale-up.

References

Abdel-Wahab, M., El-Sohaimy, S.A., Ibrahim, H.A., and Abo El-Makarem, H.S. (2020). Evaluation the efficacy of clove, sage and kiwifruit peels extracts as natural preservatives for fish fingers. *Annals of Agricultural Sciences* 65: 98–106.

Abu Yazid, N., Barrena, R., Komilis, D., and Sánchez, A. (2017). Solid-state fermentation as a novel paradigm for organic waste valorization: a review. *Sustainability* 9: 224. https://doi.org/10.3390/su9020224.

Ajila, C.M., Brar, S.K., Verma, M., and Prasada Rao, U.J.S. (2012). Sustainable solutions for agro processing waste management: an overview. In: *Environmental Protection Strategies for Sustainable Development* (eds. A. Malik and E. Grohmann), 65–109. Netherlands: Springer.

Al Khawli, F., Martí-Quijal, F.J., Ferrer, E. et al. (2020). Aquaculture and its by-products as a source of nutrients and bioactive compounds. In: *Advances in Food and Nutrition Research* (eds. J.M. Lorenzo and F.J. Barba), 1–33. Massachusetts, US: Academic Press Inc.

Al-Doury, M.K.W., Hettiarachchy, N.S., and Horax, R. (2018). Rice-endosperm and Rice-bran proteins: a review. *Journal of the American Oil Chemists' Society* 95: 943–956.

Anal, A.K., Ahmad, I., Sripinyowanich, J., and Noomhorm, A. (2013). Sustainable food grain processing. In: *Sustainable Food Processing* (eds. B. Tiwari, T. Norton and N.M. Holden), 269–294. Chichester, UK: Wiley.

Anraku, M., Gebicki, J.M., Iohara, D. et al. (2018). Antioxidant activities of chitosans and its derivatives in *in vitro* and *in vivo* studies. *Carbohydrate Polymers* 199: 141–149.

Arte, E., Rizzello, C.G., Verni, M. et al. (2015). Impact of enzymatic and microbial bioprocessing on protein modification and nutritional properties of wheat bran. *Journal of Agricultural and Food Chemistry* 63: 8685–8693.

Ashraf, S.A., Adnan, M., Patel, M. et al. (2020). Fish-based bioactives as potent nutraceuticals: exploring the therapeutic perspective of sustainable food from the sea. *Marine Drugs* 18: 265. https://doi.org/10.3390/md18050265.

Barba, F.J., Brianceau, S., Turk, M. et al. (2015). Effect of alternative physical treatments (ultrasounds, pulsed electric fields, and high-voltage electrical discharges) on selective recovery of bio-compounds from fermented grape pomace. *Food and Bioprocess Technology* 8: 1139–1148.

Barba, F.J., Zhu, Z., Koubaa, M. et al. (2016). Green alternative methods for the extraction of antioxidant bioactive compounds from winery wastes and by-products: a review. *Trends in Food Science and Technology* 49: 96–109.

Barba, F.J., Roselló-Soto, E., Brnčić, M., and Lorenzo, J.M. (2020). *Green Extraction and Valorization of by-Products from Food Processing*. Boca Raton: CRC Press.

Bartkiene, E., Mozuriene, E., Lele, V. et al. (2020). Changes of bioactive compounds in barley industry by-products during submerged and solid state fermentation with antimicrobial *Pediococcus acidilactici* strain LUHS29. *Food Science & Nutrition* 8: 340–350.

Bharat Helkar, P. and Sahoo, A. (2016). Review: food industry by-products used as a functional food ingredients. *International Journal of Waste Resources* 6: 1000248.

Bi, H., Zhao, H., Lu, F. et al. (2015). Improvement of the nutritional quality and fibrinolytic enzyme activity of soybean meal by fermentation of *Bacillus subtilis*. *Journal of Food Processing and Preservation* 39: 1235–1242.

Bordiga, M., Travaglia, F., and Locatelli, M. (2019). Valorisation of grape pomace: an approach that is increasingly reaching its maturity – a review. *International Journal of Food Science & Technology* 54: 933–942.

Borrajo, P., Pateiro, M., Barba, F.J. et al. (2019). Antioxidant and antimicrobial activity of peptides extracted from meat by-products: a review. *Food Analytical Methods* 12: 2401–2415.

Bruno, S.F., Ekorong, F.J.A.A., Karkal, S.S. et al. (2019). Green and innovative techniques for recovery of valuable compounds from seafood by-products and discards: a review. *Trends in Food Science & Technology* 85: 10–22.

Călinoiu, L.F., Mitrea, L., Precup, G. et al. (2017). Characterization of grape and apple peel wastes' bioactive compounds and their increased bioavailability after exposure to thermal process. *Bulletin of University of Agricultural Sciences and Veterinary Medicine Cluj-Napoca. Food Science and Technology* 74: 80. https://doi.org/10.15835/buasvmcn-fst:0028.

Cantatore, V., Filannino, P., Gambacorta, G. et al. (2019). Lactic acid fermentation to re-cycle apple by-products for wheat bread fortification. *Frontiers in Microbiology* 10: 1–13. https://doi.org/10.3389/fmicb.2019.02574.

Carvalho, F., Prazeres, A.R., and Rivas, J. (2013). Cheese whey wastewater: characterization and treatment. *Science of the Total Environment* 445–446: 385–396.

Castilho, N.P.A., Colombo, M., de Oliveira, L.L. et al. (2019). *Lactobacillus curvatus* UFV-NPAC1 and other lactic acid bacteria isolated from calabresa, a fermented meat product, present high bacteriocinogenic activity against *Listeria monocytogenes*. *BMC Microbiology* 19: 63. https://doi.org/10.1186/s12866-019-1436-4.

Cerda, A., Artola, A., Barrena, R. et al. (2019). Innovative production of bioproducts from organic waste through solid-state fermentation. *Frontiers in Sustainable Food Systems* 3: 63. https://doi.org/10.3389/fsufs.2019.00063.

Chang, W., Chen, Y., and Jao, C. (2007). Antifungal activity and enhancement of plant growth by *Bacillus cereus* grown on shellfish chitin wastes. *Bioresource Technology* 98: 1224–1230.

Chemat, F., Abert Vian, M., Fabiano-Tixier, A.-S. et al. (2020). A review of sustainable and intensified techniques for extraction of food and natural products. *Green Chemistry* 22: 2325–2353.

Choksawangkarn, W., Phiphattananukoon, S., Jaresitthikunchai, J., and Roytrakul, S. (2018). Antioxidative peptides from fish sauce by-product: isolation and characterization. *Agriculture and Natural Resources* 52: 460–466.

Christ-Ribeiro, A., Graça, C.S., Kupski, L. et al. (2019). Cytotoxicity, antifungal and anti mycotoxins effects of phenolic compounds from fermented rice bran and *Spirulina* sp. *Process Biochemistry* 80: 190–196.

Cilli, L.P., Contini, L.R.F., Sinnecker, P. et al. (2020). Effects of grape pomace flour on quality parameters of salmon burger. *Journal of Food Processing and Preservation* 44: 1–11. https://doi.org/10.1111/jfpp.14329.

Clément, M., Tremblay, J., Lange, M. et al. (2008). Purification and identification of bovine cheese whey fatty acids exhibiting in vitro antifungal activity. *Journal of Dairy Science* 91: 2535–2544.

Coman, V., Teleky, B.E., Mitrea, L. et al. (2020). Bioactive potential of fruit and vegetable wastes. In: *Advances in Food and Nutrition Research* (ed. F. Toldrá), 157–225. Massachusetts, US: Academic Press Inc.

Corzo, L., Fernández-Novoa, L., Carrera, I. et al. (2020). Nutrition, health, and disease: role of selected marine and vegetal nutraceuticals. *Nutrients* 12: 747.

Daley, J. (2009). Dairy derivatives. *Journal of Complementary Medicine* 8: 32–37.

Dapčević-Hadnadev, T., Hadnadev, M., and Pojić, M. (2018). The healthy components of cereal by-products and their functional properties. In: *Sustainable Recovery and Reutilization of Cereal Processing by-Products* (ed. C.M. Galanakis), 27–61. Duxford, UK: Elsevier Inc.

Das, R.K., Brar, S.K., and Verma, M. (2015). Valorization of egg shell biowaste and brewery wastewater for the enhanced production of fumaric acid. *Waste and Biomass Valorization* 6: 535–546.

de Francisco, L., Pinto, D., Rosseto, H. et al. (2018). Evaluation of radical scavenging activity, intestinal cell viability and antifungal activity of Brazilian propolis by-product. *Food Research International* 105: 537–547.

de la Fuente, B., Tornos, A., Príncep, A. et al. (2020). Scaling-up processes: patents and commercial applications. In: *Advances in Food and Nutrition Research* (eds. J.M. Lorenzo and F.J. Barba), 187–223. Massachusetts, US: Academic Press Inc.

Dei Piu, L., Tassoni, A., Isabella, D. et al. (2014). Exploitation of starch industry liquid by-product to produce bioactive peptides from rice hydrolyzed proteins. *Food Chemistry* 155: 199–206.

Denardi-Souza, T., Luz, C., Mañes, J. et al. (2018). Antifungal effect of phenolic extract of fermented rice bran with *Rhizopus oryzae* and its potential use in loaf bread shelf life extension. *Journal of the Science of Food and Agriculture* 98: 5011–5018.

Di Bernardini, R., Mullen, A.M., Bolton, D. et al. (2012). Assessment of the angiotensin-I-converting enzyme (ACE-I) inhibitory and antioxidant activities of hydrolysates of bovine brisket sarcoplasmic proteins produced by papain and characterisation of associated bioactive peptidic fractions. *Meat Science* 90: 226–235.

Díaz-Gómez, J., Twyman, R.M., Zhu, C. et al. (2017). Biofortification of crops with nutrients: factors affecting utilization and storage. *Current Opinion in Biotechnology* 44: 115–123.

Drinić, Z., Mudrić, J., Zdunić, G. et al. (2020). Effect of pomegranate peel extract on the oxidative stability of pomegranate seed oil. *Food Chemistry* 333: 127501.

Dulf, F.V., Vodnar, D.C., and Socaciu, C. (2016). Effects of solid-state fermentation with two filamentous fungi on the total phenolic contents, flavonoids, antioxidant activities and lipid fractions of plum fruit (*Prunus domestica* L.) by-products. *Food Chemistry* 209: 27–36.

Echegaray, N., Centeno, J.A., and Carballo, J. (2019). Dairy by-products as source of high added value compounds. In: *Green Extraction and Valorization of by-Products from Food Processing* (eds. F.J. Barba, E.R. Soto, M. Brncic and J.M.L. Rodriquez), 23–54. CRC Press.

European Commission (2010). Preparatory study on food waste across EU 27. doi:10.2779/85947.

Fang, B., Sun, J., Dong, P. et al. (2017). Conversion of turbot skin wastes into valuable functional substances with an eco-friendly fermentation technology. *Journal of Cleaner Production* 156: 367–377.

FAO (2018). The State of World Fisheries and Aquaculture 2018: Meeting the Sustainable Development Goals. Available online: http://www.fao.org/3/i9540en/i9540en.pdf (accessed 30 July 2020).

Franco-Arnedo, G., Buelvas-Puello, L.M., Miranda-Lasprilla, D. et al. (2020). Obtaining antioxidant extracts from tangerine (*C. reticulata* var. Arrayana) peels by modified supercritical CO_2 and their use as protective agent against the lipid oxidation of a mayonnaise. *The Journal of Supercritical Fluids* 165: 104957.

Friedman, M. (2014). Antibacterial, antiviral, and antifungal properties of wines and winery byproducts in relation to their flavonoid content. *Journal of Agricultural and Food Chemistry* 62: 6025–6042.

FUSIONS (2016). Estimates of European food waste levels. Available online: http://www.eu-fusions.org/phocadownload/Publications/Estimates%20of%20European%20food%20waste%20levels.pdf (accessed on 30 July 2020).

Gao, P., Li, L., Xia, W. et al. (2020). Valorization of Nile tilapia (*Oreochromis niloticus*) fish head for a novel fish sauce by fermentation with selected lactic acid bacteria. *LWT* 129: 109539.

Garnier, L., Penland, M., Thierry, A. et al. (2020). Antifungal activity of fermented dairy ingredients: identification of antifungal compounds. *International Journal of Food Microbiology* 322: 108574.

Granato, D., Nunes, D.S., and Barba, F.J. (2017). An integrated strategy between food chemistry, biology, nutrition, pharmacology, and statistics in the development of functional foods: a proposal. *Trends in Food Science & Technology* 62: 13–22.

Granato, D., Barba, F.J., Bursać Kovacević, D. et al. (2020). Functional foods: product development, technological trends, efficacy testing, and safety. *Annual Review of Food Science and Technology* 11: 93–118.

Gunes, R., Palabiyik, I., Toker, O.S. et al. (2019). Incorporation of defatted apple seeds in chewing gum system and phloridzin dissolution kinetics. *Journal of Food Engineering* 255: 9–14.

Hou, Y., Wu, Z., Dai, Z. et al. (2017). Protein hydrolysates in animal nutrition: industrial production, bioactive peptides, and functional significance. *Journal of Animal Science and Biotechnology* 8: 24. https://doi.org/10.1186/s40104-017-0153-9.

Hu, J., Xu, M., Hang, B. et al. (2011). Isolation and characterization of an antimicrobial peptide from bovine hemoglobin α-subunit. *World Journal of Microbiology and Biotechnology* 27: 767–771.

Hussien, A.M.S., Badr, A.N., and Naeem, M.A. (2019). Innovative nutritious biscuits limit aflatoxin contamination. *Pakistan Journal of Biological Sciences* 22: 133–142.

İnan, Ö., Özcan, M.M., and Aljuhaimi, F. (2018). Effect of location and *Citrus* species on total phenolic, antioxidant, and radical scavenging activities of some *Citrus* seed and oils. *Journal of Food Processing and Preservation* 42: e13555. https://doi.org/10.1111/jfpp.13555.

Ismail, T., Akhtar, S., Riaz, M. et al. (2016). Oxidative and microbial stability of pomegranate peel extracts and bagasse supplemented cookies. *Journal of Food Quality* 39: 658–668.

Izzo, L., Luz, C., Ritieni, A. et al. (2020). Whey fermented by using *Lactobacillus plantarum* strains: a promising approach to increase the shelf life of pita bread. *Journal of Dairy Science* 103: 5906–5915.

Jain, S. and Anal, A.K. (2017). Production and characterization of functional properties of protein hydrolysates from egg shell membranes by lactic acid bacteria fermentation. *Journal of Food Science and Technology* 54: 1062–1072.

Janiak, M.A., Renzetti, S., Noort, M., and Amarowicz, R. (2019). Antioxidant potential of wheat bran phenolics extracted with organic solvents. *Bulgarian Chemical Communications* 51: 75–78.

Jayathilakan, K., Sultana, K., Radhakrishna, K., and Bawa, A.S. (2012). Utilization of byproducts and waste materials from meat, poultry and fish processing industries: a review. *Journal of Food Science and Technology* 49: 278–293.

Jiang, W., Hu, S., Li, S., and Liu, Y. (2017). Biochemical and antioxidant properties of peptidic fraction generated from crab (*Portunus trituberculatus*) shells by enzymatic hydrolysis. *International Journal of Food Science & Technology* 52: 2479–2488.

Jiang, W., Liu, Y., Yang, X. et al. (2018). Recovery of proteins from squid by-products with enzymatic hydrolysis and increasing the hydrolysate's bioactivity by maillard reaction. *Journal of Aquatic Food Product Technology* 27: 900–911.

Kamani, M.H., Eş, I., Lorenzo, J.M. et al. (2019). Advances in plant materials, food by-products, and algae conversion into biofuels: use of environmentally friendly technologies. *Green Chemistry* 21: 3213–3231.

Khan, I.T., Nadeem, M., Imran, M. et al. (2019). Antioxidant properties of Milk and dairy products: a comprehensive review of the current knowledge. *Lipids in Health and Disease* https://doi.org/10.1186/s12944-019-0969-8.

Klempová, T., Slaný, O., Michal, Š. et al. (2020). Dual production of polyunsaturated fatty acids and beta-carotene with Mucor wosnessenskii by the process of solid-state fermentation using agro- industrial waste. *Journal of Biotechnology* 311: 1–11. https://doi.org/10.1016/j.jbiotec.2020.02.006.

König, H. and Fröhlich, J. (2017). Lactic acid bacteria. In: *Biology of Microorganisms on Grapes, in Must and in Wine* (eds. H. König, G. Unden and J. Fröhlich), 3–41. Cham: Springer International Publishing.

Kowalska, H., Czajkowska, K., Cichowska, J., and Lenart, A. (2017). What's new in biopotential of fruit and vegetable by-products applied in the food processing industry. *Trends in Food Science and Technology* 67: 150–159.

Kumar, K., Yadav, A.N., Kumar, V. et al. (2017). Food waste: a potential bioresource for extraction of nutraceuticals and bioactive compounds. *Bioresources and Bioprocessing* https://doi.org/10.1186/s40643-017-0148-6.

Lafarga, T. and Hayes, M. (2014). Bioactive peptides from meat muscle and by-products: generation, functionality and application as functional ingredients. *Meat Science* 98 (2): 227–239.

Lappa, I.K., Papadaki, A., Kachrimanidou, V. et al. (2019). Cheese whey processing: Integrated biorefinery concepts and emerging food applications. *Foods* 8: 347.

Larios, R., Juan, C., Figueroa, B. et al. (2017). Valorization of grapefruit by-products as solid support for solid-state fermentation to produce antioxidant bioactive extracts. *Waste and Biomass Valorization* 10: 763–769.

Li, B., Chen, F., Wang, X. et al. (2007). Isolation and identification of antioxidative peptides from porcine collagen hydrolysate by consecutive chromatography and electrospray ionization-mass spectrometry. *Food Chemistry* 102: 1135–1143.

Lim, J.-Y., Jang, G.-M., Garcia, C.V., and Lee, S.-P. (2017). Fortification of bioactive compounds in roasted wheat bran by solid-state fermentation using *Bacillus subtilis* HA. *Food Science and Technology Research* 23: 395–402.

Lima, K.O., da Costa de Quadros, C., da Rocha, M. et al. (2019). Bioactivity and bioaccessibility of protein hydrolyzates from industrial byproducts of Stripped weakfish (*Cynoscion guatucupa*). *LWT* 111: 408–413.

Liu, R., Wang, M., Duan, J.a. et al. (2010). Purification and identification of three novel antioxidant peptides from Cornu Bubali (water buffalo horn). *Peptides* 31: 786–793.

Liu, R., Xing, L., Fu, Q. et al. (2016). A review of antioxidant peptides derived from meat muscle and by-products. *Antioxidants* 5: 32. https://doi.org/10.3390/antiox5030032.

Londero, A., León Peláez, M.A., Diosma, G. et al. (2014). Fermented whey as poultry feed additive to prevent fungal contamination. *Journal of the Science of Food and Agriculture* 94: 3189–3194.

Lorenzo, J.M., Munekata, P.E.S., Gómez, B. et al. (2018). Bioactive peptides as natural antioxidants in food products – a review. *Trends in Food Science & Technology* 79: 136–147.

Lorenzo, J.M., Munekata, P.E., Putnik, P. et al. (2019). Sources, chemistry, and biological potential of ellagitannins and ellagic acid derivatives. In: *Studies in Natural Products Chemistry* (ed. Atta-ur-Rahman), 189–221. Elsevier.

Luo, J., Huang, W., Guo, W. et al. (2020). Novel strategy to stimulate the food wastes anaerobic fermentation performance by eggshell wastes conditioning and the underlying mechanisms. *Chemical Engineering Journal* 398: 125560.

Luz, C., Rodriguez, L., Romano, R. et al. (2020). A natural strategy to improve the shelf life of the loaf bread against toxigenic fungi: the employment of fermented whey powder. *International Journal of Dairy Technology* 73: 88–97.

Mahato, N., Sharma, K., Sinha, M., and Cho, M.H. (2018). Citrus waste derived nutra–/ pharmaceuticals for health benefits: current trends and future perspectives. *Journal of Functional Foods* 40: 307–316.

Mapelli-Brahm, P., Barba, F.J., Remize, F. et al. (2020). The impact of fermentation processes on the production, retention and bioavailability of carotenoids: an overview. *Trends in Food Science & Technology* 99: 389–401.

Martins, S., Mussatto, S.I., Martínez-Avila, G. et al. (2011). Bioactive phenolic compounds: production and extraction by solid-state fermentation. A review. *Biotechnology Advances* 29: 365–373.

Marti-Quijal, F.J., Zamuz, S., Tomašević, I. et al. (2019). Influence of different sources of vegetable, whey and microalgae proteins on the physicochemical properties and amino acid profile of fresh pork sausages. *LWT* 110: 316–323.

Marti-Quijal, F.J., Remize, F., Meca, G. et al. (2020). Fermentation in fish and by-products processing: an overview of current research and future prospects. *Current Opinion in Food Science* 31: 9–16.

Martí-Quijal, F.J., Tornos, A., Príncep, A. et al. (2020a). Impact of fermentation on the recovery of antioxidant bioactive compounds from sea bass byproducts. *Antioxidants* 9 https://doi.org/10.3390/antiox9030239.

Martí-Quijal, F.J., Príncep, A., Tornos, A. et al. (2020b). Isolation, identification and investigation of fermentative bacteria from sea bass (*Dicentrarchus labrax*): evaluation of antifungal activity of fermented fish meat and by-products broths. *Foods* 9 https://doi.org/10.3390/foods9050576.

McCarthy, T.L., Kerry, J.P., Kerry, J.F. et al. (2001). Evaluation of the antioxidant potential of natural food/plant extracts as compared with synthetic antioxidants and vitamin e in raw and cooked pork patties. *Meat Science* 58: 45–52.

McCue, P. and Shetty, K. (2003). Role of carbohydrate-cleaving enzymes in phenolic antioxidant mobilization from whole soybean fermented with *Rhizopus oligosporus*. *Food Biotechnology* 17: 27–37.

Melo, P.S., Massarioli, A.P., Denny, C. et al. (2015). Winery by-products: extraction optimization, phenolic composition and cytotoxic evaluation to act as a new source of scavenging of reactive oxygen species. *Food Chemistry* 181: 160–169.

Mendoza, L., Yañez, K., Vivanco, M. et al. (2013). Characterization of extracts from winery by-products with antifungal activity against *Botrytis cinerea*. *Industrial Crops and Products* 43: 360–364.

Moccia, F., Agustin-Salazar, S., Verotta, L. et al. (2020). Antioxidant properties of agri-food byproducts and specific boosting effects of hydrolytic treatments. *Antioxidants* 9: 438. https://doi.org/10.3390/antiox9050438.

Musina, O.N., Rashidinejad, A., Putnik, P. et al. (2018). The use of whey protein extract for manufacture of a whipped frozen dairy dessert. *Mljekarstvo* 68: 254–271.

Ng, H.S., Kee, P.E., Yim, H.S. et al. (2020). Recent advances on the sustainable approaches for conversion and reutilization of food wastes to valuable bioproducts. *Bioresource Technology* 302: 122889.

Nikmaram, N., Leong, S.Y., Koubaa, M. et al. (2017). Effect of extrusion on the anti-nutritional factors of food products: an overview. *Food Control* 79: 62–73.

Okwu, D.E., Awurum, A.N., and Okoronkwo, J.I. (2007). Phytochemical composition and *in vitro* antifungal activity screening of extracts from citrus plants against *Fusarium oxysporum* of okra plant (Hibiscus esculentus). Presented at the 8th African Crop Science Society Conference, El-Minia, Egypt, pp. 1755–1758.

Pateiro, M., Borrajo, P., Domínguez, R. et al. (2019). Extraction of valuable compounds from meat by-products. In: *Green Extraction and Valorization of by-Products from Food Processing* (eds. F.J. Barba, E.R. Soto, M. Brncic and J.M.L. Rodriquez), 55–90. CRC Press.

Peanparkdee, M. and Iwamoto, S. (2019). Bioactive compounds from by-products of rice cultivation and rice processing: extraction and application in the food and pharmaceutical industries. *Trends in Food Science & Technology* 86: 109–117.

Pescuma, M., de Valdez, G.F., and Mozzi, F. (2015). Whey-derived valuable products obtained by microbial fermentation. *Applied Microbiology and Biotechnology* 99: 6183–6196.

Przybylski, R., Firdaous, L., Châtaigné, G. et al. (2016). Production of an antimicrobial peptide derived from slaughterhouse by-product and its potential application on meat as preservative. *Food Chemistry* 211: 306–313.

Puértolas, E., Koubaa, M., and Barba, F.J. (2016). An overview of the impact of electrotechnologies for the recovery of oil and high-value compounds from vegetable oil industry: energy and economic cost implications. *Food Research International* 80: 19–26.

Putnik, P., Bursać Kovačević, D., Režek Jambrak, A. et al. (2017). Innovative "green" and novel strategies for the extraction of bioactive added value compounds from citrus wastes-a review. *Molecules* 22 https://doi.org/10.3390/molecules22050680.

Putnik, P., Lorenzo, J.M., Barba, F.J. et al. (2018). Novel food processing and extraction technologies of high-added value compounds from plant materials. *Foods* 7 https://doi.org/10.3390/foods7070106.

Putnik, P., Kresoja, Ž., Bosiljkov, T. et al. (2019). Comparing the effects of thermal and non-thermal technologies on pomegranate juice quality: a review. *Food Chemistry* 279: 150–161.

Rabetafika, H.N., Bchir, B., Blecker, C., and Richel, A. (2014). Fractionation of apple by-products as source of new ingredients: current situation and perspectives. *Trends in Food Science & Technology* 40: 99–114.

Ravindran, R. and Jaiswal, A.K. (2016). Exploitation of food industry waste for high-value products. *Trends in Biotechnology* 34: 58–69.

Reddy, M.K., Gupta, S.K., Jacob, M.R. et al. (2007). Antioxidant, antimalarial and antimicrobial activities of tannin-rich fractions, ellagitannins and phenolic acids from *Punica granatum* L. *Planta Medica* 73: 461–467.

Rocchetti, G., Lucini, L., Rodriguez, J.M.L. et al. (2019). Gluten-free flours from cereals, pseudocereals and legumes: phenolic fingerprints and in vitro antioxidant properties. *Food Chemistry* 271: 157–164.

Roselló-Soto, E., Galanakis, C.M., Brnčić, M. et al. (2015a). Clean recovery of antioxidant compounds from plant foods, by-products and algae assisted by ultrasounds processing. Modeling approaches to optimize processing conditions. *Trends in Food Science & Technology* 42: 134–149.

Roselló-Soto, E., Koubaa, M., Moubarik, A. et al. (2015b). Emerging opportunities for the effective valorization of wastes and by-products generated during olive oil production process: non-conventional methods for the recovery of high-added value compounds. *Trends in Food Science & Technology* 45: 296–310.

Roselló-Soto, E., Garcia, C., Fessard, A. et al. (2019). Nutritional and microbiological quality of tiger nut tubers (*Cyperus esculentus*), derived plant-based and lactic fermented beverages. *Fermentation* 5: 3. https://doi.org/10.3390/fermentation5010003.

Ruan, S., Li, Y., Wang, Y. et al. (2020). Analysis in protein profile, antioxidant activity and structure-activity relationship based on ultrasound-assisted liquid-state fermentation of soybean meal with Bacillus subtilis. *Ultrasonics Sonochemistry* 64: 104846.

Ruthu, Murthy, P.S., Rai, A.K., and Bhaskar, N. (2014). Fermentative recovery of lipids and proteins from freshwater fish head waste with reference to antimicrobial and antioxidant properties of protein hydrolysate. *Journal of Food Science and Technology* 51: 1884–1892.

Sadh, P.K., Kumar, S., Chawla, P., and Duhan, J.S. (2018). Fermentation: a boon for production of bioactive compounds by processing of food industries wastes (by-products). *Molecules* 23: 2560. https://doi.org/10.3390/molecules23102560.

Saiga, A., Tanabe, S., and Nishimura, T. (2003). Antioxidant activity of peptides obtained from porcine myofibrillar proteins by protease treatment. *Journal of Agricultural and Food Chemistry* 51: 3661–3667.

Schmidt, C.G., Gonçalves, L.M., Prietto, L. et al. (2014). Antioxidant activity and enzyme inhibition of phenolic acids from fermented rice bran with fungus *Rizhopus oryzae*. *Food Chemistry* 146: 371–377.

Shin, H.Y., Kim, S.M., Lee, J.H., and Lim, S.T. (2019). Solid-state fermentation of black rice bran with *Aspergillus awamori* and *Aspergillus oryzae*: effects on phenolic acid composition and antioxidant activity of bran extracts. *Food Chemistry* 272: 235–241.

Sila, A. and Bougatef, A. (2016). Antioxidant peptides from marine by-products: isolation, identification and application in food systems. A review. *Journal of Functional Foods* 21: 10–26.

Sindhu, R., Binod, P., Nair, R.B. et al. (2020). Waste to wealth: valorization of food waste for the production of fuels and chemicals. In: *Current Developments in Biotechnology and Bioengineering* (eds. S. Varjani, A. Pandey, E. Gnansounou, et al.), 181–197. Elsevier.

Smuda, S.S., Mohsen, S.M., Olsen, K., and Aly, M.H. (2018). Bioactive compounds and antioxidant activities of some cereal milling by-products. *Journal of Food Science and Technology* 55: 1134–1142.

Song, R., Zhang, K.q., and Wei, R.b. (2016). *In vitro* antioxidative activities of squid (*Ommastrephes bartrami*) viscera autolysates and identification of active peptides. *Process Biochemistry* 51: 1674–1682.

Statista (2020). Available online: https://www.statista.com/statistics/545737/distillers-dry-grains-production-volume-european-union-28/

Sultan, S., Huma, N., Butt, M.S., and Shahid, M. (2017). Antihypertensive and antioxidative potential of water soluble peptide fraction from different yoghurts. *Journal of Food Processing and Preservation* 41: e12979.

Takeda, S., Matsufuji, H., Nakade, K. et al. (2017). Investigation of lactic acid bacterial strains for meat fermentation and the product's antioxidant and angiotensin-I-converting-enzyme inhibitory activities. *Animal Science Journal* 88: 507–516.

Taranu, I., Marin, D.E., Palade, M. et al. (2019). Assessment of the efficacy of a grape seed waste in counteracting the changes induced by aflatoxin B1 contaminated diet on performance, plasma, liver and intestinal tissues of pigs after weaning. *Toxicon* 162: 24–31.

Tehranifar, A., Selahvarzi, Y., Kharrazi, M., and Bakhsh, V.J. (2011). High potential of agro-industrial by-products of pomegranate (*Punica granatum* L.) as the powerful antifungal and antioxidant substances. *Industrial Crops and Products* 34: 1523–1527.

Thomas, L., Larroche, C., and Pandey, A. (2013). Current developments in solid-state fermentation. *Biochemical Engineering Journal* 81: 146–161.

Tian, X., Liu, Y., Feng, X. et al. (2018). The effects of alcohol fermentation on the extraction of antioxidant compounds and flavonoids of pomelo peel. *LWT* 89: 763–769.

Tlais, A.Z.A., Fiorino, G.M., Polo, A. et al. (2020). High-value compounds in fruit, vegetable and cereal byproducts: an overview of potential sustainable reuse and exploitation. *Molecules* 25: 2987. https://doi.org/10.3390/molecules25132987.

Toldrá, F. (2017). The storage and preservation of meat: III-meat processing. In: *Lawrie's meat science*, 8e (ed. F. Toldrá), 265–296. Duxford, UK: Woodhead Publishing.

Tremocoldi, M.A., Rosalen, P.L., Franchin, M. et al. (2018). Exploration of avocado by-products as natural sources of bioactive compounds. *PLoS One* 13: e0192577.

Venkatesan, J., Anil, S., Kim, S.-K., and Shim, M.S. (2017). Marine fish proteins and peptides for cosmeceuticals: a review. *Marine Drugs* 15: 143. https://doi.org/10.3390/md15050143.

Verlee, A., Mincke, S., and Stevens, C.V. (2017). Recent developments in antibacterial and antifungal chitosan and its derivatives. *Carbohydrate Polymers* 164: 268–283.

Vermeirssen, V., Van Camp, J., and Verstraete, W. (2004). Bioavailability of angiotensin I converting enzyme inhibitory peptides. *British Journal of Nutrition* 92: 357–366.

Vieira, M.A., Oliveira, D.D., and Kurozawa, L.E. (2016). Production of peptides with radical scavenging activity and recovery of total carotenoids using enzymatic protein hydrolysis of shrimp waste. *Journal of Food Biochemistry* 40: 517–525.

Virtanen, T., Pihlanto, A., Akkanen, S., and Korhonen, H. (2007). Development of antioxidant activity in milk whey during fermentation with lactic acid bacteria. *Journal of Applied Microbiology* 102: 106–115.

Viuda-Martos, M., Fernández-López, J., and Pérez-Álvarez, J.A. (2019). Valuable compounds extraction from cereal waste and by-products. In: *Green Extraction and Valorization of by-Products from Food Processing* (eds. F.J. Barba, E.R. Soto, M. Brncic and J.M.L. Rodriquez), 153–186. CRC Press.

Xia, Q., Green, B.D., Zhu, Z. et al. (2019). Innovative processing techniques for altering the physicochemical properties of wholegrain brown rice (Oryza sativa L.)–opportunities for enhancing food quality and health attributes. *Critical Reviews in Food Science and Nutrition* 59: 3349–3370.

Xia, Q., Tao, H., Li, Y. et al. (2020). Characterizing physicochemical, nutritional and quality attributes of wholegrain *Oryza sativa* L. subjected to high intensity ultrasound-stimulated pre-germination. *Food Control* 108: 106827.

Yadav, A.S. (2017). LWT – food science and technology antioxidant and antimicrobial pro fi le of chicken sausages prepared after fermentation of minced chicken meat with *Lactobacillus plantarum* and with additional dextrose and starch. *LWT – Food Science and Technology* 77: 249–258.

Younes, I., Hajji, S., Frachet, V. et al. (2014). Chitin extraction from shrimp shell using enzymatic treatment. Antitumor, antioxidant and antimicrobial activities of chitosan. *International Journal of Biological Macromolecules* 69: 489–498.

Yu, H., Jing, Y., Zhang, X. et al. (2020). Comparison of intracellular glutathione and related antioxidant enzymes: impact of two glycosylated whey hydrolysates. *Process Biochemistry* 97: 80–86.

Zambrano, C., Kotogán, A., Bencsik, O. et al. (2018). Mobilization of phenolic antioxidants from grape, apple and pitahaya residues via solid state fungal fermentation and carbohydrase treatment. *LWT – Food Science and Technology* 89: 457–465.

Zamora-Sillero, J., Ramos, P., Monserrat, J.M., and Prentice, C. (2018a). Evaluation of the antioxidant activity in vitro and in hippocampal HT-22 cells system of protein hydrolysates of common carp (*Cyprinus carpio*) by-product. *Journal of Aquatic Food Product Technology* 27: 21–34.

Zamora-Sillero, J., Gharsallaoui, A., and Prentice, C. (2018b). Peptides from fish by-product protein hydrolysates and its functional properties: an overview. *Marine Biotechnology* 20: 118–130.

Zhang, Y., Liu, J., Lu, X. et al. (2014). Isolation and identification of an antioxidant peptide prepared from fermented peanut meal using *Bacillus subtilis* fermentation. *International Journal of Food Properties* 17: 1237–1253.

Zhang, H., Yun, S., Song, L. et al. (2017). The preparation and characterization of chitin and chitosan under large-scale submerged fermentation level using shrimp by-products as substrate. *International Journal of Biological Macromolecules* 96: 334–339.

Zhu, Z., Gavahian, M., Barba, F.J. et al. (2020). Valorization of waste and by-products from food industries through the use of innovative technologies. In: *Agri-Food Industry Strategies for Healthy Diets and Sustainability* (eds. F.J. Barba, P. Putnik and D.B. Kovačević), 249–266. London, UK: Elsevier.

Zotta, T., Solieri, L., Iacumin, L. et al. (2020). Valorization of cheese whey using microbial fermentations. *Applied Microbiology and Biotechnology* 104 (7): 2749–2764. https://doi.org/10.1007/s00253-020-10408-2.

Index

Fermentation Processes: Emerging and Conventional Technologies, First Edition.
Edited by Mohamed Koubaa, Francisco J. Barba, and Shahin Roohinejad.

q

r

s

www.ingramcontent.com/pod-product-compliance
Lightning Source LLC
LaVergne TN
LVHW080942021225
826312LV00014B/40